Encyclopedia
of Animal Behavior

Encyclopedia of Animal Behavior

Volume 3: R–Z

Edited by
Marc Bekoff

Foreword by
Jane Goodall

GREENWOOD PRESS
Westport, Connecticut • London

Library of Congress Cataloging-in-Publication Data

Encyclopedia of animal behavior / edited by Marc Bekoff; foreword by Jane Goodall.
    p. cm
    Includes bibliographical references (p.)
    ISBN 0-313-32745-9 (set : alk. paper)—ISBN 0-313-32746-7 (vol. 1 : alk. paper)—
    ISBN 0-313-32747-5 (vol. 2 : alk. paper)—ISBN 0-313-33294-0 (vol. 3 : alk. paper)
    1. Animal behavior—Encyclopedias. I. Bekoff, Marc.
QL750.3.E53   2004
591.5'03—dc22   2004056073

British Library Cataloguing in Publication Data is available.

Library of Congress Catalog Card Number: 2004056073
ISBN: 0-313-32745-9 (set code)
        0-313-32746-7 (Vol. 1)
        0-313-32747-5 (Vol. 2)
        0-313-33294-0 (Vol. 3)

First published in 2004

Greenwood Press, 88 Post Road West, Westport, CT 06881
An imprint of Greenwood Publishing Group, Inc.
www.greenwood.com

Printed in the United States of America

The paper used in this book complies with the
Permanent Paper Standard issued by the National
Information Standards Organization (Z39.48-1984).

10 9 8 7 6 5 4 3 2 1

*These volumes are dedicated to the memory
of one of the most amazing people
I have ever known—my mother Beatrice*

■ Contents

■ Alphabetical List of Entries

(S) indicates a side bar. It is listed under the entry within which it appears.

■ Guide to Related Topics

ANIMALS

Following is a list of entries that focus primarily on one type of animal. Most of the other entries in the *Encyclopedia of Animal Behavior* also discuss various species. Please see the index for additional references to specific animals.

Ants

Animal Architecture—*Subterranean Ant Nests*

Bats

Bats—*The Behavior of a Mysterious Mammal*
Communication—Auditory—*Bat Sonar*

Bears

Feeding Behavior—*Grizzly Foraging*
Human (Anthropogenic) Effects—*Bears: Understanding, Respecting, and Being Safe around Them*

Birds

Antipredatory Behavior—*Sentinel Behavior*
Behavioral Physiology—*Plumage Color and Vision in Birds*
Caregiving—*Brood Parasitism among Birds*
Cognition—*Grey Parrot Cognition and Communication*
Corvids—*The Crow Family*
Communication—Vocal—*Social System and Acoustic Communication in Spectacled Parrotlets*
Communication—Vocal—*Singing Birds: From Communication to Speciation*
Mimicry—*Magpies*
Play—*Birds at Play*
Reproductive Behavior—*Bowerbirds and Sexual Displays*
Tools—*Tool Use and Manufacture by Birds*
Welfare, Well-Being, and Pain—*Enrichment for Chickens*
Welfare, Well-Being, and Pain—*Feather Pecking in Birds*
Welfare, Well-Being, and Pain—*Rehabilitation of Raptors*

Cats

Cats—*Domestic Cats*
  WILD VERSUS DOMESTIC BEHAVIORS: WHEN NORMAL
  BEHAVIORS LEAD TO PROBLEMS (S)

Fish

Communication—Auditory
 NOISY HERRING (S)
Sharks
 SHARKS: MISTAKEN IDENTITY (S)
Caregiving—*Brood Parasitism in Freshwater Fish*
Caregiving—*Parental Care in Fish*
Communication—Visual—*Fish Display Behavior*
Human (Anthropogenic) Effects—*Genetically Modified Fish*
Methods
 SONIC TRACKING OF ENDANGERED ATLANTIC SALMON (S)
Reproductive Behavior—*Alternative Male Reproductive Tactics in Fish*
Reproductive Behavior—*Sex Change in Fish*
Social Organization—*Shoaling Behavior in Fish*

Frogs

Communication—Vocal—*Choruses in Frogs*

Horses

Horses—*Behavior*
 SLEEPING STANDING UP (S)
Horses—*Horse Training*

Hyenas

Development—*Spotted Hyena Development*

Insects

Animal Architecture—*Subterranean Ant Nests*
Behavioral Physiology
 COLORS—HOW DO FLOWERS AND BEES MATCH? (S)
Behavioral Physiology—*Insect Vision and Behavior*
Caregiving—*Parental Care by Insects*
Communication—*Honeybee Dance Language*
Education—Classroom Activities—*Insects in the Classroom*
Predatory Behavior—*Praying Mantids*
Reproductive Behavior—*Sexual Behavior in Fruit Flies*—Drosophila
Reproductive Behavior—*Sexual Cannibalism*

Lemurs. *See also* Primates

Lemurs—*Behavioral Ecology of Lemurs*
Lemurs—*Learning from Lemurs*

Macaques

Cognition—*Tactical Deception in Wild Bonnet Macaques*
Social Organization—*Social Knowledge in Wild Bonnet Macaques*
Tools—*Tool Manufacture by a Wild Bonnet Macaque*

Behavioral Physiology

Color Vision in Animals
    Colors—How Do Flowers and Bees Match? (S)
    Do Squid Make a Language on Their Skin? (S)
Insect Vision and Behavior
Plumage Color and Vision in Birds
Thermoregulation
Thermoregulatory Behavior
Turtle Behavior and Physiology
Visual Perception Mechanisms
    Vision, Skull Shape, and Behavior in Dogs (S)

Behaviorism

Burrowing Behavior

Caregiving

Attachment Behaviors
Brood Parasitism among Birds
Brood Parasitism in Freshwater Fish
Fostering Behavior
How Animals Care for their Young
    Helpers in Common Marmosets (S)
Incubation
Mother–Infant Relations in Chimpanzees
Non-Offspring Nursing
Parental Care
    Parental Behavior in Marsupials (S)
Parental Care and Helping Behavior
Parental Care by Insects
Parental Care in Fish
Parental Investment
    Parents Desert Newborns after Break-in! Mother Goes
    for Food as Siblings Battle to the Death! Father Eats Babies! (S)

Cats

Domestic Cats
    Wild versus Domestic Behaviors: When Normal Behaviors
    Lead to Problems (S)

Cephalopods

Octopuses, Squid, and other Mollusks
    Roving Octupuses (S)

BEHAVIOR OF ANIMALS (continued)

Cognition

Animal Consciousness
Animal Languages, Animal Minds
Audience Effect in Wolves
Behavior, Archaeology, and Cognitive Evolution
Cache Robbing
Caching Behavior
Categorization Processes in Animals
Cognitive Ethology: The Comparative Study of Animal Minds
Concept Formation
Deception
Dogs Burying Bones: Unraveling a Rich Action Sequence
Domestic Dogs Use Humans as Tools
Equivalence Relations
    FAIRNESS IN MONKEYS (S)
Food Storing
Grey Parrot Cognition and Communication
Imitation
Limited Attention and Animal Behavior
Mirror Self-Recognition
Mirror Self-Recognition and Kinesthetic–Visual Matching
Social Cognition in Primates and other Animals
Tactical Deception in Wild Bonnet Macaques
Talking Chimpanzees
Theory of Mind

Communication—Auditory

Acoustic Communication in Extreme Environments
Audition
    NOISY HERRING (S)
Bat Sonar
Long Distance Calling, the Elephant's Way
Ultrasound in Small Rodents

Communication

Electrocommunication
    ELECTRIC FISH (S)
Honeybee Dance Language
Modal Action Patterns

Communication—Olfaction

Chemical Communication
    DOG SCENTS AND "YELLOW SNOW" (S)
Mammalian Olfactory Communication

Communication—Tactile

Communication in Subterranean Animals
Vibrational Communication

Communication—Visual

Fish Display Behavior

Communication—Vocal. *See also* Communication—Auditory

Alarm Calls
Choruses in Frogs
Communication in Wolves and Dogs
Jump-Yips of Black-Tailed Prairie Dogs
Referential Communication in Prairie Dogs
Singing Birds: From Communication to Speciation
Social Communication in Dogs: The Subtleties of Silent Language
Social System and Acoustic Communication in Spectacled Parrotlets
Variation in Bird Song
Vocalizations of Northern Grasshopper Mice

Conservation and Behavior

Species Reintroduction
    Preble's Meadow Jumping Mouse and Culverts (S)
Wildlife Behavior as a Management Tool in United States National Parks

Cooperation

Corvids

The Crow Family

Coyotes

Clever Tricksters, Protean Predators

Culture

Orangutan Culture
    Culture in Our Relatives, The Orangutans (S)
Whale Culture and Conservation

Curiosity

Darwin, Charles (1809–1882)

Development

Adaptive Behavior and Physical Development
Behavioral Stages
Embryo Behavior
Intrauterine Position Effect
Spotted Hyena Development

BEHAVIOR OF ANIMALS (continued)

Domestication and Behavior

The Border Collie, a Wolf in Sheep's Clothing

Dominance

Development of Dominance Hierarchies

Emotions

Emotions and Affective Experiences
Emotions and Cognition
How Do We Know Animals Can Feel?
Laughter in Animals
Pleasure

Empathy

Exploratory Behavior

Inquisitiveness in Animals

Feeding Behavior

Grizzly Foraging
Scrounging
Social Foraging
Social Learning: Food

Friendship in Animals

Hibernation

Hormones and Behavior

Human (Anthropogenic) Effects

Bears: Understanding, Respecting, and Being Safe around Them
Edge Effects and Behavior
The Effect of Roads and Trails on Animal Movement
Environmentally Induced Behavioral Polymorphisms
Genetically Modified Fish
Human (Anthropogenic) Effects on Animal Behavior

BEHAVIOR OF ANIMALS (continued)

Parasite-Induced Behaviors

Personality and Temperament

A Comparative Perspective
Personality in Chimpanzees
Personality, Temperament, and Behavioral Assessment in Animals
Stress, Social Rank and Personality

Play

Birds at Play
Social Play Behavior and Social Morality
    DOG MINDS AND DOG PLAY (S)

Predatory Behavior

Ghost Predators and their Prey
Orb-Web Spiders
Praying Mantids
    "SEXUAL CANNIBALISM": IS IT REALLY "SEXUAL" OR IS IT JUST
    PREDATORY BEHAVIOR? (S)

Recognition

Individual Recognition
Kin Recognition

Reproductive Behavior

Adaptations and Exaptations in the Study of Behavior
Alternative Male Reproductive Tactics in Fish
Assortative Mating
Bowerbirds and Sexual Displays
Captive Breeding
Female Multiple Mating
Marine Turtle Mating and Evolution
Mate Choice
Mate Desertion
Mating Strategies in Marine Isopods
Mollusk Mating Behaviors
Monogamy
Morning Sickness
Sex Change in Fish
Sex, Gender, and Sex Roles
    FEMALE–FEMALE SEXUAL SELECTION (S)
Sexual Behavior in Fruit Flies—Drosophila
Sexual Cannibalism
Sexual Selection

Conservation and Behavior

Wildlife Behavior as a Management Tool in United States National Parks

EDUCATION IN ANIMAL BEHAVIOR

Careers

Careers in Animal Behavior Science
Significance of Animal Behavior Research

Education

Classroom Activities in Behavior
  CLASSROOM RESEARCH (S)
Classroom Activities—All About Chimpanzees!
Classroom Activities—Insects in the Classroom
Classroom Activities—"Petscope"
Classroom Activities—Planarians

History

History of Animal Behavior Studies

Zoos and Aquariums

Animal Behavior Research in Zoos
Giant Pandas in Captivity
Studying Animal Behavior in Zoos and Aquariums

HUMANS AND ANIMALS

Animal Abuse

Animal and Human Abuse
Violence to Human and Nonhuman Animals

Animal Models of Human Psychology

An Assessment of their Effectiveness

Anthropomorphism

Applied Animal Behavior

Mine-Sniffing Dogs
Scat-Sniffing dogs
Social Dynamics and Aggression in Dogs

HUMANS AND ANIMALS (continued)

Careers

Animal Behavior and the Law
Animal Tracking and Animal Behavior
Animal-Assisted Psychotherapy
Applied Animal Behavior
Careers in Animal Behavior Science
Mapping their Minds: Animals on the Other Side of the Lens
Recording Animal Behavior Sounds: The Voice of the Natural World
Significance of Animal Behavior Research
Veterinary Practice Opportunities for Ethologists
Wildlife Filmmaking
Wildlife Photography
Writing about Animal Behavior: The Animals Are Also Watching Us

Cognition

Domestic Dogs Use Humans as Tools
Grey Parrot Cognition and Communication

Conservation and Behavior

Species Reintroduction
    PREBLE'S MEADOW JUMPING MOUSE AND CULVERTS (S)
Wildlife Behavior as a Management Tool in United States National Parks

Ecopsychology

Human—Nature Interconnections

Education

Classroom Activities in Behavior
    CLASSROOM RESEARCH (S)

Education—Classroom Activities

All About Chimpanzees!
Insects in the Classroom
"Petscope"
Planarians

History

History of Animal Behavior Studies
    SOME LEADERS IN ANIMAL BEHAVIOR STUDIES (S)
Niko Tinbergen and the "Four Questions" of Ethology

Human (Anthropogenic) Effects

Bears: Understanding, Respecting, and Being Safe around Them
Edge Effects and Behavior
The Effect of Roads and Trails on Animal Movement

Tinbergen, Nikolaas (1907–1988)
> A VISIT WITH NIKO AND LIES TINBERGEN (S)

Turner, Charles Henry (1867–1923)

Washburn, Margaret Floy (1871–1939)

Whitman, Charles Otis (1842–1910)

RESEARCH, EXPLANATION, AND METHODS OF STUDY

Behaviorism

Careers
> *Significance of Animal Behavior Research*

Comparative Psychology

Development
> *Behavioral Stages*

Ecopsychology

Frisch, Karl von
> *Decoding the Language of the Bees*

History
> *A History of Animal Behavior Studies*
> *Human Views of the Great Apes from Ancient Greece to the Present*

Levels of Analysis in Animal Behavior

Lorenz, Konrad Z.
> *Analogy as a Source of Language*

Methods
> *Computer Tools for Measurement and Analysis of Behavior*
> *Deprivation Experiments*

Recognition
Individual Recognition

Imagine if all humans looked the same, with the same hairstyle, coloring, and facial features. How would you know when you passed a friend on the street? How would you distinguish between your mother and your neighbor? Human social interactions are based on the human ability to recognize each other as individuals, so living without individual recognition seems ridiculous.

There are many different types of recognition, for example, kin recognition, species recognition, nestmate recognition, mate recognition. These types of recognition are based on discriminating among classes of organisms (kin vs. non-kin, same species vs. different species, and so on). Individual recognition occurs when an organism identifies another individual according to its distinctive characteristics. Individual recognition differs from other types of recognition because it is the most specific type of recognition, based on individuals instead of on classes.

Although it is strange to imagine a world where all humans looked essentially identical, most other species are in precisely this situation. Why do individuals in many species look the same?

1. One possibility is that these animals cannot recognize each other as individuals. In fact, research suggests that most creatures do not have the ability to recognize individuals.

2. Another possibility is that these animals have individual recognition, but use other senses for this recognition.

Human are extremely visual creatures, so it makes sense that we use visual signals to recognize each other. Some animals rely more strongly on other senses, so they have evolved signals of individual identity that use sound or smell. For example, starlings and many other birds use songs to recognize each other individually. Guinea pigs and lobsters rely on chemicals in urine to smell the identity of other individuals. One thing that all signals of individual identity have in common is that they are variable. Without variability, there isn't enough information in a signal to tell different individuals apart. Imagine the example at the beginning of the paper. If all humans looked the same, there wouldn't be enough variability in appearance to allow recognition of individual humans.

Some species (like lobsters, humans, chickadees, guinea pigs) have evolved the ability to recognize individuals, but many other species have not. Current research suggests that an organism's social behavior determines whether or not it evolves the ability to recognize individuals. For

The variability in murre eggs.
Courtesy of James Dale.

example, a species like a mosquito has relatively simple social interactions, so mosquitoes don't receive any benefits by knowing the identity of other mosquitoes. Consequently, there is no selection on mosquitoes to be able to recognize other mosquitoes as individuals, and mosquitoes have not evolved individual recognition. On the other hand, some species have complex social interactions that favor the ability to recognize individuals. In these species, individual recognition provides clear benefits, so natural selection has favored individual recognition. Here are a few examples of individual recognition and the kinds of social behavior that selects for the ability to recognize individuals.

- Common murres are birds that nest in extremely dense colonies on rocky ledges near the ocean. Common murre eggs are covered in variable patterns and colors. Parents use these patterns to identify which egg is their own out of the many nearby eggs.
- When mother seals are ready to give birth, they form large "nursery" colonies with hundreds to thousands of other mothers. Mothers leave their pups alone on the beach when they hunt for fish. After returning with food, mothers must find their own pup among all the others. They do this with distinctive calls that allow mothers and pups to recognize each other as individuals.

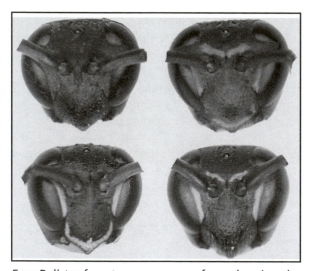

Four Polistes fuscatus *paper wasp faces showing the variation these wasps use for individual recognition.* Courtesy of Elizabeth Tibbetts.

In these two examples, mothers and offspring must find each other in busy places. It is disadvantageous for parents to invest effort in the wrong young, leaving their own without care. As a result, there is strong selection for individual recognition in situations where parents might have trouble finding their own young. In many other species, parents live alone. For example, many birds build the only nest in a tree. When these birds return to their nest, they are unlikely to find another mother's eggs (or babies) inside. As a result, there is not selection for individual recognition. In many of these solitary species, you can switch the eggs (or even babies), and the parents will happily care for the wrong young.

- Some species of paper wasps live in extremely complex societies where each individual wasp has a different role that is based on her rank in the colony. More dominant wasps lay more of the colony's eggs, while more subordinate wasps do more work. These paper wasps have variable facial markings that they use to recognize their nestmates as individuals.
- Dolphins have extremely complex social interactions, like coalitions, dominance behavior, and long term mother–offspring bonds. Dolphins use whistles to identify other dolphins as individuals.

In these examples, the species have extremely complex social networks that would be difficult to coordinate without individual recognition. It would be quite challenging to form a coalition if you couldn't keep track of the identity of your coalition partner. Therefore, the ability to recognize individuals is beneficial in species with complex social networks. This is likely the reason humans have evolved the ability to recognize individuals.

- North America bullfrog males stake out breeding territories that they defend from other males. Males use the croaks of other males to identify their neighbors as individuals.

- During the summer, pairs of chickadees form territories within which they build a nest and raise their young. Chickadees can identify other chickadees as individuals by their songs.

Chickadees and bullfrogs vigorously defend territories. They must be ready for a fight if a competitor approaches. At the same time, they are surrounded by many individuals who are not threatening. The ability to recognize individuals is beneficial because it allows them to determine who is a potentially dangerous intruder and who is not a threat. Therefore, many different types of complex social behavior can select for individual recognition.

See also Recognition—*Kin Recognition*
Cognition—*Equivalence Relations*

Further Resources

Dale, J., Lank, D. B., & Reeve, H. K. 2001. *Signaling individual identity vs. quality: A model and case studies with rugs, queleas, and house finches.* American Naturalist, 158, 75–86.

deWaal, F. B. M., & Tyack, P. L. *Animal Social Complexity: Intelligence, Culture, and Individualized Societies.* London: Harvard University Press.

Medvin, M. B., Stoddard, P. K., & Beecher, M. D. 1993. *Signals for parent–offspring recognition: A comparative analysis of the begging calls of cliff swallows and barn swallows.* Animal Behavior, 45, 841–850.

Elizabeth Tibbetts

■ Recognition
Kin Recognition

The existence of altruistic behavior, behavior that is costly to the actor but benefits another individual, seems counter to the concepts of natural selection, and it troubled Charles Darwin. For example, sterile worker bees raise the queen's young and defend the hive with suicidal stingers. How could such a situation evolve? In 1964, William D. Hamilton demonstrated that altruism can evolve if helpful behavior is directed toward relatives because relatives share many genes. In the case of bees and other social insects, an unusual genetic system causes workers and the queen, if full sisters, to share 75% of their genes by common descent. Individuals can increase their genetic representation in the next generation not only through personal reproduction, but also by helping relatives to reproduce. Therefore, even though worker bees are sterile, they can still reproduce indirectly via the queen. Cases of altruism in most other kinds of animals are not as extreme, because helpers are not sterile and full siblings share 50% of genes, but the same general principle applies. This principle also applies not only to directing altruistic behavior toward relatives, but also to limiting aggression among relatives.

Hamilton's theory, usually referred to as kin selection, was a breakthrough in evolutionary biology and led to a revolution in the study of behavior. However, an important practical problem immediately arose: In order to direct helpful behavior toward kin, individuals

Four Instructive Examples of Kin Recognition

Belding's Ground Squirrel

This species lives in colonies of female relatives, unrelated males, and their young. Females often mate with several males, and most litters contain a mixture of full siblings and maternal half siblings. Laboratory cross-fostering experiments, which involve exchanging young among litters, combined with field observations, demonstrate that female ground squirrels use a combination of association and phenotype matching to recognize their sisters. As is typical of mammals, kin are recognized by their odor. Females are less aggressive to littermates than nonlittermates, even when experimental littermates are actually nonkin and nonlittermates are actually sisters, which suggests kin recognition by association. However, although females are aggressive toward sisters reared apart, they are even more aggressive toward unrelated females reared apart. Amazingly, they are also less aggressive toward, and possibly more cooperative with, full-sister than half-sister littermates in the wild. Both of these results suggest phenotype matching, particularly self-referent phenotype matching, since in the laboratory experiment the other possible referents were often unrelated littermates and foster mothers. This is one of the few examples of kin recognition that demonstrates fine-scale discrimination among classes of close kin, in this case full sisters (who share on average 50% of genes by descent) versus half sisters (who share 25%).

Peacock

Peacocks form *leks*, areas where males gather to display their impressive plumage to females. The males tend to display near their genetic brothers, even after cross-fostering. The possibility that display-site preferences are inherited was ruled out because sons did not use their father's old sites. Therefore, peacocks are able to identify brothers whom they have never before encountered. As with the ground squirrels, self-referent phenotype matching is the most likely explanation. Unlike the ground squirrels, where association probably has a larger impact on behavior than phenotype matching, association plays no role in peacock kin recognition. Peacocks do not display near their unrelated nestmates, instead preferring to display near siblings even when they are complete strangers.

Common Tern

Terns nest on the ground in colonies where hundreds of nests are present in a small area. When chicks reach 4–5 days old, they begin to occasionally wander from the nest site or scatter when a predator (or biologist) enters the colony. Chicks need a way to locate the correct nest to return home. Away from the nest they lose food

A common tern chick is shown approaching its nestmate in a laboratory test of kin recognition.
Courtesy of B. G . Palestis.

to their siblings and are exposed to the elements and to aggression from adult terns. A field experiment demonstrated that tern chicks can locate their nests more easily if siblings are present in the nest than if the nest is empty, or if nonsiblings are present instead. Occasionally they even approached the wrong nest if it contained siblings. Both laboratory and field experiments demonstrated that sibling recognition, based on a combination of vocal and visual cues, develops at approximately 4–5 days of age, and thus corresponds to the time when mixing between broods first becomes possible. This association between the development of kin discrimination and the development of mobility also occurs in other colonial birds and mammals, such as swallows and ground squirrels.

Botryllus schlosseri

Larvae of the sea squirt *Botryllus schlosseri* settle near kin and avoid settling near non-kin, and adults form clonal colonies that will fuse only with kin. Sea squirts can recognize kin even though they lack brains, using a mechanism analogous to the rejection of incompatible tissue transplants. These organisms distinguish kin by chemically comparing alleles at one histocompatibility gene. Non-kin bred to share the same histocompatibility allele are preferred over kin bred to have different alleles. Although *Botryllus* kin recognition is similar to the green beard effect, it probably does not qualify, because there is no evidence suggesting that production of the chemical cue is genetically linked to recognition of and response to the cue.

would usually need some way to discriminate relatives from nonrelatives or close relatives from distant relatives. There had been previous work demonstrating parent–offspring recognition, largely in colonial seabirds, but none demonstrating recognition of nondescendent kin. Today the situation is much different—since the late 1970s numerous examples of the ability of animals to recognize kin have been found across diverse taxa, including mammals, birds, amphibians, fish, social insects, and sea squirts. Although this review will focus entirely on kin recognition in animals, examples have also been found in plants.

One note on terminology must be mentioned since it has caused some confusion in the kin recognition literature. In this chapter I use the term "kin recognition" broadly, as have most authors, not explicitly separating it from "kin discrimination." However, technically "recognition" refers to internal neural processing, whereas "discrimination" refers to an overt behavioral act. In other words, it is possible to recognize kin without acting on this knowledge and actually discriminating kin from non-kin.

In most cases kin recognition has been studied in the context of kin-directed altruism, as described above, but kin recognition can have other functions. One important function of kin recognition is inbreeding avoidance because mating among relatives can allow harmful recessive mutations to be expressed and it also decreases genetic variability. Most animals avoid mating with kin, and in many cases this requires the ability to discriminate kin from non-kin. On the other hand, mating with an individual that is too genetically different can also be costly because favorable combinations of genes may be broken up. Therefore, there can be selection to optimize the balance between inbreeding and outbreeding—mating with individuals not too similar, but also not too different. Bateson showed that Japanese quail, reared only in the presence of siblings, prefer to associate with (and presumably mate with) first cousins over familiar and unfamiliar siblings, third cousins, and non-kin.

Kin recognition can also give survival advantages in certain situations. The young of birds that nest on the ground in colonies can easily get lost, and can use siblings (and parents) as a cue to the location of the correct nest site. Another survival advantage to recognizing kin is in the avoidance of disease. Diseases may spread more easily among relatives, since relatives have similar immune systems. The presence of a deadly bacterium may partly explain why cannibalistic salamander tadpoles avoid eating relatives. Of course, kin selection could also contribute to an inhibition against eating kin.

Kin recognition is widespread in the animal kingdom, but how do animals recognize their relatives? Classically, the mechanisms of kin recognition were divided as follows: spatial distribution, association, phenotype matching, and recognition alleles. *Spatial distribution* is the simplest way to identify relatives: If kin live in predictable locations, individuals in a particular place (such as a nest) can be assumed to be kin. *Association*, probably the most common mechanism of kin recognition, involves learning the identity of relatives when encountered in predictable locations or situations. These individuals can be treated as kin later, when they are encountered again in a different location. Unlike association, *phenotype matching* allows the recognition of kin that were never previously encountered. At first an individual learns the characteristics of its relatives through association or learns its own characteristics. When a stranger is encountered, it can be compared to this learned template.

All three mechanisms discussed above are prone to occasional recognition errors because they depend on kin being reliably located in a particular place or being reliably encountered in particular contexts. One exception is if self-referent phenotype matching occurs. Richard Dawkins has called this mechanism the "armpit effect," because an individual can use its own odor (or other characteristics) to judge the relatedness of others. An individual "knows" that it is related to itself, but may not be able to safely assume that all nestmates or littermates are kin, because adoption of unrelated young does occur occasionally. Even if all nestmates are siblings, they may often consist of a mix of full and half-siblings due to promiscuous mating. Another possible advantage of self-referencing is that it may not require memorizing a template, since the self is always present when a comparison is needed. Possible examples of self-referent phenotype matching in peacocks and ground squirrels are described in the sidebar to this essay.

Hamilton suggested the possibility of innate recognition based on (at the time hypothetical) *recognition alleles*, a process that Dawkins dubbed the "green beard effect." Recognition alleles would cause an individual to express a particular trait (such as a green beard), to recognize this trait in others, and to direct helpful behavior to those sharing the trait. Although examples of green beard genes have been found recently in fire ants and slime molds, among others, recognition alleles are best thought of as an alternative to kin recognition rather than as a mechanism of kin recognition. One way to ensure that altruism is usually directed toward other altruists is to help relatives, who are likely to share genes *for* altruism. Another way is to recognize fellow altruists directly by their green beards (or by their altruistic behavior), regardless of whether they are relatives.

A general classification of kin recognition mechanisms into two categories, direct and indirect kin recognition, is often useful. *Indirect recognition* relies on contextual cues of time or place, and thus is similar to the spatial distribution mechanism described above. *Direct recognition* is based on the traits of the organisms being discriminated among, as in association and phenotype matching, and involves three components: production of a cue, perception of the cue and comparison to a template, and action based on the similarity between the cue and the template and on the context of the encounter. The cues used in direct recognition vary among species, and can have an olfactory, vocal, or visual basis. These cues can be genetically determined or can be acquired from the environment, as long as they are reliable indicators of relatedness. For example, paper wasps acquire a colony-specific odor that allows them to discriminate relatives returning home from unrelated intruders.

Regardless of whether cues are genetically or environmentally determined, they are almost always learned. Mice base kin recognition on odors that are genetically determined by the major histocompatibility complex (MHC). The *MHC* is a set of highly variable genes that allows the immune system to discriminate the body's own cells from foreign cells, and relatives have more similar MHC genes than do nonrelatives. Mice of both sexes prefer to mate with individuals whose MHC's differ from their own, to avoid inbreeding. Female mice prefer to nest communally with female kin, whose MHC-based odors are similar to their own. However, even in a case like this one, where the recognition cue is based in the genes, recognition is not innate. Mice learn their familial odor from exposure to the odor of their parents (or experimental foster parents), an example of phenotype matching. In addition, even the production of this genetically based cue is not without environmental influences since diet can alter the odor.

An important consideration in the study of kin recognition is the relative cost of the two kinds of recognition errors, acceptance and rejection errors. *Acceptance errors* involve treating non-kin as kin, and *rejection errors* involve treating kin as non-kin. The relative cost of these two kinds of errors varies among species and situations, and plays a role in the action component of recognition. Parent birds and mammals will often assume that any individuals in their nests or burrows are their own young, an example of indirect recognition. Acceptance of non-kin in the nest occurs even in species where direct parent–offspring recognition has been demonstrated to occur in other contexts. This is probably because having an additional mouth to feed, due to an acceptance error, is not as costly as mistakenly ejecting one's own offspring, due to a rejection error. Once the offspring reach an age at which they are mobile enough to mix frequently with members of other broods, parents become more discriminating.

Kin recognition is an important component of the evolution of altruistic behavior and plays a key role in inbreeding avoidance. Biologists have found kin recognition nearly everywhere they have looked, in taxonomically and ecologically diverse organisms, and more examples are sure to come. More research is needed on the functions of kin recognition. Often researchers have studied kin recognition in the laboratory without demonstrating its adaptive value in the field. Additionally, although a large number of papers have been written on the mechanisms of kin recognition, more work is also needed here. As Mark Hauber and Paul Sherman recently pointed out, the neural basis of kin recognition remains a black box. This is particularly true for the perception component of recognition and for the "decision rules" underlying the action component. The study of kin recognition has been a highly productive area of research for the past quarter century, and will likely remain so well into the future.

See also Cognition—*Equivalence Relations*
Hamilton, William D. (1936–2000)
Recognition—*Individual Recognition*
Sociobiology

Further Resources

Fletcher, D. J. C. & Michener, C. D. (Eds.) 1987. *Kin Recognition in Animals*. New York: John Wiley & Sons.

Hauber, M. E. & Sherman, P. W. 2001. *Self-referent phenotype matching: Theoretical considerations and empirical evidence*. Trends in Neurosciences, 24, 609–616.

Hepper, P. G. (Ed.) 1991. *Kin Recognition*. Cambridge: Cambridge University Press.

Nakagawa, S. & Waas, J. R. 2004. *O brother, where art thou? A review of avian sibling recognition by referring to the mammalian literature*. Biological Reviews, 79, 101–119.

Nakagawa, S., & Waas, J. R. 2004. *O singling, where are thou? A review of avian sibling recognition with respect to the mammalian literature*. Biological Reviews, 79, 11–119.

Pfennig, D. W. & Sherman, P. W. 1995. *Kin recognition*. Scientific American, 272, 98–103.

Sherman, P. W., Reeve, H. K. & Pfennig, D.W. 1997. *Recognition systems*. In: *Behavioural Ecology: An Evolutionary Approach* (Ed. by J. R. Krebs & N. B. Davies), pp. 69–96. Oxford: Blackwell Science.

Brian Palestis

■|Religion and Animal Behavior

Nonhuman animals play significant and varied roles in many different religions, ranging from locally-based indigenous traditions to world religions such as Christianity and Buddhism. Animals appear in religious stories, art, rituals, and morality as sacred figures, deserving of respect, love, awe, or fear; as embodiments of virtues or vices, as objects of sacrifice and as subjects of reverence. Interpretations of animals in different religions reflect not only attitudes toward particular species and nonhumans in general, but also human understandings of their own societies, of the natural world, of rules for proper human behavior, and of where and how to draw the line between humans and (particular) nonhumans. These issues are seen in relation to several different religions, including Native American, Christian, and Buddhist traditions.

A Benin bracelet from Nigeria depicts Anansi, a recurring spider character in the mythology.
© The Art Archive / Antenna Gallery Dakar Senegal / Dagli Orti.

It is dangerous to generalize about "religion" in general or about clusters of religious traditions, because generalizations tend to miss the great diversity and complexity of religious thought, practice, and values. Not only do religions such as Christianity, Islam, and Buddhism differ from each other, within each of these traditions there are many diverse, often conflicting schools of thought about important topics, including conceptions of and attitudes about nonhuman animals. With this warning in mind, it is possible to identify some significant—though never absolute—differences among various religions and even among the three broad categories of indigenous, Western, and Asian traditions.

Many indigenous religions, in the Americas, Africa, and elsewhere, have been and still are closely tied to local ecological conditions and natural histories. Most of these cultures have long traditions of living in close relationship with wild and/or domestic animals: They have made their livings primarily as hunters, fishers, and gatherers or as small-scale farmers and pastoralists. There is no linear relationship between cultures' livelihoods and their religious life, but often a group's closeness to, knowledge of, and dependence upon particular animals have contributed to a prominent role for the animals in the culture's religious stories, rituals, and moral guidelines.

Many (but not all) indigenous religions are polytheistic, meaning that they envision a cosmos with multiple gods and goddesses, many linked to natural processes and places. Some are pantheistic, meaning that they find sacred value not only in divine beings but in natural objects, such as animals, trees, bodies of water, or mountains. In many cases, these cultures do not establish a sharp dividing line between all humans and all nonhumans. Instead, they perceive a series of different types of distinctions and relationships among them. The religious importance of a particular kind of animal rests on many factors, including local abundance, economic significance, and perceived behavioral traits. Some locally abundant species may be religiously insignificant, for example, while less common ones may play central roles, as creators, tricksters, enemies, or allies of humans. Sometimes timing matters: In many Native American cultures, sacred stories and rituals exclude animals that are not native to the area or that were introduced after the arrival of Europeans (the horse is a major exception for many groups). When hunting is a major source of livelihood for a group, its religious stories and rituals often emphasize guidelines for treating prey animals. For pastoral or agricultural societies, domestic animals may take on religious and moral significance. For example, sheep, a major source of livelihood, are conceptualized as "mothers" of humans in Navajo culture, their significance another exception to the generally minor role accorded European-introduced animals. Religious importance does not always correlate with economic importance, however. Although wild pigs (peccaries) are a major prey source for the Huaorani people of the Brazilian Amazon, they are not especially significant in Huaorani religion and myth, as Laura Rival has noted. Further, sometimes animals with little material importance are central religious figures, as are ravens or coyotes in many North American cultures.

Brides and grooms sit on elephants representing Ganesha during their wedding ceremony in the northern Thai province of Ayutthaya.
© Reuters / Corbis.

Despite the great diversity among indigenous cultures, they do share a common attention to and interest in nonhuman animals and a sense that at least some animals are spiritually important. While we should not overgeneralize, it is possible to identify a qualitative difference between many native traditions and the three Western religions of Judaism, Christianity, and Islam. These monotheistic faiths often assert that humans are both unique and superior to other animals. This claim rests on the idea that humans alone have an eternal soul, the "image of God" which divides them from the rest of creation which is ultimately lacking in spiritual value. For Judaism and Christianity, claims about human uniqueness begin with the Biblical narrative of human origins in Genesis 1:26–28, according to which:

And God said, "Let us make man in our image, after our likeness, and let them have dominion over the fish of the sea, and over the birds of the air, and over the cattle, and over all the earth, and over every creeping thing that creeps upon the earth."

So God created man in His own image, in the image of God he created him; male and female he created them.

> And God blessed them, and God said to them, "Be fruitful, and multiply, and fill the earth and subdue it; and have dominion over the fish of the sea and over the birds of the air and over every living thing that moves upon the earth."

While various interpretations of this passage are possible, the most common one, especially in Christianity, takes it as a basis for asserting humans' spiritual superiority over other animals, which in turn justifies their domination. There are, of course, many alternative schools of thought within Christianity and the other Western monotheistic traditions. Perhaps the most influential dissenter is St. Francis of Assisi, who, following some earlier theologians, suggested that human salvation is tied to the spiritual renewal of all of creation. Francis viewed God's creation inclusively, refusing to make a sharp separation between the spiritual and the material or to attribute ultimate value only to the former.

Asian traditions, including Buddhism, Hinduism, and the Chinese traditional religions of Confucianism and Taoism, differ significantly from each other and also internally. It is safe to say, however, that most Asian religions differ from Western monotheistic traditions in that the former do not make an absolute division between spirit and matter, or between humans and other animals. Both Buddhism and Taoism, for example, assert that there is substantial continuity between humans and the rest of life. However, they do not deny a special place to humans. For Buddhism, only humans can reach the ultimate fulfillment of becoming a Buddha by practicing meditation and becoming enlightened. Despite the special value of being born as a human, Buddhism sees all sentient beings as fundamentally similar, in their urge to avoid pain and experience well-being and, perhaps most important, in their connectedness through the core Buddhist doctrine of interdependence. This interconnectedness and the importance of compassion to nonhuman animals are frequent themes in Buddhist teachings and stories (Badiner 1990). Traditional Chinese religions such as Taoism also assert both continuity among all life and a certain distinctive value for humans. In many Asian traditions, humans' unique talents bring not only special privileges but also special responsibilities in regard to other creatures.

While it is possible to make some broad distinctions among different religions in their attitudes toward nonhuman animals, it is also possible to identify a number of common themes, both in the roles animals play in different religions and in religiously inspired conceptions of animal behavior. In a number of religions, especially in indigenous cultures, nonhuman animals possess great sacred power: An animal may even be creator of the universe, of humans, or of other life forms, as the raven is for many cultures in northern North America. Different species may also embody particular character traits, good or bad, that are shared with humans: They may be tricksters, storytellers, good or bad parents, powerful hunters, or healers, for example. Some figures, such as coyote in North America or the clever spider Anansi in West African traditions, are recurring characters in stories that both entertain and pass on important cultural values and moral lessons.

Such characterizations, and the relative significance of different species, depend both on the animals' importance in a particular area and on human perceptions of animal behavior. The Huaorani, for example, use the same word to describe both woolly monkey social groups and human groups because they perceive similarities in the two species' forms of social organization (Rival 1996). Some cultures prohibit killing, consumption, or capture of certain species because perceived character traits are related to religious and moral significance. The Hindu reverence for cows and subsequent prohibition on eating beef is well known, for example. From a different angle, both Islam and Judaism ban the consumption of pork, not out of reverence but because pigs are believed to be unclean.

These examples suggest, not only that there are a wide range of ways that animals can be religiously important in different traditions, but also that religious significance itself has many possible consequences for treatment of nonhuman animals. Important animals may be protected from harm and killing, as cattle are in Hindu areas. On the other hand, significant animals may be singled out for captivity, killing, or consumption. The totem animal of an indigenous group, for example, may be hunted in order to provide feathers or fur for ritual clothing, or an animal believed to possess special powers or value might be subject to sacrificial killing, as guinea pigs and llamas are in some Andean cultures. And sometimes animals play an important religious role, for example as the objects of ritual sacrifice, even though they are not seen as carrying special spiritual power. (It is important to note, however, that animals used in religious sacrifices, such as chickens killed in rituals of the Afro-Caribbean religion Santería, are not necessarily subject to any worse treatment—and many times have more space and freedom to move—than livestock bred for industrial agriculture.)

A blessing of the animals taking place in New York City for Saint Francis Day in 2002.
© *SETBOUN / Corbis.*

The examples and discussion above underline, first of all, that although animals are religiously significant in many different cultures, the perceptions of and value accorded to different species, or even to nonhuman animals in general, vary greatly. Not all cultures draw a single line between humans and nonhumans. Many, probably most, have a series of distinctions among nonhumans and between humans and different nonhuman species. This is clear, for example, in the Huaorani understanding of woolly monkeys as similar to humans in ways that other animals, even other primates, are not. Even modern Western societies and religions, which often seem to draw the sharpest line between the human and nonhuman realms, actually understand different species in very different ways. For example, some are acceptable as food, others are not—because they are sacred, unclean, economically significant, or emotionally tied to humans.

Finally, understandings of nonhuman animals are closely related to the ways religions and cultures define human nature. In other words, the meaning and value of human life are understood in relation to nonhuman creatures. Many traditions define human nature in part through what we have in common with other animals. Thus, some Native North American cultures have linked human traits to the intelligence of ravens, the social cohesion of wolves, or the trickiness of coyotes. And many traditions define human nature by what distinguishes us from nonhumans. For example, most Christian theologies assert that "we" have an eternal soul and "they" do not.

It is important, in this discussion, to distinguish between difference and value. Making a distinction (between human and nonhuman or between two particular species) does not necessarily entail making a value judgment. In other words, we might attribute a certain characteristic to humans that is lacking in all other species, but it is not inevitable, on that basis, that we assign humans a higher value than all other species. This theoretical possibility may not have been realized with great frequency, although there are some historical examples, especially from indigenous cultures. Especially in the modern West, both religious and secular philosophies have tended to link difference and value quite strongly: What makes us distinctive, be it a soul or rational language, is also understood as something that makes us qualitatively better than all the other creatures who lack this trait. By shedding light on the distinctive capacities and characteristics of various species, the study of animal behavior can contribute, perhaps, to a broader vision of what makes a species valuable and how humans are related to, as well as different from, countless nonhuman species.

Further Resources

Badiner, A. H. 1990. *Dharma Gaia: A Harvest of Essays in Buddhism and Ecology.* Berkeley: Parallax Press.

Rival, L. 1996. *Blowpipes and spears: The social significance of Huaorani technological choices.* In *Nature and Society: Anthropological Perspectives* (Ed. by P. Descola & G. Palsson). London and New York: Routledge.

Santmire, P. 1985. *The Travail of Nature: The Ambiguous Promise of Christian Theology.* Philadelphia: Fortress Press.

Sorrell, R. D. 1988. *St Francis of Assisi and Nature: Tradition and Innovation in Western Christian Attitudes toward the Environment.* New York: Oxford University Press.

White, L. Jr. 1967. *The Historical Roots of Our Ecologic Crisis.* Science, 155, 1203–1207.

Anna L. Peterson

■ Reproductive Behavior
Adaptations and Exaptations in the Study of Behavior

In the study of evolution, *adaptation* (the process of evolutionary change resulting from natural selection and improving the efficiency with which organisms survive and reproduce in a given environment) was certainly the process that claimed most interest and intellectual commitment of biologists, which is easy to understand.

After the revolutionary work of Charles Darwin, biologists became aware that natural selection was a process capable of producing highly integrated sets of traits astonishingly well suited to the needs of an organism, without any purposefulness or preexisting plan. It is not a bit surprising that these processes that we call adaptations attracted the attention of innumerable biologists. Finding new adaptations became a challenging task to all those interested in evolution. Trying to find adaptive meaning in traits that at first sight seemed to contradict the evolutionary theory (like "altruistic behavior"), is doubtless an exiting adventure to the intellect.

However, not everything in living organisms is adaptive. To imagine that everything is an adaptation would be to transform one of the most fundamental concepts of biology—natural selection—into a caricature because then we would have to ascribe to selection all the attributes of a supernatural entity, with unlimited intelligence and power. Evolution is an historical process. Each organism presents traits that evolved in very .different times along the history of its lineage, often when its ancestors were quite different and lived in environments that differed much from the present one.

Some traits persist apparently because natural selection was unable to eliminate them, although it seems very likely that it would be advantageous to the organism to get rid of them. There is a lineage of European newts to which marbled and crested newts belong. All the species of this lineage have a genetic make-up that causes 50% of the embryos to die at early stages of development. Although for other species, biologists invested much thought and experimental work to test the hypothesis that minute details of the biology of the organism tend to maximize their fitness, this is pointless with this 50% loss of embryos in newts. Some structural arrangements of chromosomes suppress recombination in a segment of the chromosome, so that harmful and beneficial genes are inherited as a block, and selection has little chance to separate favorable and unfavorable genes and the traits they cause. This seems to have been the case in the evolution of newt species. This example illustrates one of the possible limitations (and there are many others) of natural selection, that will cause many disadvantageous genes and the traits that are their phenotypic expressions to persist in evolution.

Natural selection, as François Jacob and Stephen J. Gould argued so eloquently, does not create new complex traits in a vacuum. Faced with new environmental changes, populations will respond by increasing the frequency of the phenotypes that best fit the new needs of the organism. Selection will benefit phenotypes that were already present as part of the natural variation of the population. Thus, every adaptation will also bear the marks of the species' past history, and often the raw material with which natural selection operates in a particular environment has a much older origin.

The term *exaptation* was coined to express this historical characteristic of the evolutionary process. We use the term exaptation to characterize traits that evolved for a function different from the one they serve in the present. To identify exaptations, it is of course necessary to have sufficient information on the phylogenetic history of the studied species and of the sequence of structural, physiological and behavioral changes that occurred in the history of that species.

Most of us remember the painful experience caused by the sting of a bee or wasp. Living wasps, bees, and ants are members of a group of insects called the Hymenoptera that possess a specialized ovipositor which helps the females to lay their eggs in favorable conditions. There is a fascinating variety of behaviors that explore the ovipositor in the evolution of different lineages of Hymenoptera. Some of these insects insert their eggs into the new buds of trees (like oaks) forcing the bud to develop into a tumor that affords nourishment and protection to the larvae. Others insert the eggs into living insect larvae, injecting secretions that paralyze them, although leaving them alive. After they hatch, the hymenopteran larvae will devour the paralyzed host. Still others use the ovipositor and associated secretions to paralyze insect larvae that they carry to the nest where they lay their own eggs. They form thus a reserve of paralyzed, but living food, that will be ready to be eaten by their young. The same ovipositor is used by sterile workers of many bees, ants, and wasps to sting other animals as a defense mechanism, already independent of its primitive usage in reproduction. The use of the sting in defense is clearly an exaptation originated from a structure and behavior primitively evolved for a different function.

Another obvious example is the frontal position of the eyes in humans. The structure of our face and the location of the eyes are of paramount importance in human communication. Everyone knows this by his/her own experience, but it would be wrong to say that the frontalization of the eyes was an adaptation to our social needs. Frontal eyes are very widespread in primates regardless of their social organization. They probably evolved in an ancestral primate for which having good binocular vision improved the perception of depth and computation of distance of objects. This capability is likely to have evolved as an adaptation to the needs of arboreal animals that had to make precise movements and find inconspicuous food on trees, and must have subsequently been explored by evolution in the context of communication.

The old concept of ritualization, so important in the history of ethology, expressed, several decades in advance, the notion of exaptation applied to the study of behavioral evolution. The term ritualization refers to the evolution of signals used in communication from preexisting behaviors that served different functions. To scratch the ground with the feet is apparently a threat signal in some mammals like bulls. It probably originated as a by-product of high tension in an animal in a conflict on whether to attack or avoid an opponent. During courtship, female chaffinches may display the soliciting behavior used by chickens to beg food from their parents. Apparently, a behavior that originally evolved in young to elicit parental care was subsequently incorporated in the behavior of females in a sexual context.

Fish of many different families independently evolved lineages that successfully colonized rocky habitats between tide marks. Gobies, blennies, and clingfish provide many examples. The true residents in these habitats are species that, after settlement (from the plankton at the end of their larval life), grow and breed in rock intertidal habitats. Almost all these fish show parental care and protect the eggs until hatching. Parental care is, however, not a true adaptation to life in the intertidal zone. Each family possessing resident intertidal fish contains many more species in other habitats (coral reefs, estuaries, etc.) that exhibit similar forms of parental care. It is unlikely that the ancestors of all these fish were intertidal, and parental care, being so widespread in these groups, must have evolved before some of the members of each family colonized the intertidal zone. Why did only groups showing parental care succeed in becoming residents in the rocky intertidal habitat? In marine fish, parental care is especially common in small species with reduced mobility, like those that hover and hide in the crevices and interstices of coral reefs. Rock intertidal fish are typically small (often less than 10 cm; 4 in) and show reduced swimming (possibly a useful trait in an environment where wave action is a major physical factor). The prevalence of parental care in the rock intertidal may thus be linked to the life history of these fish, because being small sized and poor swimmers, they derive the same advantage from parental care as their relatives do in other environments. Egg protection, however, may be especially important in this environment. Many fish spawn in open water, and their eggs drift in the plankton for days or weeks. For fish that live under strong wave action, swimming to the water column to spawn may be quite risky. The eggs would also face the danger of being smashed against rocks, captured by the many predatory invertebrates (like sea anemones), or being swept ashore. Regardless of the current utility of parental care for survival in intertidal conditions and its importance in the colonization of this habitat, it certainly is an exaptation that evolved in ancestors that lived in other environments.

The concept of exaptation is important for the study of evolution because it calls our attention to its historical nature and the way selection tinkers with available variation to build new adaptations from the scraps left by past history. Exaptations clearly show that we have to distinguish carefully between the selective pressures, if any, which operated at the origin of a character, and those that maintain it in present populations.

See also Behavioral Phylogeny—*The Evolutionary Origins*
    of Behavior
      Communication—Vocal—*Singing Birds: From*
        Communication to Speciation
      Levels of Analysis in Animal Behavior
      Sociobiology

Further Resources

Almada, V. C. & Santos, R. S. 1995. *Parental care in the rocky intertidal: A case study of adaptation and exaptation in Mediterranean and Atlantic blennies*. Reviews in Fish Biology and Fisheries, 5, 23–37.

Gould, S. J. & Lewontin, R. C. 1979. *The spandrels of San Marco and the Panglossian Paradigm: A critique of the Adaptationist Programme*. Proceedings of the Royal Society of London—Series B, 205, 581–598.

Gould, S. J. & Vrba, E. S. 1982. *Exaptation—A missing term in the science of form*. Paleobiology, 8(1), 5–15.

Jacob, F. 1994. *The Possible and the Actual*. Seattle: University of Washington Press.

Vitor Almada

■ Reproductive Behavior
Alternative Male Reproductive Tactics in Fish

We are used to thinking that each sex has a typical set of behavioral, morphological, and physiological traits that characterize its role in reproduction. In recent years, however, as studies of reproductive behavior became more and more detailed, we became increasingly aware that, for many animal species, there may be several different ways through which males and females contribute to the next generation. This is specially so in males, in which competition for access to the gametes of the opposite sex is usually more intense than in females.

Some males may be more attractive to females than others. Males may compete aggressively for access to females or for the control of nest sites, food resources, or shelter used by them. Both situations frequently lead to large variation among males in their breeding success. Some males are not very attractive; others are too small or too weak to be successful in aggressive encounters with superior competitors. Both scenarios may favor the evolution of Alternative Male Reproductive Tactics (AMRT). When we speak of tactics, we are not assuming that the animals are making purposeful and intentional decisions. We simply mean that, in many species, males that are not very attractive to females or are not able to compete successfully with rival males may display complexes of behavioral, morphological, and physiological traits that provide them with alternative routes to reproduction.

In fish, alternative reproductive tactics have been found in a very wide spectrum of families. Fish groups so diverse as minnows, salmons, pupfish, toadfish, blennies, gobies, cichlids, and wrasses include species with AMRT, but the list of known cases is much longer. In many of these groups, AMRTs must have originated independently in their evolution.

There are some basic characteristics of the biology of fish that make the evolution of AMRTs especially likely. These include the prevalence of external fertilization and the fact that most fish show indeterminate growth.

External fertilization creates many more opportunities for sperm competition than internal fertilization, because the sperm of many males ejaculating almost simultaneously may be present around the eggs at the time of spawning, making the control of fertilization by a male much more difficult. A large male salmon that is mating with a female may be unable to avoid the very quick approach of smaller males that swim rapidly towards the spawning pair, ejaculate near the eggs and escape before the large male has time to attack them. For the large male, leaving the female to chase other males may be costly. Many small males may take advantage of the occasions when the large males are driving others away to approach the spawning female and ejaculate.

Indeterminate growth means that an organism continues to grow after reaching sexual maturity. This type of life history tends to generate a large variation in the size of adults. This discrepancy in size will also favor the evolution of AMRTs, as smaller males would be at a disadvantage in species where the large dominant individuals aggressively control the access to females and/or nest sites.

Sneakers

The example of salmon illustrates what are called sneaker males: small-sized quick intruders that avoid confrontation with large males and attempt to fertilize some eggs when an opportunity arises. These sneakers differ from large aggressive males in traits other than behavior. In species where the large males have conspicuous ornaments (like crests on the head, striking color patterns, and so on) or modified structures to fight, sneakers typically lack these conspicuous masculine traits and are often very difficult to distinguish from females or juveniles. They also differ in their physiology. Typically, they have testes that are proportionately much larger in relation to their body mass than those of large males. In summary, sneakers typically display an integrated complex of traits (behavioral, morphological, and physiological) which together allow them to achieve an efficient alternative way to reproduce. They are inconspicuous, reducing the risk of being detected as rivals by larger males. They invest strongly in their gonads producing a comparatively larger amount of sperm. They are effective at avoiding confrontations and detecting opportunities to move in and ejaculate when eggs become available for fertilization.

Female Mimicry

In some species the males that "steal" fertilizations go a step further. They display the behavior of receptive females and are sometimes courted by a territorial male, thus being able to approach the nest in the presence of its owner, with reduced risk of attack. This female mimicry will work if the visits of females to the nest are sufficiently frequent. In these circumstances the males that mimic their behavior may be able to fertilize some of the eggs. In the peacock blenny, a marine inshore fish from southern Europe, the biggest males control the available nests, and, when they have sufficient eggs, they no longer court females. The females, in turn, have to compete for the limited space available to deposit eggs, and they display specialized courtship patterns involving color changes and specific body movements. Small parasitic males approach the nest sites by mimicking this female behavior and actively court the nest owners.

Satellites

Another AMRT is displayed by "satellites." Frequently, satellites are of intermediate size between territorials and sneakers, and stay in the vicinity of the nest of a larger territorial male. They defend the area against other males, while at the same time attempting to fertilize eggs. The territory owner may be moderate in his aggression toward the satellite.

These situations are difficult to interpret functionally. If the satellite fertilizes some of the eggs obtained by the territory owner, it will be imposing a cost on the territorial male. However, it may happen that the action of the satellite, by helping to patrol the area and drive away other intruders, may be sufficiently beneficial to compensate for the fertilizations he "steals." Thus, this phenomenon is a challenging topic for evolutionary biologists.

Indeed, minute changes in the details of the situation may transform a scenario of parasitism into one of cooperation.

Helpers

Another situation where the boundary between conflict and cooperation may be quite tenuous is when helpers associate with a nest. In substratum spawning cichlids, one female (or several in some species) associates with a male to defend a territory and care for the eggs and the young until the juveniles are fully formed. In some species, the juveniles stay in the nest for such a long period that they reach sexual maturity while still there. They help to defend the nest site, thus affording added protection to younger broods. At the same time, however, the younger females may spawn their own eggs and the younger males may fertilize some of the eggs of the adult females. What is the net cost/benefit balance of the presence of these helpers? To complicate matters, we must remember that they are relatives of the larger fish, raising the possibility of kin selection. In addition, if the nest sites are scarce, they may benefit by "inheriting" the territories of the large fish when they die or are no longer able to defend them.

Territory Takeovers

Finally, there are large males, of a size similar to territory owners, that avoid the costs of fighting for a territory and/or building a nest. They move around the spawning grounds, and when they find a suitable nest site they mount a full-scale attack on the owner and sometimes succeed in taking over the territory. Afterwards they may behave as territorials, courting females and caring for the eggs.

These "pirates" may be fish that were less successful in establishing territories at the beginning of the spawning season when males were at peak physical condition. They may have continued to grow while others were already breeding. They may test the strengths of several males until they find some less capable opponents that have been weakened by continued reproduction.

Females of several fish species seem to prefer to spawn in nests that already contain eggs. In these circumstances, it may even be beneficial to males that took over territories to refrain from eating all the eggs fathered by the previous territory owner, and engage in aloparental care as a means to attract new females to spawn. This is another example of how difficult it may be to differentiate between conflict and cooperation.

In many species, there may be more than two alternative tactics in the same population. Territorial males, satellites, and sneakers, sometimes performing female mimicry, may be all attempting to get eggs fertilized in their own ways. Sometimes, a territorial male may even leave his nest to intrude into the territory of a neighbor where spawning is taking place.

There are species (like Pacific salmon and bluegill sunfish) in which each male may be committed to a single alternative mode of reproduction for life (Pacific salmon typically spawn once and die). In many others, there is an ontogenetic shift from sneakers or helpers to satellites or territorials as the fish grow larger and stronger. In some cases, the fish keeps a plastic set of behavioral capabilities, adopting different alternatives depending on his social position in a given breeding event.

Regardless the degree of plasticity of male behavior, their alternative reproductive tactics may vary greatly in the role of genes and environmental factors underlying them.

With the advent of molecular techniques based in the study of very small amounts of DNA, it is now possible to fingerprint large numbers of individuals, for instance, an entire sunfish breeding colony. These new tools are beginning to open fascinating avenues for research. We may be able to determine who are the father and mother of whom and what is the relative success of males showing alternative reproductive tactics. These molecular techniques will help us to answer one major unsolved question: Are parasitic alternative reproductive tactics, like sneaking "inferior" options, adopted by males that are poor competitors and must make "the best of a bad job," or are they alternatives with equivalent success? Up to now, these studies have already demonstrated that sneakers or satellites actually fertilize some of the eggs in the territories of larger males, in various proportions, depending on species and populations. As more data accumulate, we will be able to get a broader base from which to access the functional importance of different AMRTs in distinct fish groups and ecological contexts.

See also Behavioral Plasticity

        Caregiving—*Brood Parasitism in Freshwater Fish*

Further Resources

Colbourne, J. K., Neff, B. D., Wright, J. M. & Gross, M. R. 1996. *DNA fingerprinting of bluegill sunfish* (Lepomis macrochirus) *using (GT)n microsatellites and its potential for assessment of mating success.* Canadian Journal of Fisheries and Aquatic Science, 53, 342–349.

Dominey, W. J. 1981. *Maintenance of female mimicry as a reproductive strategy in bluegill sunfish* (Lepomis macrochirus). Environmental Biology of Fish, 6, 59–64.

Gonçalves, E. J., Almada, V. C., Oliveira, R. F. & Santos, A. J. 1996. *Female mimicry as a mating tactic in males of the Blenniid fish* Salaria pavo. Journal of the Marine Biological Association of the UK, 76, 529–538.

Gross, M. R. 1984. *Sunfish, salmon, and the evolution of alternative reproductive strategies and tactics in fishes.* In: *Fish Reproduction: Strategies and Tactics* (Ed. by G. Potts & R. J. Wootton), pp. 55–75. London: Academic Press.

Jones, A. G., Östlund-Nilsson, S. & Avise, J. C. 1998. *A microsatellite assessment of sneaked fertilizations and egg thievery in the fifteenspine stickleback.* Evolution, 52, 848–858.

Kodric-Brown, A. 1986. *Satellites and sneakers: Opportunistic male breeding tactics in pupfish* (Cyprinodon pecosensis). Behavioural Ecology and Sociobiology, 19, 425–432.

Santos, R. S. 1995. *Allopaternal Care in the Redlip Blenny.* Journal of Fish Biology, 47, 350–353.

Taborsky, M. 1994. *Sneakers, satellites, and helpers: Parasitic and cooperative behavior in fish reproduction.* Advances in the Study of Behavior, 23, 1–100.

Taborsky, M. 1999. *Conflict or cooperation: What determines optimal solutions to competition in fish reproduction?* In: *Behaviour and Conservation of Littoral Fishes* (Ed. by R. Oliveira, V. Almada & E. Gonçalves), pp. 301–349. Lisbon: ISPA.

Taborsky, M. 2001. *The evolution of bourgeois, parasitic, and cooperative reproductive behaviors in fishes.* The American Genetic Association, 92, 100–110.

Vitor Almada

■ Reproductive Behavior
Assortative Mating

Mating is rarely, if ever, random in natural populations. *Assortative mating*, or *homogamy*, occurs when individuals of similar *phenotypes* (the characteristics of an organism as they result from the joint influence of the organism's genotype and its environment) mate with a greater frequency than predicted by random pairings. Assortative mating is well documented in both invertebrates and vertebrates, including human beings, and is probably the most widely described mating pattern. It can occur with respect to one or several traits with continuous (such as age, body size, weight), or discrete (such as color morph, parasitic prevalence, or social status) variation. The study of assortative mating is of particular interest since it may have profound consequences on the genetic structure of populations and, eventually, lead to or reinforce *sympatric speciation* (i.e., the formation of two distinct species starting with the division of a single population into two distinct and reproductively isolated populations). It is important to note that assortative mating is somewhat ambiguous because it refers to both a pattern of mating and the process of mate acquisition underlying the observed pattern. Because different processes can lead to the same pattern of assortative pairing, it is most often difficult to infer the process from knowledge of the pattern.

Patterns of Assortative Mating

Let's consider assortative pairing for a continuous character such as age or size. In a large range of species, individuals tend to mate with conspecifics of similar size or age. The easiest way to visualize assortative mating is simply to plot the size or the age of females against that of males for all pairs that can be sampled in a population. However, significant correlations can be obtained even if the pattern of mating is not truly size-assortative. Graph "a" in the first figure describes a mating pattern in which small males tend to pair with females of all sizes, whereas large males pair only with large females. The reverse is shown in graph "b." Only graph "c" corresponds to "true assortative mating," although a significant correlation between the sizes of males and females would be computed for each graph. Graphs "a" and "b" correspond only to "apparent assortative mating." A simple inspection of the distribution of points on a two-dimensional graph is therefore a prerequisite before concluding that size-assortative mating does occur in a population. In the case of a character with discrete variation, the pattern of pairing can be assessed from the analysis of a contingency table (see table on p. 878). Statistical tests are available to assess precisely to which extent a pairing pattern conforms to true homogamy.

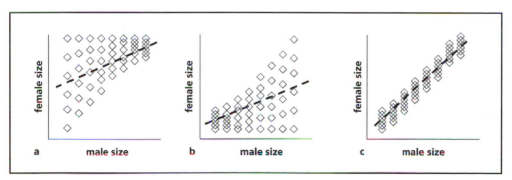

Patterns of size-assortative mating when size is measured as a continuous variable.
a and b: apparent assortative mating, c: true assortative mating (n = 48 in each case).
Courtesy of Frank Cézilly.

		female size								
		Small	Medium	Large	Small	Medium	Large	Small	Medium	Large
male size	Small	10	10	10	30	0	0	30	0	0
	Medium	0	15	15	15	15	0	0	30	0
	Large	0	0	30	10	10	10	0	0	30
		a			b			c		

Patterns of size-assortative mating when size is measured as a discrete variable, with three levels: small, medium, large. a and b: apparent assortative mating, c: true assortative mating (n = 60 pairs in each case).
Courtesy of Frank Cézilly.

Underlying Processes

Assortative mating can occur passively, for instance through temporal or spatial segregation of different phenotypes during the process of pair formation. In long-lived seabirds, such as common terns, *Sterna hirundo*, or kittiwakes, *Rissa tridactyla*, for instance, age-assortative mating can occur passively if individuals start breeding at a similar age and show both high mate fidelity and high adult survival. If individuals rarely change mates, the initial similarity in age between pair members is preserved over time, and a strong correlation between ages is observed. On the other hand, spatial segregation between phenotypes may result from habitat heterogeneity. In the aquatic crustacean amphipod, *Gammarus pulex*, pairs form several hours or days before insemination, with the male carrying the female underneath his body and performing swimming movements. Pairs in so-called "precopula" show a strong homogamy for size that results primarily from male–male competition. However, size-assortative pairing can be reinforced by the spatial segregation of individuals of different sizes that tend to distribute themselves in the river stream according to the susbtrate. Large individuals tend to be found in the vegetation or in areas of coarse gravel, whereas the smallest individuals occur predominantly in fine gravel. Pooling samples taken from different microhabitats significantly increases the correlation between the sizes of males and females in precopula, indicating a potential role for spatial heterogeneity in the observed patterns of size-assortative pairing in *G. pulex*. In territorial birds, such as raptors, territory quality varies largely between pairs. Age of mates is often correlated with territory quality, with territories of lower quality being occupied by immature birds. Long-term data from a Spanish population of imperial eagles, *Aquila adalberti*, indicate that variation in territory quality provides the most parsimonious explanation for the observed pattern of age-assortative mating in that species.

Assortative pairing does not necessarily involve mutual preferences in males and females. In *G. pulex*, the fecundity of females increases exponentially with body size. This means that a small difference in female size may yield susbtantial benefits in terms of eggs that are fertilized by a male. Accordingly, males prefer to mate with large females and compete among themselves to gain access to them. However, the energetic cost of forming a precopula pair increases with female size. It has been shown that larger males are better able to sustain the cost of precopula with large females and are also able to take over smaller males that engaged in precopula with large females. Female behavior seems to play little role, if any, in the formation of precopula pairs. Because of their competitive advantage, large males tend to monopolize large females, and small males are left to pair with small females. In this case, size-assortative mating results essentially from size-related variation in female fecundity combined with a competitive advantage for large males. The same combination of phenomena seems to underlie size-assortative mating in a majority of invertebrate species.

"Apparent" assortative mating can occur when the degree of mate selectivity in one sex varies with phenotypic variation in that sex. Size-assortative mating in red-sided garter snakes, *Thamnophis sirtalis parietalis*, provides an excellent illustration of this phenomenon. Probably because they are limited in their ability to court and inseminate several females, males are highly selective with respect to mate choice. They tend to prefer large females over smaller ones, but this degree of preference shifts with male body size. Smaller (younger) males court both small and large females, whereas large males restrict their courtship and mating essentially to large females. Actually, males rely on both visual and pheromonal cues to assess female body size, and large and small males differ in their responses to these cues. Large males require both visual and pheromonal evidence of large female body size, whereas smaller males tend to react to visual or pheromonal cues from females of all sizes. Such a variation in male preference for female size results in the pattern of assortative mating shown in graph "a." Although smaller male garter snakes still court stimuli that do not elicit reaction from larger males in the absence of larger rivals, it is possible that size-based divergence in mate-recognition systems has initially evolved to reduce intrasexual competition between small and large males.

Finally, active assortative mating for a trait can occur through a mutual preference linked to that trait or to some another attribute closely associated with the trait. Active choice underlying assortative mating can correspond to either homotypic preference (i.e., a mutual preference for mates of similar phenotypes), or for a mutual directional preference for individuals expressing a more extreme phenotype. In the latter case, individuals of extreme phenotypes (either the largest individuals or the smallest ones in the case of size-assortative pairing, either the oldest individuals or the youngest ones in the case of age-assortative pairing), should mate first, leaving individuals of less extreme phenotypes to mate among themselves. Although both processes lead to patterns of assortative pairing, they may be separated on the basis of the dynamics of pair formation. In the case of homotypic preference, the timing of pair formation should not be influenced by the phenotypes of individuals, whereas in the case of directional preference, it is predicted that pairings between individuals of the most valued phenotype should occur early in the breeding season. Age-assortative mating in birds provides examples of both homotypic and directional preferences. Homotypic preference has been evidenced in the barnacle goose, *Branta leucopsis*. In this species with long-time partnership, divorce occurs rarely (about 2% of pairs annually), and most mate changes are due to the death of one partner. Divorce seems to be accidental, but on some occasions pair members can switch mates if better alternatives are available. Forming a new pair, however, takes time and energy. Divorced or widowed birds may spend one or more breeding seasons without finding a mate. Birds that mate with an individual much older than themselves have a risk of becoming single following the death of their partner. Choosing a mate of similar age then seems to be a near-optimal strategy to avoid the risk of becoming a widow. The situation differs markedly in the greater flamingo, *Phoenicopterus ruber roseus*. The species shows absolutely no mate fidelity, and new pairs form in each breeding season. Still, as observed in the Camargue (southern France) population, pairing is strongly size-assortative. The study of displays and pair formation revealed that, as expected in the case of directional preference, older flamingos stopped displaying and formed pairs earlier in the breeding season than their younger conspecifics. Additional evidence for the existence of a directional preference for older individuals in birds comes from studies of captive feral pigeons, *Columbia livia*, and black-billed magpies, *Pica pica*. Why individuals of both sexes should prefer older mates is not totally clear. Presumably, older individuals have more experience and have proven their ability to resist diseases and predators.

Disassortative Mating

Although less frequent, *disassortative mating* (or negative assortative mating) occurs when individuals of dissimilar phenotypes mate more often between them than expected by chance. Disassortative mating has been reported, for instance, in some insects for body size and in some bird species for color morph (i.e., pattern of color). In the white-throated sparrow, *Zonotrichia albicollis*, two morphs coexist, and mating is dissassortative by morph. For example, females of both morphs tend to prefer tan-striped individuals over white-striped ones when allowed to interact with males, whereas in the same conditions males of both morphs show no preference for females of either morph. During within-sex competition trials, white-striped individuals were dominant over tan-striped individuals of the same sex. Thus, disassortative mating in white-throated sparrows seems to result from white-striped females outcompeting tan-striped ones for access to the preferred tan-striped males, rather than from mutual heterotypic preference (i.e., preference for a mate of dissimilar phenotype).

Mate choice for genetic dissimilarity can lead to disassortative mating. The genetic benefits of mating are highly dependent upon the compatibility of the sperm and egg genotypes. Genetic incompatibility results in developmental problems and poor viability of offspring. More specifically, survival and vigor of offspring might depend on its degree of *heterozigosity* (the presence of two different alleles at a genetic locus) at the *major histocompatibility* (MHC) (i.e., tissue compatibility) complex. MHC genes play a central role in immunocompetence, that is, the ability of organisms to resist diseases and infections. MHC genes come in many alleles, with no single allele predominating. Therefore, different individuals tend to have different MHC genotypes. Offspring that are heterozygotes for MHC genes may have higher immunocompetence compared to that of homozygotes, or benefit from better viability. Individuals should then be selected to prefer reproductive partners whose MHC genotype best complements their own. There is some evidence that MHC genotypes provide a stimulus for sexual imprinting that results in disassortative mating. Several independent studies have found that house mice, *Mus musculus domesticus*, prefer mates dissimilar, rather than similar, to them at MHC loci and that such MHC-disassortative mating preference is mediated by olfaction. Cross-fostering experiments have indeed revealed that mice impregnate on the MHC identity of their family early in life and later use it as a referent to which they compare the MHC type of potential mates. Presumably, mice benefit from mating disassortatively with respect to MHC type through confering better immunocompetence to their offspring. Some evidence exists that mate selection in human beings is partly mediated by odor preference, possibly in association with MHC types.

See also Reproductive Behavior—*Mate Choices*

Further Resources

Arnquist, G., Rowe, L., Krupa, J. & Sih, A. 1996. *Assortative mating by size: A meta-analysis of mating patterns in water striders*. Evolutionary Ecology, 10, 265–284.

Bollache, L., Gambade, G. & Cézilly, F. 2000. *The influence of micro-habitat segregation on size-assortative pairing in* Gammarus pulex (L.) (Crustacea, Amphipoda). Archives für Hydrobiologie, 147, 547–558.

Burley, N. 1983. *The meaning of assortative mating*. Ethology and Sociobiology, 4, 191–203.

Cézilly, F., Boy, V., Tourenq, C. & Johnson, A. R. 1997. *Age-assortative pairing in the Greater Flamingo*, Phoenicopterus ruber roseus. Ibis, 139, 331–336.

Crespi, B. J. 1989. *Causes of assortative mating in arthropods.* Animal Behaviour, 38, 980–1000.

Ferrer, M. & Penteriani, V. 2003. *A process of pair formation leading to assortative mating: Passive age-assortative mating by habitat heterogeneity.* Animal Behaviour, 66, 137–143.

Houtman, A. M. & Falls, J. B. 1994. *Negative assortative mating in the white-throated sparrow, Zonotrichia albicollis: The role of mate chocie and intra-sexual competition.* Animal Behaviour, 48, 377–383.

Penn, D. & Potts, W. 1999. *The evolution of mating preferences and major histocompatibility complex genes.* American Naturalist, 153, 145–164.

Plaistow, S. J., Bollache, L. & Cézilly, F. 2003. *Energetically costly precopulatory mate guarding in the amphipod.* Gammarus pulex: *Causes and consequences.* Animal Behaviour, 65, 683–691.

Shine, R., Phillips, B., Waye, LeMaster, M. & Mason, R. T. 2003. *The lexicon of love: What cues cause size-assortative courtship by male garter snakes?* Behavioral Ecology and Sociobiology, 53, 234–237.

Frank Cézilly

■ Reproductive Behavior
Bowerbirds and Sexual Displays

Naturalists have long marveled at the bowerbirds' intricately decorated display courts and bowers. The stick bowers of the Vogelkop bowerbirds are hut-like structures built around moss-covered saplings that are commonly 1.5 m (5 ft) tall and 2 m (6.5 ft) in diameter and decorated with large piles of different, bright-colored decorations. They are found only in Australia and New Guinea. The eighteenth century explorer Beccari thought that these were human-built huts, and the artfulness of these displays has led some to claim that bowerbirds are the most mentally advanced birds. Recently, detailed behavioral and genetic studies have replaced speculative claims about these extraordinary creatures with no less interesting details of how their unique displays are central to a complex and highly evolved process of mate selection.

There are 20 species in the family Ptilonorhynchidae among which 17 clear a display court and 15 of these build bowers where courtship and copulation occurs. The bower, a stick structure, is built on or adjacent to the display court. Bowers and the decorations used on the court are unique to each of these species. Male court-building bowerbirds provide females with sperm only and are not involved in parental care. Bowerbirds are long-lived, with males showing a prolonged juvenile period (up to 6 years) during which young males engage in same-sex practice courtships.

A male satin bowerbird (Ptilinorhynchus violaceus) *displays to a female at a bower at an eastern Australian rainforest.*
© *Michael and Patricia Fogden.*

Male court and copulate with females at the bower. The opportunity to use cameras to monitor key behaviors at individual bowers, to individually mark a large proportion of males and females in our study population, together with the ability to manipulate display elements such as bower decorations, makes bowerbirds an ideal system for studying sexual selection. In satin bowerbirds several features of male display have been found to be important in affecting male attractiveness: bower decorations, bower quality, courtship intensity, parasite infection, and the quality of vocal display. Males steal decorations and destroy the bowers of their neighbors, both reducing the quality of their competitor's displays and providing a major source of key and otherwise rare decorations. Females search the bowers of multiple males before nest building and then return to visit a smaller set of males in later rounds of mate choice, typically mating with a single male. Males show a high skew in mating success, with some mating with as many as 25 females in a single year. Females mated to successful males the previous year visit bowers of fewer males and tend to remate with their mate from the previous year. Females who previously mated with lower quality males visit more males and mate with a higher quality males than their mate from the previous year.

In satin and spotted bowerbirds, females prefer intense male displays, but these displays can also threaten females. Male displays are aggressive, and males in these species show adaptations to reduce the threat from display. Some populations of spotted bowerbirds have especially intense displays and have altered how they display: Instead of courting the female through the bower entrance, the bower wall is modified so that the male courts through the see-through wall which acts as a protective barrier that reduces threat to females. In satin bowerbirds, females signal their level of comfort for high intensity displays by their speed of crouching, and male ability to read and react to these signals is important for their mating success. These high intensity displays and the ability of females to signal comfort appear to differ with female age. Young females choose males primarily with regard to blue decorations on bowers, but older females use blue decorations early in the mate choice process and then focus on behavioral displays later in courtship sequences. This polymorphism in female preferences may explain multifaceted male displays; males need to have different display types to appeal to different classes of females.

Males tend to repeat as bower holders over multiple years and, unlike other species with extreme skews in mating success, male mating success is correlated across years. This leads to a very high lifetime variability in male mating success and suggests that male displays are not extremely costly. This finding is consistent with the observation that male displays are learned over a long juvenile period, and thus some critical honest indicators of male quality are inexpensively renewable across years. Long-term studies of parasite infection show that males who had low infection as juveniles tend to become bower holders. This result suggests that effects of parasites on the ability to produce sexual displays, as considered in the bright male hypothesis, must be considered across individual lifetimes and not just for its effects on adults. Lastly, the complexity of male displays, the use of different components of male displays by different age classes of females, and the complex signaling between sexes suggest a very dynamic and highly adapted mate-choice process. This is in contrast to the suggestion from sensory exploitation and runaway models that male displays are the unselected by-products of other sometimes arbitrary processes.

An important issue in sexual selection studies is how new display traits originate. Several key male display traits in bowerbirds appear to be the product of cooption. This is most clearly demonstrated in spotted bowerbirds in which males of the genus *Chlamydera* have incorporated aggressive display as part of their courtship calls. In this group, courtship and aggressive calls are nearly identical and closely match the aggressive calls used by males in other species across this family. Other likely instances of cooption include the modification in bower walls to serve as a screen and reduce the threat from high intensity male courtship displays in spotted bowerbirds. In one of the two main clades (a group of species descended from common ancestors) of bower-building bowerbirds, males hold decorations in their beak for threat reduction during courtship. With the development of the aggressive displays in spotted bowerbirds, males violently throw decorations rather than holding them, suggesting that their function has changed from threat reduction to an indicator of display intensity. In satin bowerbirds males copy the calls of at least five different other species for use in their courtship display, and the number of these calls used predicts male mating success. The bower appears evolved as a device that protects females from forced copulations from the courting males, and it has taken on a secondary function such that differences in bower quality are used by females to assess male quality. These multiple instances of trait borrowing/cooption suggest that it is an important mechanism affecting the development of sexually selected traits. Cooption of already existing display elements may help explain the rapid divergence in display seen among even closely-related bowerbird species.

See also Reproductive Behavior—*Female–Female Sexual
Selection*
Reproductive Behavior—*Female Multiple Mating*
Reproductive Behavior—*Mate Choice*
Reproductive Behavior—*Sex, Gender, and Sex Roles*
Reproductive Behavior—*Sexual Selection*

Further Resources

Borgia, G. 1995. *Why do bowerbirds build bowers?* American Scientist, 83, 542–547.
Borgia, G. & Coleman, S. 2000. *Co-option of male courtship signals from aggressive display in bowerbirds.* Proceedings of the Royal Society of London Series B, 267, 869–874.
Borgia, G. & Presgraves, D. 1998. *Coevolution of elaborated male display traits in the spotted bowerbird: An experimental test of the threat reduction hypothesis.* Animal Behaviour, 56, 1121–1128.
Gilliard, E. T. 1969. *Birds of Paradise and Bowerbirds.* London: Weidenfield and Nicholson.
Humphries, N. & Ruxton, G. D. 1999. *Bower-building: coevolution of display traits in response to the costs of female choice?* Ecology Letters, 2, 404–413.
Kusmierski, R., Borgia, G. Uy, A. & Crozier, R. 1997. *Molecular information on bowerbird phylogeny and the evolution of exaggerated male characters.* Proceedings of the Royal Society of London, 264, 307–313.
Patricelli, G. L., Uy, J. A. C., Walsh, G. & Borgia, G. 2002. *Sexual selection: Males adjust displays in response to female signals.* Nature, 415, 279–280.
Uy, J. A. C., Patricelli, G. & Borgia, G. 2001. *Loss of preferred mates forces female satin bowerbirds (Ptilonorhynchus violaceus) to increase mate searching.* Proceedings of the Royal Society of London B., 268, 633–638.

Gerald Borgia

■ Reproductive Behavior
Captive Breeding

Why bring animals into captivity? What roles should zoos play and how can animal confinement be justified in an age of animal ethics? No longer just menageries, the modern zoo pursues three goals: education, conservation and research. Zoos occupy a unique niche in the conservation community, reaching many urbanites that might not otherwise embrace conservation. The challenge for zoos is to educate this populace and to recruit supporters of nature conservation in the wild. Zoos also have an opportunity—and responsibility—to better understand endangered species through scientific investigation. Scientists often cannot study many aspects of a species' biology in the wild because the animal is rare and elusive or because they cannot conduct a proper experiment. Zoos now play an increasing role in conservation science by taking their research programs simultaneously to both field and captive settings.

Another major impetus for captive breeding is to create a viable self-sustaining population that can serve as a genetic reservoir in case things take a turn for the worse in the wild. According to the International Union for the Conservation of Nature, captive breeding programs should be established *before* the wild population is so small that further removals will exacerbate its decline. Years of study are often required to establish a captive breeding program. Take the case of the Sumatran rhinoceros. The population was already highly endangered when animals were first removed to establish a captive safeguard against extinction. This attempt failed dismally because the species' behavior and biology remained too poorly understood to keep them alive in captivity, much less to get them to reproduce. The captive population has all but died out, and more removals from the wild cannot be risked. Clearly, conservation cannot rely on last-ditch efforts in captivity. Indeed, many question whether these efforts are worthwhile at all, given the enormous expense and high failure rate.

Without doubt, conservation should focus primarily on wild populations, but recent scientific advancements suggest a brighter future for captive breeding. Behavioral research by zoo biologists can claim a prominent role in this renewed optimism. California condors are once again sailing over California skies, golden-lion tamarins scramble along networks of branches in Brazilian tropical dry forest, and black-footed ferrets scamper over Wyoming prairies in part because of behavioral research. Critics once believed that the giant panda, popular icon of the worldwide conservation effort, was so reluctant to breed in captivity that a self-sustaining captive population would never be realized. Today, with much of the mystery erased by numerous research programs, an increasing number of pandas are mating and rearing their young in captivity, while the Chinese government struggles to preserve what remains of its natural environment.

These successes hold out promise for a small role for captive breeding in the larger conservation effort. But why is captive breeding so difficult? Why can't we just place two animals together and let nature take its course? The answer lies in the captive environment, which often is a poor imitation of nature. The challenge is to understand the biologically relevant dimensions for each species, including foraging strategies, social organization, mating systems, parental care systems, habitat needs, and much more—all this, and with compromised research tools. Zoo biologists often have few animals to work with, and the goals of experimental design often conflict with the goals of animal management. For example, consider the ideal experiment to evaluate the role of scent signals in mating. The unencumbered researcher would exchange social odors for 10 pairs, but deprive another 10 pairs of olfactory communication as an experimental control. Captive breeding programs for endangered species cannot afford to risk the aggression and reproductive failure that may accompany this experimental control. It is also likely that while behaviorists are studying one factor important to breeding, managers are also making changes to husbandry, veterinary care, and nutrition, making it difficult to tease out cause and effect if reproduction improves. Researchers in these environments must find a way over or around these obstacles to gain sound scientific understanding. Despite these constraints, behavioral research has significantly enhanced our understanding of many species' needs in captivity.

All animals have basic needs that go well beyond the simple needs for food and shelter. The "ethological needs" model of motivation proposes that animals are motivated not only to obtain important biological resources such as food and shelter, but they are also motivated to *perform* the behaviors that throughout the species' evolutionary history have been used to obtain these resources. Similarly, animals may have motivational needs to gather information or have some control over their environment. An animal given a bowl of processed food cannot search for, extract, handle and process food resources. Studies show that some animals prefer to work for food, a phenomenon amusingly referred to as "contrafreeloading." For example, European starlings will pass over a bowl of mealie worms to search for the same food hidden under plastic flaps, a task mimicking their natural foraging strategy in leaf litter.

Unlike their wild counterparts, captive animals often exercise little control over their environment. Too often they cannot do anything to change what, when, and how they are fed, nor can they remove themselves from aversive stimuli such as heat, light, human "predators," noise, and so forth. Behavior, which connects the animal to its environment, has become irrelevant, and with detrimental consequences. The animal may show signs of stress and may perform a variety of abnormal behaviors, especially "stereotypies." Stereotypic behaviors are highly repetitive behaviors, invariant in form, and have no apparent function or goal. Some common examples include pacing back and forth, tossing the head, self-biting, and exaggerated spinning when an animal turns a corner.

Students of animal behavior have played a major role in addressing these problems. Field studies suggest potential behavioral needs and guide experimentation with environmental enrichment. *Enrichment* is the behaviorally relevant modification of captive environments to optimize physical and psychological well-being. Simply enlarging pens is not enough. Complexity, challenge, and change must be incorporated into captive environments. Types of permanent additions to enclosures include substrate such as loose dirt to roll in; live vegetation for shade, hiding, and eating; water pools for play, bathing, and cooling off; and trees or climbing structures for play, exercise, and resting. Food can be scattered, hidden, placed in puzzle containers, or provided in a way that allows the animal to use natural feeding behaviors. Novel objects can be given to encourage play, exploration, and cognitive exercise. Species-appropriate social interaction is one of the best forms of enrichment. All of these enrichments give back some control to the animal, and allow the animal to choose among different options.

It is important to articulate the specific goals of enrichment and to collect data to determine whether enrichment is effective at reaching those goals. Here behaviorists can claim many successes. When leopard cats living next to lions were given hiding places, they explored more, paced less, and had reduced corticoid hormone levels, suggesting reduced stress. Giant pandas given opportunities to work for food and explore novel objects showed more diverse natural behaviors and fewer abnormal behaviors. Scattering food treats in straw, pebbles, and other substrates can keep animals occupied in search behavior for hours and reduce stereotypic behavior in species as varied as monkeys, bears, cats, and walruses. In other species more amenable to scientific investigation, such as the laboratory rat, enrichment is known to have more long-lasting effects. Animals reared in enriched environments have increased brain weight and neuronal synapses, enhanced learning abilities, and are better prepared to cope with change. These cognitive improvements serve the animal well when dealing with important changes, such as an introduction to a potential breeding partner or the birth of offspring. Endangered black-footed ferrets raised in enriched environments proved to be better predators and were more likely to survive when reintroduced from captivity to the wild. Enrichment also aids in reproduction by alleviating stress, which is known to reduce immune function and suppress reproductive hormones. It is clear that meeting these basic behavioral needs is a prerequisite to getting animals "to do what comes naturally."

The social environment is one of the most important aspects of captive breeding. The first lesson that zoos learned is that anthropomorphism often leads down the wrong path. Following the biblical Noah's ark format, zoos typically housed animals in monogamous pairs, though only about 3% of mammals are monogamous. Today's zoos look to nature and seek to emulate species-specific social arrangements. Animals with solitary tendencies are usually kept in separate enclosures, except during mating. More social species are held in groups. However, further study in the field often reveals flexibility in social systems, and animals may switch from one to another depending on prevailing circumstances, especially resource distribution. For example, hummingbirds will change from territoriality to a dominance organization when resources are highly clumped: When intruder pressure increases beyond the level that an individual can economically defend the resource patch, the territory holder has no choice but to tolerate other birds. Understanding the mechanisms of social flexibility has important implications for captive breeding because it provides managers with a framework for accommodating different social arrangements in limited space.

Attention to finer distinctions in social organization often pays off. Although wolves live in large social groups in nature, closer study revealed that the group is comprised of a breeding pair and its adult offspring. Thus, it is better to start off a captive population with a single breeding pair rather than a group of unrelated individuals, which may fight excessively. In colonial nesting, Humboldt's penguins, increasing group size from 5 to 10 pairs doubles chick production per pair. In nature, female cheetahs are solitary, yet many zoos housed them in pairs. In an elegantly controlled study, Nadja Wielebnowski and colleagues demonstrated that this social arrangement causes female cheetahs to shut down ovarian activity, leading to reproductive failure. In several highly social species, such as cotton-top tamarins, wolves, and naked mole-rats, parents suppress reproduction in adult offspring, which care for younger siblings while waiting for the parents' or a neighboring territory to become vacant, rather than risk the dangers of dispersal. Thus, removal of such offspring from the parents can increase reproduction.

Relatively asocial species typically pose the greatest challenge for captive breeding. These animals must overcome their tendencies toward aggression and avoidance just long enough to come together for mating. In nature, they have infinite space and elaborate communication behavior to negotiate this process. In captivity, members of these species too often respond to the opposite sex with indifference or escalated aggression. Male clouded leopards and black-footed ferrets may kill the female, the normally docile giant panda may inflict serious injury, and cheetahs may have nothing to do with each other. One goal of communication in solitary species is to attain an optimal level of familiarity with individuals of the opposite sex. Too much familiarity may "breed contempt." Keeping a pair of white rhinoceros together is a guaranteed recipe for reproductive failure, but give the females more space and force the male to distribute his attentions between many females, and females start producing calves. Using this arrangement, the San Diego Wild Animal Park has had more than 80 births in the last 30 years. Keeping kangaroo rats separately works well if the male and female are allowed to communicate and become familiar with each other. The notorious giant panda, so often reluctant to mate in captivity, becomes sexually aroused when given access to the odors of the opposite sex. Sexual motivation in the male is further enhanced by exposure to the odors of the female when she is in estrus. A few days of such olfactory exchange before placing the pair together in a mating introduction appears to reduce aggression and promote mating.

Sometimes the problem is simpler. Several species of prosimian copulate while suspended upside down, and artificial surfaces may be too slick to allow animals to hold on while mating! Adding a few natural branches may be all that is required. In some species, such as the black-footed ferrets, cheetahs, and gorillas, improper orientation while mounting is a primary cause of reproductive failure. In these species it may be important for the male to gain sexual experience with females. Alternatively, deficient sexual motivation may be the cause of improper mounting, requiring further study of motivational processes.

Captive breeding involves much more than just getting animals to mate. Equally important is getting them to mate with the right partner, because conservation is as much about genetics as it is about numbers of individuals. When individuals mate randomly in a population, maintenance of genetic diversity is maximized, but if only a small proportion of individuals breed, the *effective* population size can be one fifth of the actual population size. One implication is that if many females share preferences for the same males, the population will have to be much larger to maintain genetic diversity, which is essential for the population to adapt to environmental change such as rapidly evolving viruses and bacteria. To avoid these problems, managers pay close attention to breeding patterns, and pair only those individuals that are a good match from a genetic standpoint.

But what if female partner preferences run counter to the goals of genetic management? Managers must either fall back on artificial methods or encourage the female to choose the right male. When choosing males, females often use traits that are strongly affected by the male's physical condition, which indicate male quality. One function of such choosiness might be to obtain "good genes" to pass on to offspring. Examples include the colorful exaggerated tail of the peacock, comb size in jungle fowl, and antler size in deer. The challenge facing behaviorists, then, is to manipulate male traits that females use for mate choice. One option might be to lengthen the tail artificially. Or if color is dependent on certain nutrients, the male's diet could be altered to make his colors brighter. Some female mammals use olfactory cues related to male testosterone to select mates, so perhaps "spiking" the male's odors with the right chemicals will work. Studies with the threatened pygmy loris show that females reliably choose the male whose odors are most familiar, probably because frequent encounters with a male's scent indicates that he is a territorial male capable of excluding competitors. Thus, all managers need to do is transfer "scent postcards" from the right male to the female for a few weeks and the female's choice is made for her, albeit deviously.

For most zoo species, inadequate parental care means almost certain death of offspring. In the majority of mammal species, the mother provides all care for the offspring, whereas most birds have biparental care. In captivity several bear and cat species kill their offspring if conditions are not conducive to maternal care, whereas many primate mothers abandon their newborn. Other mothers may attempt to care for young, but are ineffective or accidentally crush the young. Although hand-rearing by humans is an option, it often leads to abnormal development and subsequent reproductive incompetence as adults. Hand-reared female primates, for example, rarely know how to care for their young.

Again, insight from the wild, as well as experimentation in captivity, is crucial for promoting parental care. The first obstacle is determining whether the female is pregnant and when she will give birth. Without advance warning, animal caretakers will not be able to prepare an area suitable for rearing offspring. Behavioral monitoring can play a critical role when it is unsafe to capture and stress the animal for abdominal palpation or other tests. In some cases hormonal monitoring is predictive, but samples can be difficult to obtain and expensive to process. Some species experience a pseudopregnancy that is difficult to distinguish from true pregnancy because of similar hormonal, morphological, and behavioral profiles. However, even in these difficult cases, behavioral data can prove invaluable, as Melissa Rodden and her colleagues at the National Zoo have shown for maned wolves. By examining a suite of behaviors during the breeding season, these investigators were able to predict which females were most likely to become pregnant.

Once pregnancy is determined, caretakers must provide the appropriate environment for the birth. Complete isolation from other animals and human disturbance is required for some species, and many will need a den or nesting materials. Cheetahs and other carnivores often move their young to another den when disturbed. If only one den is provided, these mothers may carry their young excessively until they die of exposure or bite wounds. Tree shrew mothers typically leave their young to sleep with the male, and if two nestboxes are not provided, the female will kill the offspring. Some females may respond fearfully to the birth of young (a truly novel event!); systematically exposing such females to the odors and vocalizations of young may reduce this fear and encourage maternal care.

In sum, captive breeding requires a holistic approach, relying heavily on research into many aspects of behavior, as well as across many other disciplines, such as physiology, nutrition, veterinary care and basic husbandry.

See also Welfare, Well-Being, and Pain—*Sanctuaries*
Zoos and Aquariums—*Animal Behavior Research in Zoos*
Zoos and Aquariums—*Giant Pandas in Captivity*
Zoos and Aquariums—*Studying Animal Behavior in Zoos and Aquariums*

Further Resources

Fisher, H. S., Swaisgood, R. R. & Fitch-Snyder, H. 2003. *Countermarking by male pygmy lorises (Nycticebus pygmaeus): Do females use odor cues to select mates with high competitive ability?* Behavioral Ecology and Sociobiology, 53, 123–130.
Kleiman, D. G., Allen, M. E., Thompson, K. V. & Lumpkin, S. 1996. *Wild Mammals in Captivity*. Chicago: University of Chicago Press.
Lindburg, D. G. & Fitch-Snyder, H. 1994. *Use of behavior to evaluate reproductive problems in captive mammals*. Zoo Biology, 13, 433–445.
Shepherdson, D., Mellen, J. & Hutchins, M. 1998. *Second Nature: Environmental Enrichment for Captive Animals*. Washington: Smithsonian Institution Press.
Swaisgood, R. R., Zhou, X., Zhang, G., Lindburg, D. G. & Zhang, H. 2003. *Application of behavioral knowledge to giant panda conservation*. International Journal of Comparative Psychology, 16, 65–84.
Wedekind, C. 2002. *Sexual selection and life-history decisions: Implications for supportive breeding and the management of captive populations*. Conservation Biology, 16, 1204–1211.
Wielebnowski, N. 1998. *Contributions of behavioral studies to captive management and breeding of rare and endangered mammals*. In: *Behavioral Ecology and Conservation Biology* (Ed. by T. Caro), pp. 130–162. Oxford: Oxford University Press.
Wielebnowski, N., Ziegler, K., Wildt, D. E., Lukas, J. & Brown, J. L. 2002. *Impact of social management on reproductive, adrenal and behavioural activity in the cheetah* (Acinonyx jubatus). Animal Conservation, 5, 291–301.

Ronald R. Swaisgood

■ Reproductive Behavior
Female Multiple Mating

A major revolution in the study of animal behavior in the last 20 years has been the discovery that females often mate with several males during a single reproductive cycle. The researchers most surprised by this were those studying birds. For decades birds have been held up as models of monogamy. Today, thanks to molecular paternity testing techniques like DNA fingerprinting, we know that females in at least 85% of passerine songbird species produce chicks sired by a male other than their social partner. This multiple paternity could simply be a consequence of extrapair males forcing females to copulate with them. Numerous studies have shown, however, that females sneak away from their mates and fly into the territories of neighboring males to mate. In Australian fairy wrens, for example, females sneak away from their social partners shortly before dawn and fly up to five territories away to mate with another male on his territory. Multiple paternity is therefore often a result of an active decision by females to mate multiple males in so-called "extrapair copulations." Multiple mating is by no means confined to birds though. It is extremely common in most species with internal fertilization including insects, spiders, reptiles, and mammals. Honeybee queens, for example, often mate with over 20 males before establishing a new colony. Species where a female only mates with a single male to fertilize a set of eggs now appear to be the exception to the rule.

It is easy to see why males engage in multiple mating, but less clear why females do the same. Generally speaking, male lifetime reproductive success depends on the number of females he copulates with. Access to females is the main constraint on male reproductive success, so the more females a male mates with the greater the total number of offspring he sires. But what benefits will a female gain from mating with several different males, especially when the potential cost of multiple mating is high? There is, for example, the risk of acquiring a sexually transmitted disease. Alternatively, if a female delays fertilization and egg-laying/giving birth until she has mated with several males, she increases her risk of dying before breeding. Finally, in species with male parental care, there is always the risk that a female's partner will assist less in caring for offspring if he is unsure whether or not he has actually fathered them.

To compensate for the costs of multiple mating there must be benefits for females. Two main types have been proposed. First, females may gain direct material benefits by multiple mating. These benefits increase the total number of offspring a female can produce and rear to independence. Second, females may gain genetic benefits that increase the quality of their offspring.

There is abundant evidence that females sometimes gain material benefits from multiple mating. In insects, for example, a single male often fails to provide sufficient sperm to fertilize all of a female's eggs. Females must therefore mate again to replenish depleted sperm supplies. In insects there is also good evidence that males transfer chemicals and nutrients in their ejaculates that allow a female to produce more eggs. Alternatively, males may provide females with nutritious nuptial gifts, the consumption of which increases a female's egg-production capabilities. In some cases, especially in birds, females that mate with two males may be able to convince both to contribute to the care of offspring. In group-living mammals, female multiple mating may be a way for females to hide the paternity of infants. This may be a strategy that decreases the risk that any given male will commit infanticide because he cannot be sure that an infant is not his own offspring. In all these examples, a female, by mating with several males, will increase the total number of offspring she produces and rears to maturity. Multiple mating for direct material benefits is relatively uncontroversial.

The real dilemma for biologists is to explain why females engage in multiple mating when there are no obvious material benefits. To return to an earlier example, what possible benefit does a female fairywren gain from sneaking off at dawn to mate with a distant male who provides her with nothing more than sperm? Sperm depletion is an improbable explanation because she can always mate again with her social mate; and, if she is worried about male sterility, why not just mate with the male next door rather than flying five territories away? The answer seems to lie in the genetic quality of extrapair males. In birds, females usually choose to engage in extrapair copulations with males with more attractive plumage, songs or sexual ornaments than their own social mate. Consequently, attractive males gain in two ways. They are less likely to be cheated on by their own mate, and are more likely to achieve extrapair copulations with other females. This may largely explain why in many socially monogamous birds, males are more ornamented than females. Female choice of additional mating partners favors males with brighter plumage, longer tails, or more complex songs because these characteistics increase a male's total reproductive output.

An obvious question is why don't females simply mate with the most attractive male in the first place? The answer is probably that female mate choice is constrained. In many birds, two parents are essential to ensure that offspring are successfully reared to independence. If there are similar numbers of adult males and females, some females have no choice other than to pair with a less attractive male so that they have a mate to help with parental duties. These females, however, can then improve the quality of some of their offspring by sneaking off to mate with a more attractive male.

The genetic benefits of engaging in extrapair mating with attractive males are of two main sorts. First, it allows females to produce sexier sons, who have above average mating success. Breeding studies have shown that male traits favored by female choice tend to have a heritable genetic basis so that sexy fathers sire sexy sons. Second, male attractiveness may be related to health. If there is a heritable genetic basis to healthiness then offspring sired by extrapair males may be stronger than those sired by their social mate. In a few birds, this has been directly tested. Researchers have compared growth rates or immune responses of offspring sired by the social mate with those sired by an extrapair male. This is a powerful test because it is possible to compare offspring reared under identical conditions (in the same nest) that share the same mother. The only difference between the two types of offspring is therefore the genes they inherited from their fathers. The available evidence suggests that extrapair offspring have better immune systems and higher growth rates.

In some species female multiple mating does not appear to be a strategy to improve on an earlier mating. Instead, females mate with several males in a genuinely indiscriminate manner. Tiny female pseduoscorpions, for example, show an active preference for mating with a new male rather than a previous mate. Jeanne and David Zeh have offered a novel explanation for this behavior. They argue that females differ in their ideal mating partner because of various sorts of genetic incompatibilities. The simplest example of a genetic incompatibility is that associated with inbreeding. Mating with a close relative, like a brother or father, often leads to the production of stillborn or weak offspring. The Zehs argued that, in general, females might struggle to detect which males they are genetically more compatible with simply by inspecting potential mates. They then suggested that females may mate with several males and allow subsequent chemical interactions between sperm and the female reproductive tract or even sperm and eggs to ensure compatible sperm are more likely to fertilize their eggs.

Several studies have now confirmed that multiple mating can reduce the likelihood of eggs being fertilized by genetically incompatibile sperm. In Swedish populations of European adders, for example, females that mate with multiple males give birth to fewer stillborn off-spring than those that mate singly. Stillbirth is known to occur when the parents of offspring are close relatives. The beneficial effect of multiple mating suggests that females have some way to ensure that sperm from less closely related males are more likely to fertilize their eggs. Subsequent work on another Swedish reptile, a sand lizard, has confirmed this claim. Mats Olsson and his colleagues used DNA fingerprinting techniques to examine offspring pater-nity and show that when a female mated with several males, the less closely related the male was to the female, the greater the proportion of her offspring he sired.

Female multiple mating has enormous implications for the evolution of both male and female mating behavior, physiology, and morphology. From a male perspective, the main consequence is that mating does not guarantee reproductive success. Not only must a male mate, he must also ensure that his sperm, rather than that of a rival, ends up fertilizing the female's eggs. This leads to males trying to prevent females from mating with other males by, for example, mate guarding or producing mating plugs that seal a female's reproductive tract. Males may also release chemicals in their semen that cause females to lay more eggs shortly after mating than would otherwise be the case (increasing the likelihood a female will fertilize eggs using his sperm). In addition, because females often succeed in multiple mating, males must then invest in traits that increase the chances that their sperm will be used when there is "sperm competition." There is a strong pattern that males in species where females tend to mate with many males per breeding cycle have larger testes (so that they can produce more sperm) than in species where females are less promiscuous. From a female perspective, multiple mating means that females must have mechanisms of "cryptic choice" if they want to increase the likelihood that the sperm of specific males fertilize their eggs.

The most dramatic consequence of female multiple mating is, however, that it creates a "battle of the sexes." When males and females are truly monogamous, their reproductive in-terests are identical. The more offspring a female produces, the more offspring her mate sires. With multiple mating, however, male and female reproductive interests only partly coincide. If a female remates, her previous mate's share of paternity decreases because he must now share fatherhood with another male. So a male may benefit for example, by pro-ducing chemicals that increase a female's daily egg-laying rate while she is mainly using his sperm to fertilize eggs, even if this reduces her lifespan. It does not matter to a male whether a female dies, if her offspring are no longer his. For a male it is often better to in-crease his reproductive success in the short term even if the long-term effect is detrimental to the female.

In sum, due to material and genetic benefits there may be gains to individual females in a population that mate with multiple males compared to those that do not. The evolu-tionary outcome, however, is that the greater the level of female multiple mating in a species, the more male–female conflict there is over reproduction. Unless females evolve measures to combat males pursuing their own selfish interests, we therefore expect to find that the act of mating has a more negative effect on females in species where females prac-tice multiple mating.

See also Reproductive Behavior—*Female–Female Sexual Selection*
Reproductive Behavior—*Mate Choice*
Reproductive Behavior—*Monogamy*

Reproductive Behavior—*Sex, Gender, and Sex Roles*
Reproductive Behavior—*Sexual Selection*

Further Resources

Birkhead, T. R. 2000. *Promiscuity: An Evolutionary History of Sperm Competition*. Cambridge, MA: Harvard University Press.

Diamond, J. 1998. *Why Is Sex Fun? The evolution of Human Sexuality*. New York: Basic Books.

Eberhard, W. G. 1996. *Female control: Sexual selection by cryptic female choice*. Princeton, NJ: Princeton University Press.

Gould, J. L. & Gould, C. G. 1989. *Sexual Selection (Scientific American Library, No 29)*. New York: Freeman & Company.

Hrdy, S. B. 1999. *Mother Nature: Natural Selection and the Female of the Species*. London: Chatto & Windus.

Judson, O. 2002. *Dr. Tatiana's Sex Advice to All Creation*. New York: Owl Books.

Olsson, M., Madsen, T., Ujvari, B., et al. 2004. *Fecundity and MHC affects ejaculation tactics and paternity bias in sand lizards*. Evolution, 58, 906–909.

Olsson, M., Shine, R., Madsen, T., et al. 1996. *Sperm selection by females*. Nature, 383, 585.

Ridley, M. 1995. *The Red Queen: Sex and the Evolution of Human Nature*. New York: Penguin.

Simmons, L. W. 2002. *Sperm Competition and its Evolutionary Consequences in the Insects*. Princeton, NJ: Princeton University Press.

Michael Jennions

■ Reproductive Behavior
Marine Turtle Mating and Evolution

Marine turtles have long been a successful group. They shared the world with dinosaurs and survived the mass extinction of so many other reptiles at the end of the Cretaceous period. Through time, there have been dozens of species (including the amazing 15-foot *Archelon*), but for the last 20 million years or more there have only been six species, each with specialized feeding habitats, home ranges, and nesting locations.

What has fascinated biologists, is that some marine turtle species have successfully mated with each other on occasion, including green and loggerhead turtles, which have been distinct for at least 50 million years. This may be the record for hybridization between distant species for any vertebrates, and it fascinating also because green turtles are primarily herbivores in contrast to the carnivorous loggerheads, which use their massive heads to crush molluscs and other invertebrates.

Scientists have speculated on whether this ability to hybridize has been important in the evolution of marine turtles. And if so, it would depend in part upon turtle behavior; especially where and when they mate and how selective they are about choosing mates. Observations of male sea turtles during the mating season, mounted on floating logs or harassing scuba divers, makes one wonder why hybrids aren't any more common than they are.

In contrast, how might new species of marine turtles originate and how might behaviors evolve in ways that would lead to the formation of new species? By physically tagging turtles and conducting population genetic studies, biologists have found that even though marine turtles are highly migratory, they form distinct breeding populations. Thus, even though different populations might share the same feeding grounds, when it comes time to mate, turtles travel back to home breeding grounds, thus reducing the exchange of genes between them. If this isolation is strong enough, then, over time, populations could diverge and possibly form a new species.

The formation of new species, known as *speciation*, can occur through many processes, one of which is known as *sympatric* (meaning "same home") speciation. In this process, a population evolves into a new species within the range of the parent species, rather than from a physically isolated population. For this process to occur, the population must somehow be intrinsically separated from the other populations, at least during mating. In green turtles in Australia, we see such an example of intrinsic separation due to recent behavioral differences that have evolved in that last few hundred turtle generations.

In Australia, the most prolific marine turtle is the green turtle, *Chelonia mydas*, which nests at specific mainland and island beaches around the northern half of the continent. In the center of this distribution is a unique population that nests in the Gulf of Carpentaria; unique, because these turtles nest primarily in the austral winter, whereas the rest of Australia's green turtles nest primarily in the austral summer. Why is this, and what might be the implications of such changed behavior?

Up until about 8,000 years ago, the Gulf of Carpentaria was not available to green turtles as a nesting area. Instead, the sea level was too low, and the area contained a freshwater lake, Lake Carpentaria. With rising sea levels, the shallow Gulf of Carpentaria was inundated, and genetic studies indicate that turtles from the northwest colonized the area. Given the shallow waters and high summer temperatures, it is hypothesized that sand temperatures were too hot in the summer and reduced the successful development of hatchlings. Thus, turtles that nested in the middle of the summer had less success in producing hatchlings, whereas those nesting in early or late summer had better success. This selective pressure would have resulted in a shift in the timing of nesting, to favor females who nested in winter.

That might explain the females' behavior, but what about the males whose hormones trigger breeding for only a few months in a given year? Genetic studies of the males that were breeding with these females identified the males as also having originated from the Gulf of Carpentaria population. Thus, in the midst of all the other green turtle populations in Australia, exists a population in which males and females are mating with each other in the winter while the rest of the turtles are busy eating.

Although isolated for now, changing temperatures and sea levels over time are likely to shift the amount of genetic exchange among populations, including changes brought about by human-influenced global warming.

Further Resources

Bolten, A. B. & Witherington, B. E. (Eds.). 2003. *Loggerhead Sea Turtles*. Washington D. C., Smithsonian Books.

FitzSimmons, N. N., Goldizen, A. R., Norman, J. A., Moritz, C., Miller, J. D. & Limpus, C. J. 1997. *Philopatry of male marine turtles inferred from mitochondrial markers*. Proceedings of the National Academy of Science, 94, 8912–8917.

Lutz, P. A. & Musick, J. A. (Eds.). 1997. *The Biology of Sea Turtles*. CRC Marine Science Series. Boca Raton, FL: CRC Press,

Perrine, D. 2003. *Sea Turtles of the World*. Stillwater, MN: Voyageur Press, Inc.

Sea Turtle Research Web site: http://www.turtles.org

Nancy N. FitzSimmons

■ Reproductive Behavior
Mate Choice

Animals sometimes act in ways or possess characteristics that reduce their lifespans. For example, the bright colors of male guppies attract predatory fish, and the incessant courtship calls of male frogs consume vast amounts of energy. These life-threatening traits are usually more prominently expressed in males. They are thought to exist because they increase a male's reproductive success when competing with rivals for breeding opportunities. The most easily explained sexually selected characters are weapon-like traits such as horns, tusks and greater body size or musculature that increase fighting success. These traits are clearly advantageous when males engage in direct physical contests with each other for access to fertile females.

Charles Darwin, however, also suggested that sexual selection leads to the evolution of male traits that circumvent the need for direct competition with rivals. Why not appeal directly to the customer? Darwin argued that males have evolved traits that act like advertisements or seductive devices. If females find these traits attractive then males that express them—be they elaborate courtship, gaudy plumage, or even complex genitalia—will mate successfully with more females and produce more offspring. Studies over the last 20 years have extensively confirmed Darwin's initially controversial claim that females are choosy. In most species, males with brighter plumage, bigger ornaments, and louder, longer, or more complex courtship displays have higher mating success than the average male. For example, in fiddler crabs, males with bigger claws or males that wave their claws more vigorously are more likely to entice a passing female to mate with them.

The long tail of this widowbird indicates good genes and attracts females.
© Joe McDonald / Visuals Unlimited.

Some have argued that certain males may have high mating success, not because these males are preferred by females, but because they win male–male fights and females are constrained to mate with winners because victors prevent losers from mating. Direct experimental manipulations of male characteristics have confirmed, however, that even when the effects of male–male competition are removed, females still show a mating preference for more ornamented or energetically courting males. In African long-tailed widowbirds, for example, Malte Andersson performed a "scissors and glue" experiment in which males were randomly assigned to one of two experimental treatments. Males in each treatment were therefore presumably of equal average fighting ability. In one treatment, males' tails were experimentally shortened. In the other treatment, tails were experimentally extended. Andersson showed that males with experimentally lengthened tails attracted more females onto their territories than those with shortened tails (without any obvious effect of changing tail length on a male's ability to win contests with neighboring males). Females therefore prefer males with longer tails.

In most species, females, and sometimes males, are choosy when deciding on a mating partner. If females are choosy, it is easy to see why males invest in costly displays and characters that make them more attractive because it increases their reproductive success. Despite this, though, there is still a genuine difficulty with Darwin's theory that continues to stimulate debate to this day. The real problem is that being choosy is costly. A choosy female must invest time and energy into sampling males before she finds an appropriate partner. She will expose herself to predators as she moves between males' territories, and, if her sampling regime involves close inspection of males, may increase her risk of picking up parasites or diseases. All of these costs will reduce her survival and lifetime *fecundity* (number of offspring produced) compared to a female that mates with the first male encountered that is of the same species as she is. The key question is therefore: What benefit do females gain from being choosy?

One answer is that females are choosy because this actually increases their total lifetime fecundity. For example, females that avoid mating with diseased or heavily parasitized males reduce the risk of becoming infected and probably live longer, thereby producing more sets of offspring. Males may also vary in their fertility such that some are better than others at fertilizing eggs. The extent of such variation in wild animals is poorly known, but male sterility is a problem in humans and has been detected in some mammals involved in captive breeding programs. Females that know how to detect sterile males will be at an obvious advantage.

In many species, males also provide material resources to females. In insects and spiders, these are often in the form of nuptial gifts of prey items, specially manufactured food packages, or even the male's own body. Consuming larger nuptial gifts often leads to females laying more eggs. In other insects, there are nutrients or protective chemicals in male ejaculates, and larger males tend to provide more of these chemicals, so females that prefer to mate with larger males are at an advantage. In other animal groups, especially birds and mammals, some males may provide females with access to better feeding territories and nesting sites, or even assist more with the rearing of offspring. All these diverse direct material resources are beneficial because they increase the ability of choosy females to produce and rear offspring successfully.

If males vary in their ability to provide material resources, then choosy females that preferentially mate with males offering more resources will be favored by natural selection. There are, however, a few problems with this scenario.

In some cases, females may be able to inspect the resource directly. For example, they can spend some time foraging on a male's territory to test whether or not it has plentiful food. But when females can directly assess the material benefits of mating with a male, why would he expend energy on a costly advertisement of what he has to offer? All he needs to do is produce a minimal signal that allows females to locate him. Resources that females can directly inspect are, therefore, less likely to explain the evolution of highly extravagant male displays, coloration, or body features. In other cases, females cannot directly measure the alleged benefits a male will provide. This may be because the benefit is in the future ("I will be a good and caring father") or because there is no way to physically assess the resource without actually mating ("My ejaculate is full of useful chemicals you can absorb and use to protect yourself and your eggs from predators"). This raises a more general question we will return to shortly: How do females ensure that males are advertising honestly?

A second explanation for female choosiness differs from the first because it concedes that choosiness might decrease a female's lifetime fecundity. The benefit of choosiness in this case is thought to be genetic rather than material gains. Specifically, attractive male traits may provide females with reliable information about a male's genetic quality. Choosy females are prepared to incur the costs of reduced lifespan or fecundity because mating with attractive males leads to the production of better quality offspring. So, even though

This wild turkey displays his in order to attract a mate.
© Anthony F. Chiffolo.

choosy females produce fewer sons and daughters, in the long run they gain because their higher quality offspring will produce more grandchildren. There are two main ways in which the offspring of choosy females might be better than those of less choosy females.

First, the characteristics that make males attractive may be heritable. If fathers with sexually attractive characters produce more attractive sons, these sons will, as a consequence, have higher than average mating success. So choosy females, because they are more likely to mate with an attractive male, will produce sexier sons. There is good evidence that sexually selected characteristics, like high calling rate in crickets, brighter color in fish, and longer tail length in birds, are heritable because fathers pass on to their sons genetic coding for stronger expression of these characteristics.

Second, male attractiveness may also signal to a female that he carries genes that will increase the survival or fecundity of her offspring. For example, sexual traits that females find more attractive may signal that a male has genes for increased disease resistance, better body condition, faster growth, the ability to evade predators, and so on. If offspring inherit these genes, then choosy females will produce sons and daughters with the potential to live longer and produce more grandchildren. This second "genetic" explanation for choosiness brings us back to the issue of honesty.

What ensures that a male honestly signals that he is a good parent or will pass on genes for greater offspring survival? Females only gain by choosing if attractive male traits, at least on average, are honest signals of the benefits males provide to females. Something must prevent a low-quality male from lying and signaling that he is high-quality. The answer, perhaps unsurprisingly, requires that male signals are costly. After all, if attractive signals are cheap to produce then every male will produce the biggest, most attractive signal possible. Costliness alone might seem sufficient to explain honest signaling because only a high-quality male can afford to pay the price of producing a big and costly sexual signal. As we all know though, it is sometimes possible for a conartist to deceive people into thinking he is wealthy by driving a flashy car—even if this leaves him with no cash in the bank.

The problem with using costly behavior as a signal is that it is becomes unreliable if the behavior *totally* uses up the very resource it is meant to signal. In a biological context, for example, just think how a cumbersome elongated male tail will reduce a male's ability to feed and care for offspring. An attractive long tail therefore obliterates the very thing it is suppose to signal. When signaling genetic benefits, the scenario is a little trickier. The resource of interest is, say, "genes that allow the bearer to forage more efficiently and therefore fly faster to avoid predators." Now a high-quality male with these genes might actually be able to signal that he has them by diverting all the extra energy they provide into growing a long tail (in the same way that a rich person can show their absolute wealth by spending everything they own on an ultrafancy car). The problem though is that the genetic benefit the female is interested in is how her offspring will perform. If the father transfers genes for flying faster to his sons, but also transfers genes that make his sons put all their extra energy into tail growth, as he did, then they, too, will not fly faster. There is therefore no net benefit.

Amotz Zahavi provided an elegant solution to the dilemma of how honest indicators of male quality can evolve, called the *handicap theory of signaling*. Surprisingly, it not only retains the idea that a signal must be costly but, in fact, emphasizes that the signal *must* consume the resource it is designed to indicate. It adds, however, one small, but essential, element to the above argument. If the expression of a sexual trait is to increase with male quality (so that it honestly indicates quality) then the marginal cost of producing a bigger trait must be greater for low-quality individuals. Marginal cost is simply the increase in cost per unit of investment. To return to our analogy, this would require that buying a slightly fancier car adds more to the bill of a poor person than to that of a rich one. (Perhaps car salesmen are concerned that a poor person is more likely to suffer a sudden financial crisis, and therefore they demand extra payment). A rich person, therefore, pays less than a poor person to buy the same flashy car.

In biological terms, the handicap principle requires that the costs of greater investment in a signal increase more slowly for a high-quality than a low-quality male. The beauty of this is that the optimal level of signaling does not fully consume the resource being signaled. The low-quality male starts running out of resources sooner than the rich one and therefore has to stop investing in the signal. To return to our analogy, the rich person ends up with a better car than the poor person, but still has more money in the bank. The flashiness of a car can therefore be used to judge the size of a person's bank balance.

The handicap principle is often phrased in terms of male body condition. (The unstated assumption is that, on average, higher-quality males achieve better condition than low-quality males). The larger a sexual trait, the greater the mating benefit to a male. However, a male in good condition will pay a smaller cost to produce a slightly larger sexual signal than a male in poor condition. For example, if you are healthy you can afford to carry around a heavier set of antlers. Consequently, the optimal investment (the difference between the mating benefits and energetic/predation costs) for a male in good condition is to put more into sexual advertisement than a male in poorer condition. This is indeed what we generally find. There is a positive relationship between a male's body condition and the size of his sexual traits or intensity of his courtship displays in most species. Bright coloration, for example, is usually a good way for a farmer (or a female) to pick a healthy male bird or fish. In addition, when biologists experimentally provide males with extra food, these males tend to develop brighter colors, larger sexual characters and court more vigorously.

In summary, females may choose males because this increases their lifetime fecundity, either through greater fecundity per breeding event or through increased longevity. This is favored by natural selection. In addition, choosiness may be sexually selected because choosy females increase the fitness of their offspring if attractive males sire sexier sons and/or sire offspring with higher than average lifetime reproductive success. Female choosiness will only persist, however, if males honestly signal the benefits they offer. This requires that the costs of producing a slightly more attractive signal increase more rapidly for a low-quality male than they do for a high-quality one.

See also Reproductive Behavior—*Female–Female Sexual Selection(s)*
   Reproductive Behavior—*Female Multiple Mating*
   Reproductive Behavior—*Monogamy*
   Reproductive Behavior—*Sex, Gender, and Sex Roles*
   Reproductive Behavior—*Sexual Selection*

Further Resources

Andersson, M.B. 1994. *Sexual Selection*. Princeton, NJ: Princeton University Press.

Birkhead, T.R. 2000. *Promiscuity: An Evolutionary History of Sperm Competition*. Cambridge, MA: Harvard University Press.

Cronin, H. 1991. *The Ant and the Peacock*. Cambridge: Cambridge University Press.

Diamond, J. 1998. *Why is Sex Fun? The Evolution of Human Sexuality*. New York: Basic Books

Gould, J. L. & Gould, C. G. 1989. *Sexual Selection (Scientific American Library, No 29)*. New York: Freeman & Company.

Judson, O. 2002. *Dr. Tatiana's Sex Advice to All Creation*. New York: Owl Books.

Ridley, M. 1995. *The Red Queen: Sex and the Evolution of Human Nature*. New York: Penguin.

Zahavi, A. & Zahavi, A. 1999. *The Handicap Principle: A Missing Piece of Darwin's Puzzle*. Oxford: Oxford University Press.

Zuk, M. 2002. *Sexual Selection: What We Can and Can't Learn about Sex from Animals*. Berkeley: University of California Press.

Michael Jennions

■ Reproductive Behavior
Mate Desertion

Mate desertion occurs when one parent leaves their partner before the period of parental care has ended. Brood desertion then occurs if the remaining parent decides to abandon the offspring rather than care alone. Many studies of desertion ask why some individuals in a population provide more care than others. For example, in a Panamanian cichlid fish, females guard their fry from predators for 5–6 weeks until they can fend for themselves. In some breeding pairs, the male also stays and guards the brood until the fry reach independence. In other pairs, however, the male deserts when the fry are only a few days old, leaving the female to guard alone. Although it is worthwhile to explain differences within a single species in the timing of desertion between the sexes or among parents of the same sex, the identical question can be asked at a much broader level. Why do some species mainly show female-only care, others male-only care, yet others biparental care, and still others no parental care at all? These four patterns of care clearly depend on desertion decisions, specifically whether males, females, neither, or both sexes desert.

Despite variation in desertion patterns among species, there is one clear trend: If only one parent deserts, it is usually the male. So, to rephrase the question slightly, why do females care more than males? To answer this we need to know whether the benefits and costs of deserting differ for males and females. The main advantage of desertion is that the deserter can breed again sooner. Deserting male Panamanian cichlids, for example, often court another unpaired female while they are still with their current mate so that, shortly after desertion, they breed again with this new female. The main downside to deserting is that, when only one parents provides care, this decreases the likelihood that offspring will survive to independence.

The defining feature that distinguishes males from females is the size of their gametes (reproductive cells). Males always produce smaller gametes than females (sperm versus eggs). This difference in gamete size is known as *anisogamy*. The explanation most textbooks offer to explain why females care more than males do runs as follows. Sperm are small and relatively cheap to produce. Consequently, the reproductive success of males is limited by how many females they mate with. In contrast, females produce large eggs that are far more costly to manufacture. As such, females gain a smaller benefit by deserting be-cause it takes them longer to generate a new set of eggs and re-enter the mating market. Therefore males benefit more from desertion than females because they have a higher potential rate of reproduction. Males "miss out" on more mating opportunities when they care. However, as David Queller has noted, this widely promoted argument is fundamen-tally flawed. If the population has equal numbers of each sex, then males cannot lose more future mating opportunities than females, because every mating requires both a male and a female. The average male and female must therefore mate at the same rate.

Queller noted two ways in which the traditional explanation can be salvaged. First, ani-sogamy may have the effect of making the adult sex ratio such that there are more females than males. If true, this will increase the benefits of desertion for males because, on average, each male will be a parent in more future broods than each female. For example, if there are 5 males for every 10 females, then the average male will be involved in twice as many mat-ings. A male is therefore likely to mate again twice as fast as a female if he deserts. But why would adult males tend to be consistently less common than females? One obvious reason is that it is a consequence of *sexual selection* via female *mate choice* or direct male–male contests, which leads to some males consistently failing to mate. When males compete for females, some males are more likely to succeed than others. This is either because females find these males more attractive or because they win more fights for direct access to fertile females. Although the average adult male and female must have the same rate of reproduction when the total adult sex ratio is one-to-one, sexy/dominant males have a higher than average likeli-hood of mating again in the future. Desertion therefore becomes a more profitable option for these superior males, and only males that mate get to decide whether or not to desert. Another way to think about this is to realize that, even if the total adult sex ratio is equal, some males are so ugly that choosy females will not mate with them. So, because of sexual selection, the *effective* adult sex ratio (which excludes males that have no hope of mating) be-comes female-biased. Due to male–male competition, only a subset of males have a chance of mating. These males therefore have a higher mating rate than females.

Queller's second explanation is that the cost paid when deserting (due to lower offspring survival when there is reduced parental care) is smaller for males than females. Why would this be so? The simplest answer is that males are usually less certain than females of the parentage of offspring. In species with external fertilization, like most marine fish, this occurs because several males may simultaneously release sperm when a female is spawning. There is also abundant evidence from most species with internal fertilization (like birds, mammals, and insects) that females copulate with more than one male per breeding cycle. Paternity is therefore often shared among several males. If a male delays desertion and provides parental care, some of the offspring he cares for are not his own. (In human populations, estimates of the percentage of fathers caring for children they mistakenly think are their own range from less than 2% to up to 30%). The average benefit of caring is therefore smaller for males than females because fathers sometimes waste energy caring for another male's offspring. The cost of desertion due to reduced paternal care is therefore smaller for males than females.

Queller's first explanation still requires that we understand why females are usually choosier than males and why males compete more strongly for females than the reverse. A truly satisfying explanation from first principle must be based solely on anisogamy. We cannot start by simply assuming that females provide more postmating care for offspring than males. This, after all, is the very fact we wish to explain. The most likely explanation is that anisogamy allows males to invest less than a female prior to mating by releasing just enough sperm to fer-tilize the female's eggs. Consequently, a female tends to invest more per mating than does a male so that it takes her longer to return to a state where she can breed again. This is true even if she provides no postmating parental care and deserts at exactly the same time as her mate.

There are two implications of females taking longer to re-enter a state when they are ready to mate. First, females are less likely to mate again before the breeding season ends or before they die. As such, every mating is of greater value to a female. It is more likely to be her last. If the opportunity arises to choose between two potential males, and some males are better fathers than others, the cost of choosing, which may result in forfeiting breeding opportunities in the future, is consequently smaller for a female than a male. Females can therefore afford to be choosier.

Second, because males take less time after mating to produce enough gametes to breed again, the ratio of sexually mature males to fertile females is male-biased. This ratio is known as the *operational sex ratio* (or OSR for short). A male-biased OSR is why males compete more strongly for access to females than the reverse. Every time a female becomes fertile there are several males waiting to compete to mate with her.

Finally, it is important to distinguish between the OSR and the effective sex ratio. The OSR includes every male that competes to mate. It therefore includes some males that are unlikely to succeed. After all, there are no winners without losers. In contrast, the effective sex ratio only includes those males that mate. Only a male that mates has the luxury of deciding whether to care or desert. The male-biased OSR due to anisogamy leads to stronger sexual selection on males for traits to help them compete successfully for females. And, due to stronger sexual selection on males, the effective sex ratio becomes female-biased which favors male desertion.

See also Reproductive Behavior—*Female Multiple Mating*
Reproductive Behavior—*Mate Choice*
Reproductive Behavior—*Sexual Selection*

Further Resources

Andersson, M. B. 1994. *Sexual Selection*. Princeton, NJ: Princeton University Press.
Clutton-Brock, T. H. *The Evolution of Parental Care*. Princeton, NJ: Princeton University Press.
Gould, J. L. & Gould, C. G. 1989. *Sexual Selection (Scientific American Library, No 29)*. New York: Freeman & Company.
Jennions, M. D. & Polakow, D. *The effect of partial brood loss on male desertion in a cichlid fish: An experimental test*. Behavioral Ecology, 12, 84–92.
Queller, D. C. 1997. *Why do females care more than males?* Proceeding of the Royal Society of London Series B, 264, 1555–1557.
Ridley, M. 1995. *The Red Queen: Sex and the Evolution of Human Nature*. New York: Penguin.
Short R. V. & Balaban, E. (Eds.). 1994. *The Difference between the Sexes*. Cambridge: Cambridge University Press.

Michael Jennions

■ Reproductive Behavior
Mating Strategies in Marine Isopods

Paracerceis sculpta is a sea-dwelling isopod crustacean. Their terrestrial isopod relatives, pill bugs or "roly-polies," are familiar to most gardeners and turners of stones. Pill bugs thrive in dark, protected places, and are named for their ability to roll themselves into a tight ball. *Paracerceis* isopods also perform this trick, and like pill bugs, they thrive in dark, protected places. For *P. sculpta*, such places are spongocoels, cavities within sponges inhabiting the Sea of Cortez. At the tip of this finger of ocean separating Baja, California, from Mexico's mainland, isopods use spongocoels as breeding sites. Here, alternative mating strategies have evolved.

Alternative mating strategies represent *polymorphisms*, distinct forms of reproductive behavior, life history or body type that persist within a single population. Such polymorphisms exist in most species in which competition for mates occurs. In *P. sculpta*, females are similar in appearance. Males, however, exhibit three distinct "morphs" that each display different repertoires of mating behavior. Alpha males are largest, although they are less than a centimeter long. Like tiny sultans, α-males (alpha males) attempt to sequester harems of females within spongocoels, using their robust bodies and horn-like posterior limbs to defend or usurp these sites. Many fail to mate at all. Smaller β-males (beta males) are about half as long as α-males. They invade harems by looking and acting like females. Tiny γ-males (gamma males) are less than 3 mm (.35 in) long. They invade harems by moving fast and hiding within spongocoels.

All immature isopods in this species feed and grow on algae before making their way to sponges. Alpha males establish spongocoel territories and wait for females to arrive, whereas β- and γ-males wait for harems to become established before beginning their mating attempts. Females spend their preadult lives storing fat and developing eggs. They swim to sponges to find mates. Courtship is brief because females discriminate neither α-male nor spongocoel characteristics. Instead, females prefer spongocoels containing harems, perhaps because the existence of these established sites predicts their relative safety. Once inside, females molt, become receptive for one day, then shed their mouthparts and enter a 3-week pregnancy. Juvenile isopods leave their spent mothers and settle on algae. Spent mothers leave spongocoels and die.

What conditions allow the three male morphs to coexist? Alpha males grow slowly but may live 6 months. Beta males have intermediate growth and survival rates. Gamma males mature fast but die young. Because male reproductive tenure varies inversely with maturation time, life history differences that could influence the relative contributions of each morph to population numbers appear to cancel. Alpha males devote about 20% of their bodies to sperm production. In β-males, this amount exceeds 50%, and as much as 80% or more of γ-male bodies consist of sperm! Yet within spongocoels, the relative fertilization success of each male morph varies, with the density of receptive females, and with the frequency of other male morphs. In terms of total offspring numbers, the average fitnesses of the three male morphs are equivalent.

Such conditions are considered sufficient to maintain a genetic polymorphism, and indeed, the three male morphs are distinct at a single gene. The number of females that enter spongocoels varies throughout the year and is

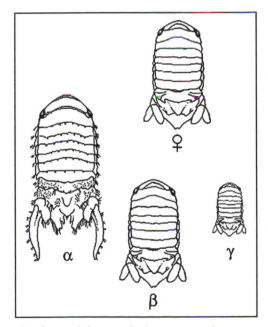

The four adult morphs in Paracerceis sculpta; females are monomorphic; α-males are largest, possess elongated uropods, and comprise 81% of aggregate male population samples; β-males are smaller than α-males, resemble females in their behavior and external morphology (although β-males are slightly smaller than females), invade spongocoels by mimicking female behavior, and comprise 4% of aggregate male population samples; γ-males are smallest, comprise 15% of aggregate male population, and use their small size and rapid movements to invade spongocoels.

Courtesy of Stephen M. Shuster.

unpredictable at any given time, causing the success of each male morph to oscillate. Thus, like the prows of three shrimp boats moored together in the nearby harbor, the relative frequencies of α-, β- and γ-males, rise and fall, always in motion, but remaining constant over time.

Further Resources

Emlen, D. J. 1997. *Alternative reproductive tactics and male-dimorphism in the horned beetle* Onthophagus acuminatus (*Coleoptera: Scarabaeidae*). Behavioral Ecology and Sociobiology, 41, 335–341.
Gross, M. R. 1985. *Disruptive selection for alternative life histories in salmon.* Nature, 313, 47–48.
Shuster, S. M. 1990. *Courtship and female mate selection in a semelparous isopod crustacean* (Paracerceis sculpta). Animal Behaviour, 40, 390–399.
Shuster, S. M. 1992. *The reproductive behaviour of α-, β- and γ-males in* Paracerceis sculpta, *a marine isopod crustacean.* Behaviour, 121, 231–258.
Shuster, S. M. 2002. *Mating strategies, alternative.* In: *Encyclopedia of Evolution* (Ed. by M. Pagel, et al), pp. 688–693. Oxford, UK: Oxford University Press.
Shuster, S. M. & M. J. Wade. 2003. *Mating Systems and Strategies.* Princeton, NJ: Princeton University Press.
Sinervo, B. & C. M. Lively. 1996. *The rock-paper-scissors game and the evolution of alternative male strategies.* Nature, 380, 40–243.

Stephen M. Shuster

■ Reproductive Behavior
Mollusk Mating Behaviors

As exciting as it may be for us to think about and participate in, our mammalian mating customs are rather dull compared to that of some other animals. Humans may have elaborate courtship rituals, but the actual mating act is remarkably similar and relatively brief for most mammals. There are variations on the theme among humans of various cultures, but it is a recurrent theme. Other vertebrates such as birds, reptiles, and fish may not even have external sexual organs. They have to mate by pressing their genital openings together for the transfer of sperm into the female so that the female's eggs can be fertilized in what some have called the "cloacal kiss" (a *cloaca* is an undifferentiated combined anal and genital opening). Such a coupling is necessarily brief and undistinguished. But the sexual coupling behaviors in other groups of animals—the *invertebrates*, animals without backbones)—are far more complex and interesting, some might even say bizarre, than our own human sexual behavior.

The phylum Mollusca is a good group of animals in which to look at different methods and behaviors of mating. Mollusks are mostly shellfish—they are animals most people are familiar with, having eaten them, or having seen them on the seashore or in public aquariums. This *phylum* (a major animal grouping) is remarkably large and diverse. Animals in it are almost all marine with the exception of the slugs and snails, which have managed the transition from the salty sea to freshwater and even to dry land. The phylum is composed of seven subgroupings (called *classes*) of which three are rare and obscure. The rest are the *chitons* (slug-like animals with eight plates or shells on their backs), *bivalves* (clams and oysters), *gastropods* (slugs and snails), and *cephalopods* (octopuses and squids).

The mollusks have been around for an extraordinarily long time, as evidenced by the fossil record. Having shells to protect their soft bodies has allowed fossil mollusks to be well-documented because shells fossilize readily. Some fossil ancestors of mollusks have been found in billion-year-old shales of the Burgess rock formation in Canada. In contrast, the first vertebrate fossils showed up only 300 million years ago and the first mammals appeared about 65 million years ago. Having such a long lineage has given mollusks a longer time to evolve. Their behavior styles have evolved over this long time as much as their body shapes have, and it is fascinating to study, for example, how a snail enclosed in a hard shell has sex.

But before considering mollusk sex, we need to examine why there is sex at all. Why do mollusks (and humans) have two sexes? Why not three or four? Why do we couple together to exchange sexual products? We need to take a brief excursion into genetics to answer these questions.

When life arose on earth several billion years ago, and evolution began to shape it, there needed to be a mechanism for reproduction. Being able to reproduce is part of the basic definition of life itself. Organisms die and hence must have a method for replacing themselves. One of the simplest methods of reproduction is to split and divide oneself into two animals. Students of basic biology classes will learn of amoebas, simple single-celled animals that divide down the middle to become two smaller animals. Higher animals than amoebas also practice this method of reproduction—animals that frequently live in colonies, such as sea anemones and tunicates. Many other animals also split by budding, where a piece of themselves breaks off, either by natural or unnatural means, and becomes a new individual. Flatworms, annelid worms, and sponges use this classic method of reproduction.

Another way of reproduction without use of sex is by *cloning*, which is production of fertile eggs without the benefit of sex (without contribution of sperm to a developing egg). This process of cloning animals (and there has been much controversy lately over possible cloning of humans) has been in the news lately. But contrary to what the media or the "Star Wars" saga would have people believe, the process of cloning animals has been around for hundreds of millions of years. When environmental conditions are beneficial, aphids on your garden roses all reproduce *parthenogenetically* (without contribution of a male), as do brine shrimp ("sea monkeys") that are fed to tropical fish. There are over 500 species of bdelloid rotifers (tiny wheel animals that live marshy areas or mud puddles) for which males have never been found—*all* bdelloid rotifers are female.

The above methods of reproduction (splitting, budding, and cloning) share a common result: The offspring are all alike. They have exactly the same genetic makeup as the "parents." They are unable to evolve (or at least unable to evolve rapidly in response to evolutionary pressures) since they have no method for introducing mutations into their populations. Consider what would happen to a colony of cloned sea anemones if the sea water warmed up, perhaps in response to an El Niño event or global warming. *All* the sea anemones are alike, *all* cannot withstand the warmer water, *all* die. But if there is some variation or diversity between individual sea anemones, some may survive the warm water event and live to reproduce and reestablish the colony. Sex is thus a method of ensuring genetic variety within a population or species of animals, and having genetic diversity is a means of evolution.

During sexual reproduction, each parent contributes half its genes to the offspring. A male human contributes 23 chromosomes, and a woman has 23 in the egg she contributes, for a total of 46. This sexual mixing is what drives evolution.

In animals that use sex as a means of reproduction, each individual is different from its parents since each individual receives only half of the parents' genes. An offspring is thus a mix of its parents' genes. This contributes to having a great diversity in the population. Consider, for example, humans. We all don't look alike because we all have different genetic make-ups, since we received half our genes from our mother and half from our father. Such diversity guaranteed that a least a few hardy individuals will survive under evolutionary pressures such as changing environments, increased predation or increased parasite load. The fact that animals have two genders reflects the fact that this method is the simplest way to achieve diversity—it only takes two. Having more sexes would increase the cost and complexity of producing diverse offspring. (Imagine how more difficult and expensive it would be to arrange to have take three or more to "dinner and a movie!") The fact that most animals have two sexes shows that this system evolved at the very beginning of life itself and has worked adequately ever since. This system is evident across all the modern phyla, most of which diverged more than a billion years ago, indicating the ancient age of the system.

Other than responses to environmental changes, animals need to evolve and adapt to other pressures. They need to evolve better defenses against their predators and their parasites, otherwise the predators would wipe them out. At the same time, the predators and parasites are also evolving better methods to attack their prey. This system of coevolution has been called the *Red Queen Theory*, from a scene in the book *Alice Through the Looking Glass* by Lewis Carroll. Alice and the Red Queen are running as fast as they can, but they never get anywhere, since the scenery and other players are running just as fast. Although they are running fast, they just maintain where they are. Related practical examples are the "Arms Race," the "Rat Race," and "keeping up with the Joneses." As a snail develops a thick shell to prevent being eaten by a crab, the crab evolves thicker, stronger pincers to break the snail's shell. Again, this evolutionary arms race is driven by sex, which allows mutations and diversity, thus allowing some individuals to have better survival rates and to be able to pass on their improved genes to their offspring. Note that they *must* reproduce in order for this to be effective.

Evolution has worked its magic on the mating event, the method by which genes are mixed and get passed to the offspring, especially among the mollusks which have some very unusual methods of reproduction.

Sexual Change Artists—Mollusk Switch Hitters

Oysters

Consider the oyster. The oyster of the genus *Crassostrea* is a *bivalve* (a mollusk with two shells). It sits on its left shell in the mud of an estuary or attaches its left shell to a rock or another oyster. An oyster is not the Einstein of the mollusk world. It doesn't even have a brain. It doesn't need one because it only does about three things: It filters food from the water, it releases eggs or sperm into the water, and it closes up in response to some environmental stimuli, such as low tide (no water), bad water, or touch by a predator such as a sea star or human. But an oyster does do one thing that is interesting: It changes sex, sometimes frequently and sometimes rapidly.

The common oyster starts off as a swimming larva drifting in the plankton. When it gives up its nomadic planktonic life and settles down to the bottom, it

An oyster bed in Maine.
© Gary Carter / Visuals Unlimited.

begins its boring phase of life of filtering food. It develops into a functional male, then slowly turns into a female. At one stage it may have both functional ovaries and testes. This process of beginning life as a male and changing into a female is known as *protandry* (as opposed to *protogyny*, where females become males). Once an oyster has become a functioning female, it spawns its millions of eggs into the water where they are fertilized by sperm from male oysters. It then rather quickly over several weeks changes back into a male, when it emits sperm into the water.

It is logical that the change from male to female would be longer than the change from female to male. Eggs are much larger than sperm and contain nutrients for the developing oyster larvae. Sperm are tiny and are little more than a swimming genetic package. Therefore, it takes fewer resources (food) to become a male and produce tiny sperm than it does to become female and produce large yolky eggs. It thus makes sense that protandrous mollusks first become males, since they can do this at a smaller size. As they grow, they can turn into females, or in the case of an oyster, switch back and forth.

Since it takes more resources to become a female, the frequency of the sex changes depend on environmental factors. It has been shown that oysters in the colder waters of Norway may only go through one sex change a year, whereas the same species in the warm waters of the bays of France go through two or three changes each year.

Slipper Shells

Slipper shell limpets—which are not true limpets—only go through one sex change in their lives. Slipper shells are small snails with a bowl-shaped shell with a partial shelf going across the inside of the shell. Their scientific name is *Crepidula fornicata*, but the species name has nothing to do with its odd mating habits—it is named for the arched *fornix* of its shell. They live on oyster shells, other slipper shells, other snails, or inside other, empty dead shells. They filter food from sea water to make a living, and since they have very limited mobility, they usually attach to the same spot for life.

Larvae of slipper shells settle preferentially onto the shells of other slipper shells and form a pyramid of up to 30 individuals. Newly settled limpets develop into functional males which eventually turn into females, passing through a hermaphroditic stage, after 2–3 years. Thus the pyramid always has males and females, the smaller males on top of the chain and larger females on the bottom.

Although there may be pairings of only two individuals, one lower female with a male on top of her, in longer chains of individuals the males may be separated from the females by some distance, especially since there may be transitional individuals in between. This poses a problem for the males since they don't move and since the males have to inseminate the females by internal

Sperm shoots out of a limpet on the sea bed 3 m (10 ft) deep near Signy Island, Antarctica. The slightest water current, such as a hand wafting past, triggers spawning. Spawning occurs only a few days a year.

© Rick Price / Corbis.

fertilization. They need to get their sperm into the female somehow. The males have overcome this challenge by evolving a very long penis. The male slipper shell's penis is at least as long as its body and is tremendously prehensile and active. Since these animals have little or no eyesight, they can't see a possible mate to find it. They can't see where to place their penis and they have to feel for the females (but remember, they know a female is down there somewhere!). Once they are in the right general vicinity of a receptive female, they may find the proper receptacle for their sperm by touch or by chemo-reception. As yet, we know little of the mating systems of the slipper limpet—whether the female has to be receptive, whether there is courtship, or whether there is competition among the males to mate females.

The advantages of such a mating system is that the males and females are always together. Males don't have to go looking for females, and females always have a ready source of males when they are ready to mate. Females deposit egg clusters on the hard surface beneath them. Eggs develop under the protective cover of the female's shell and hatch into free-swimming larvae that eventually settle onto other shells.

Hermaphrodites

What is a *hermaphrodite*? All of the sea slugs and pulmonate gastropods (land snails and slugs) are hermaphrodites. They have fully functional organs of both sexes, and they produce eggs and sperm simultaneously. But they don't have to act as both male and female simultaneously during mating (although many do). They might have elaborate fights—as some slugs do—to be male, since males need fewer resources to mate. Some land slugs fight with their penises, which are greatly enlarged, sparring with them to prove who is the biggest and best and "most male."

One species of banana slug on the American west coast that indulges in this type of fighting is called *Ariolimax dolichophallus*. The species name means "large penis." This slug engages in mutual mating, but during mating their male organs are greatly enlarged and may get caught inside its partners female organs. Although the partner may just crawl off with the "male" still lodged inside it, dragging behind, this slug has evolved a novel solution to the problem—it turns around and chews off the other's male organ and eats it. There are two advantages to this odd behavior other than just getting rid of an annoying hanger-on. The gnawer gains some nutritive value from eating the other's appendage, but there is also a genetic advantage to making one's partner lose its member. The partner that has become *apophallated* (lost its penis) then has to act as a female and again has to spend the resources in producing large eggs, while the victorious slug can act as a male with other slugs, spreading its genes while using fewer resources (time and energy). Some scientists have argued that development of such gigantism in this slug may lead it toward extinction, as we have seen in other giant animals in the past (some dinosaurs, saber-toothed tigers, wooly mammoths).

Sea slug nudibranchs often mate with multiple partners at the same time. The sea hare *Aplysia* (not a true nudibranch, but closely related) is a good example of this. If there are just two slugs mating, the slugs align themselves so that the slug in back acts as a male to the slug in front, and the one in front acts as a female to the one behind it. When many sea slugs mate together (up to 20 or more), each acts as a male to one in front. Sometimes the end slug bends around and meets the slug at the other end of the chain, thus forming a circle.

Other sea slugs and many freshwater snails fertilize themselves, a process known in scientific language as *selfing*. There are some snails where mating has never been observed; they all self-fertilize. From a genetic and evolutionary point of view, such a process might appear to be primitive, but it works for these snails, and it is obviously convenient if there are no other snails around.

The hermaphroditic great gray slug *Limax maximus*, introduced to many gardens around the world, exhibits yet another bizarre mollusk mating behavior. During their courtship behavior, two slugs writhe around each other, almost as in a knot. They both produce copious amounts of thick mucus as they rub each other all along their bodies. During the course of this courtship behavior, they crawl out onto a limb of a shrub or low tree and suspend themselves in the air from their tails by a thin thread of tough mucus. It is here that they mutually mate.

Yet another very bizarre mating ritual among the mollusks is that of the garden snail *Helix aspera*. This is a common species that has been introduced to vegetable gardens in many parts of the world. It is a true hermaphrodite with working reproductive organs of both sexes at the same time. Like most hermaphrodites, it doesn't fertilize itself but cross-fertilizes with another snail, thus increasing genetic diversity and providing yet another example in support of the Red Queen Theory.

But before it mates with another snail and trades sperm with it, it shoots a calcareous dart into the body of the other snail and shortly afterward gets shot in turn. The snails then engage in mutual mating that may last 4–6 hours. The dart is made of calcium carbonate, the material most shells are made of, and is usually toothpick-shaped, about a centimeter (.4 in) long. These tiny spears are known as "love darts." This behavior has been known for centuries and may have even spawned the myth of Cupid's arrows.

Why would these snails develop this habit? The darts are costly to produce and possibly harmful to the recipient. Several theories have been proposed to account for this odd mating behavior. One was that it stimulated the recipient to be more receptive to mating with the shooter, but many times these snails miss or hit the shell and mating occurs anyway, so this theory has been discounted. Another theory suggested the dart may be a gift of calcium to the recipient, since egg-laying uses much calcium from the parent, but this theory has also been proven wrong. The most recent theory is that the love darts have mucus on them that contain hormones. These hormones stimulate enlargement of the sperm receptacle in the female organs of the snail, so that more sperm is stored in the dart receiver and consequently used when eggs are laid. If mating occurs without dart penetration, most of the sperm is simply digested by the recipient and not used for reproduction. Thus dart shooting is yet another way of making sure one's genes are passed on to the next generation.

Octopuses and Squid

Sexes are always separate in the cephalopods, that is, they are *dioecious*. Although males and females may look alike superficially, there are usually some external differences (*dimorphism*) between males and females. Males usually have one or more modified arms that are used in mating to pass their sperm to the female.

Squid use a number of different mating behaviors. Caribbean reef squid *Sepioteuthis sepioidea* go through elaborate courtship rituals, perhaps used to determine mate assessment and readiness to mate over hours and days, but the actual mating takes a fraction of a second. A male deposits a sperm packet (*spermatophore*) onto the arms of a female. In a successful mating she then takes the spermatophore with her arms and

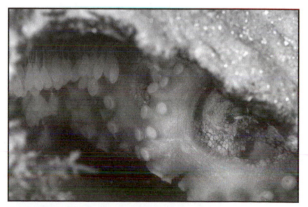

Female pallid octopus brooding eggs off the coast of Australia.
© Mark Norman / Visuals Unlimited.

places it inside her body to be used in fertilizing her eggs. She may somehow determine that the male is not suitable and rid her arms of the spermatophore by blowing it away with her funnel.

Members of other squid species come together in giant mating gatherings. The California market squid, *Loligo opalescens*, cluster in huge groups of up to hundreds of thousands when they mate. A male grasps a female head to head with his arms and deposits his sperm packets into the female over the course of several minutes. As he does so, he produces marvelous color patterns of purple and brown on his head and arms. Within hours, the female lays her eggs; then both male and female die. There is no such thing as "safe sex" for cephalopods. They die after reproducing, just as salmon do.

These two cuttlefish are mating. Cuttlefish mate head to head. The male is on the right and the female is on the left. *Courtesy of James B. Wood.*

Male octopuses have a modified third right arm, known as a *hectocotylus*, for mating. At the end of this hectocotylized arm is the *ligula*, a grooved, spatulate organ without suckers on it. A spermatophore is passed along this arm in a groove, and the ligula places the spermatophore into the oviduct of a female octopus. Since this arm is moveable, flexible, and extendible, it allows a male octopus to mate at a distance. Depending on species, a male octopus is able to extend his hectocotylized arm a considerable distance to mate a female, up to a meter (3 ft) away. He is able to stay outside a female's den, assess her readiness and willingness to mate, and extend his hectocotylized arm into the den and into the female to accomplish mating. Matings may take up to 4 hours, depending on the species.

Mating behaviors of most squids and octopuses are unknown to science. Most cephalopods have not been studied yet in detail in the wild or in captivity. Even the technology of keeping them in captivity has not been determined yet. Many of them live in mid-water and avoid trawl nets, or live on the abyssal plains of the ocean floor where it is impossible to observe them for more than a few hours from inside a submersible or using an ROV (remotely operated vehicle). We count ourselves lucky to see these deep water animals briefly and even luckier still to see the few matings that have been observed. We have never even seen the notorious giant squid alive (except as a tiny juvenile), much less learned anything about its mating behavior. Squid researchers would be ecstatic to see a giant squid embracing another in the ocean's murky depths.

These molluskan examples show that there is nothing new under the sun. Human teenagers grappling in the back seat of a car have nothing on the mollusks. Mollusks have done it all millions of years ago. Nowadays we practice safe sex, but for a mollusk it may not be safe: They run the risk of falling from trees, getting shot with darts, having one's sexual member chewed off, or even dying. This study of the mollusks' mating behavior will provide interesting work for animal behaviorists for many years to come.

See also Cephalopods

Further Resources

Forsyth, A. 2001. *A Natural History of Sex*. New York: Firefly Books.
Gordon, D. G. 1994. *Field Guide to the Slug*. Seattle: Sasquatch Press.
Lutz, R. A. & Voight, J. R. 1994. *Close encounter in the deep*. Nature, 371, 563.
Morton, J. E. 1967. *Molluscs*. London: Hutchinson University Library.
Ridley, M. 1994. *The Red Queen: Sex and the Evolution of Human Behavior*. New York: Macmillan.

Roland C. Anderson

■ Reproductive Behavior
Monogamy

Monogamy traditionally corresponds to the mating of a single male with a single female. As such, it denotes both a social organization and its genetic consequences, and is only relevant to *gonochoric species* (i.e., species in which male and female gametes are produced by different individuals). *Pair bonding*, the strong affiliation that exists between reproductive partners, usually characterized by mutual familiarity, cooperative behavior, and efforts to maintain proximity, is often considered to be the essence of monogamy. However, the strength and duration of this affiliation varies among monogamous species. Social monogamy can be continuous over the whole year, or parttime when partners spend only part of the year together, usually during the breeding season. Pair bonds may last for just one reproductive attempt or continue until the death of one partner. Although monogamy is supposed to imply sexual exclusivity between pair members, strict genetic monogamy appears to be the exception rather than the rule in the animal kingdom. Indeed, evidence obtained from using genetic markers over the last 15 years has shown that in most monogamous species, pair bonding does not necessarily go hand in hand with sexual exclusivity. When a male and a female herring gull, *Larus argentatus*, apparently cooperate in parental duties such as incubation and delivering food to the brood, there is no absolute guarantee of genetic monogamy. The female may have engaged in extrapair copulations that eventually resulted in extrapair paternity. Similarly, the male, although assisting only one female in parental duties, may attempt to copulate with many others at the same time. Therefore it seems appropriate to consider monogamy more broadly as the association of one male and one female in breeding, without any implication on the degree of sexual exclusivity.

Taxonomic Distribution of Monogamy

Monogamy, in its broad sense, is a rare mating system that is, however, widely distributed across phylogenetic groups. It is rare in invertebrates, amphibians, and reptiles, occurs at low frequencies in fish and mammals, and only predominates in birds where about 90% of species are considered to be socially monogamous. However, monogamy should not be taken as a monotypic form of social organization. Instead, it refers to a diverse and taxonomically varied set of associations between one male and one female, that differ largely with respect to grouping patterns, duration of pair bonds, male and female behavioral profiles, and phylogenetic history.

Among invertebrates, the most extreme form of monogamy probably corresponds to what is observed in termites, where a male and a female typically associate and remain for life in the same nest. In other arthropod species, cooperation between a male and a female to raise a brood is observed in some beetle species (Coleoptera: Scarabaeidae). In burying beetles (genus *Nicrophorus*) males and females cooperate to bury carrion which becomes the food source for their larvae and is used to rear young. Similarly, in dung beetles (genus *Copris*) males and females cooperate in excavating a nest underneath a dung pad and filling it with fresh dung from the pad. The female then converts the dung into separate brood balls and lays an egg in each of them. Monogamy also occurs in some crustacean species, as in snapping shrimps (genus *Alpheus*) where male and female of a reproductive pair jointly defend a territory, or in the starfish-eating shrimp, *Hymenocera picta*, where pair members cooperate in catching prey.

Among freshwater fish, only a few species are known to display monogamy with bi-parental care in bony tongues, catfish, and snakeheads. The cichlids are an important exception. The Cichlidae is a large family consisting of about 1,000 species in which monogamy with biparental care is predominant. An important feature of monogamy in freshwater fish is that caretaking generally is not limited to the protection of eggs, but extends to protecting free-swimming young. In addition, monogamous pairs possess breeding territories. Monogamy also occurs in some coral-reef fish, such as butterflyfish, filefish, and surgeonfish, despite the absence of biparental care of free-swimming young. Individuals of such species usually form closely coordinated heterosexual pairs that cooperate to defend a territory. In syngnathids (pipefish and seahorses) males provide all postzygotic care by carrying the eggs on their ventral surface or in a specialized brood pouch. Some pipefish and all seahorses so far documented are monogamous. The formation of pair bonds has been documented by direct observations in several species of seahorses, with pairs performing daily greetings throughout the breeding season. In some seahorse species and at least one pipefish, pair bonds can be maintained over successive broods or over breeding seasons.

Among invertebrates, the most extreme form of monogamy probably corresponds to what is observed in termites where a male and a female typically remain in the same nest for life.

© *Brian Rogers / Visuals Unlimited.*

Social monogamy is virtually unknown in amphibians, although some results suggest that it may occur in the territorial red-backed salamander, *Plethodon cinereus*. In reptiles, short-term monogamous relationships have been reported for a few lizard species, with males defending an area occupied by a single female or both pair members cooperating to defend a territory. A noticeable exception is the case of the Australian sleepy lizard, *Tiliqua rugosa*. This nonterritorial lizard forms monogamous pair bonds for about 8 weeks before mating each year. Although pair members separate after mating, they usually reunite the next year.

Although most bird species engage in social monogamy, only 21% of the 159 bird families show long-term partnerships. Pair stability can be extreme in some species, such as barnacle geese, *Branta leucopsis*, where pair members stay together each day, often for life. Indeed, most geese have only one partner during their life. On the reverse, in the greater flamingo, *Phoenicoptreus ruber roseus*, pairs typically split after breeding is completed and do not reunite in successive years.

About 5% of mammal species are considered to be socially monogamous, although this figure may underestimate the true value, given that many mammal species are nocturnal and difficult to study. Monogamy has been reported in some marsupials, in shrews, in several rodent species, including the beaver, *Castor fibex*, and the African porcupine, *Hystrix cristata*, in bats, in antelopes, and in several carnivores, including the wolf, *Canis lupus*, the coyote, *Canis latrans*, and the badger, *Meles meles*. About 10–15% of primate species have been classified as socially monogamous. However, they differ extensively in their grouping patterns and ecology. Indeed, only 24 primate species occur exclusively or primarily in two-adult groups throughout their range. Finally, although in some human cultures the main social unit corresponds to the association of an adult male, an adult female, and their offspring, there is no evidence that monogamy is the predominant mating system of *Homo sapiens*.

The Evolution of Social Monogamy

Based on its taxonomic distribution, it is clear that social monogamy evolved many times in independent lineages. This may seem surprising because, to a certain extent, monogamy is an evolutionary puzzle. From a theoretical point of view, monogamy is expected to be rare. Basically, the asymmetry between males and females in gamete production (males produce cheap sperm in large quantities, whereas females produce only a few costly eggs) promotes contrasted mating strategies between the two sexes. Females are expected to be coy and choosy, whereas males are supposed to be promiscuous and seek multiple matings. Why then do males engage in monogamy in several species?

Actually, the selective forces that favor social monogamy vary largely between species. In the absence of biparental care, at least three different and nonexclusive hypotheses can explain the evolution of social monogamy. First, males can be constrained to monogamy because of their inability to locate, attract, or provide resources to more than one female. Females generally distribute themselves in relation to the distribution of limiting resources, and males map onto the distribution of females. If resources are sparse but uniformly distributed, the large distance between females may prevent males from monopolizing more than one mate. In the desert spider, *Agelenopsis aperta*, monogamy is the predominant mating system, although about 10% of males are polygamous. Studies conducted in the laboratory have shown that male spiders readily accept matings with several females if given the opportunity. In the field, however, high travel costs to males limit their potential for polygyny. In some rodent species, adult males end up in monogamous pairs in years of low food availability just because they are unable to defend a territory large enough to accommodate more than one female. Second, males and females may benefit from monogamous pairing if a solitary individual is unable to defend a territory alone. In snapping shrimps, for instance, territorial cooperation between a male and a female reduces the risk of eviction from the territory. Males and females benefit from monogamy through dividing the labor of territorial defense and maintenance. Third, monogamy can result from mate guarding and intrasexual aggression. In several crustacean species, females are constrained in sexual receptivity to a short period of time after molting, whereas males are available for mating most of the time during their molt cycle. This results in strong competition between males for access to receptive females. In this situation, mate guarding of prereceptive females is selectively advantageous for males. In conditions that make searching for females too costly for males, temporary mate guarding can transform into prolonged heterosexual pairing as observed, for instance, in snapping shrimps. In some other species, both males and females may enforce monogamy though reciprocal mate guarding. This is likely to occur when there is extensive variation in mate quality in both sexes. For instance, larger females might be more fecund, and larger males might be better able to defend a territory. In such species, each pair member attempts to monopolize its mate through being aggressive toward individuals of its own sex. It is important to note that all three hypotheses can combine to promote social monogamy without biparental care, as observed in the coral-reef goby, *Valenciannea strigata*. This fish species occurs in monogamous pairs that share an all-purpose territory with several burrows used for nests and refuge. Pairs remain together over several episodes of reproduction, mate exclusively, and only males perform parental care. In this species, uniformly distributed resources lead to monogamy through lowering the costs of reciprocal mate guarding, while monogamous pairing is beneficial to both sexes in terms of shared maintenance of burrows and food provisioning.

The evolution of male parental care is thought to be determined by a male's ability to increase offspring survival and competitive ability and by the trade-off between enhancing offspring survival and reducing future reproductive success. Males are faced with a sort of dilemma. Assisting a female in incubation duties and brood care may contribute to increase offspring fitness. However, time and energy spent in parental duties could also be invested in searching for and copulating with additional females. From a theoretical point of view, mate desertion should occur whenever benefits of remating outweigh those of caring for young. This logic of mate desertion may help to explain why the frequency of monogamous pairing differs so markedly between birds and mammals. In mammals, gestation is internal and therefore females are indispensable for the development of young. By contrast, the contribution of males can be limited to egg fertilization, since females are physiologically capable of providing for their own offspring before and after birth. In the absence of ecological constraints favoring group living or limiting access to females, male mammals benefit from deserting their mate soon after fertilization. The situation is quite different in birds. Males have an opportunity to increase their own fitness through incubating eggs and feeding nestlings. In addition, a male that deserts his female is exposed to a risk of retaliation: If females desert the clutch as well, then there will be no benefit at all for the male. Female mammals just do not have the same latitude. As soon as they become pregnant, they are forced to invest in their progeny. Abandoning the young after birth would incur an important loss in terms of past investment.

Under particular ecological circumstances, however, the additional offspring that survive to maturity as a direct consequence of paternal care may compensate for the loss of additional mating opportunities. Susceptibility of young to predation risk may, for instance, be reduced when both reproductive partners tend their progeny. It is also possible that in some species, paternal care first evolved in populations where males were constrained to monogamy because of limited opportunities to mate polygynously. Reduced aggression between a male and a female sharing a common territory may have then facilitated attendance of the progeny by the male. The fitness benefits accruing from paternal care may have, in turn, positively selected male genotypes that engaged in some sort of paternal care. In the long run, males may have developed particular physiological and behavioral features that canalized the species in social monogamy. As a result, in some species, male contribution to parental care has become essential for young survival and development. In several monogamous fish species, males incubate the young in their mouth, or in a special brood pouch, as observed in seahorses. The size of the buccal cavity or of the brood pouch limits the number of young a single male can raise to the clutch laid by one female, and therefore makes mating with additional females of little value. In birds, experimental removal of males tends to have the most dramatic effect in species where males both incubate and feed the young.

Interspecific Variation in Mate Fidelity

In several monogamous species, particularly vertebrates, individuals live long enough to engage in more than one reproductive attempt. Thus, pair members regularly face the same alternative, that is, keep the same mate or switch partners. Mate fidelity occurs when pair members remain paired or reunite over successive breeding attempts. The term *divorce* has been coined to denote the separation of former mates when both survive from one reproductive attempt to another, whether this is due to failure to reunite or to re-pairing of at least one individual with a new mate. Although the term divorce is somehow fraught with some anthropomorphic connotation, it has been extensively used in the literature in preference to alternatives such as mate switching or pair splitting.

In most invertebrate species, monogamy is only seasonal; that is, pair bonds generally do not extend beyond one reproductive attempt. The situation is more complex in vertebrates, although only a few experimental studies have directly examined the factors that determine the strength and duration of pair bonds. In addition, data on pair-bond duration in vertebrates essentially come from bird studies, with only a few studies conducted in other groups. In the sleepy lizard, for instance, 79% of females were found to keep the same mate over consecutive breeding seasons, and one pair is known to have remained together over 10 consecutive years. It is suspected that a large proportion of mate changes may actually be due to the death of one partner. In addition, empirical evidence suggests that mate switching is particularly frequent among young males that may need some years before establishing a more stable partnership. In the Western Australian seahorse, *Hippocampus subelongatus*, genetic data obtained from 14 males and their broods revealed that eight males remated with the same female over two consecutive breeding attempts, whereas six males mated with a new female. However, the proximate determinants of divorce in seahorses remain unknown.

As noted above, there is extensive variation in pair-bond duration in birds. Several hypotheses have been put forward to explain the occurrence of divorce in birds. One simple possibility is that variation in mate fidelity among birds directly reflects differences in ecological constraints affecting the possibility of mate reuniting. It has been therefore suggested that mate fidelity might simply be a by-product of fidelity to nest site. In several bird species, individuals are extremely faithful to their nesting place, often returning to nest in the same location or very close to it. A recent comparative study in Ciconiiformes (i.e., wading birds and seabirds) has shown that, indeed, a significant correlation exists between the degree of mate fidelity and the degree of site fidelity. Furthermore, it appeared that in this group site fidelity evolved before mate fidelity, suggesting that the latter might have first evolved as a consequence of the former. However, evolutionary scenarios recapitulating the coevolution between the two traits also indicated that in more recent evolutionary times mate fidelity and site fidelity have evolved independently of each other. Thus, other factors might influence the interspecific variation in mate fidelity in birds.

According to the "incompatibility hypothesis" introduced by the British ornithologist John Coulson, pair fidelity depends on breeding success, which is in turn determined by compatibility between mates independently of their intrinsic qualities. Incompatible pairs are more likely to fail to fledge young. Mate switching would allow former pair members to acquire new, compatible mates. This hypothesis predicts that divorce should be more frequent among young individuals, and empirical evidence in birds supports this view to a certain extent. However, the incompatibility hypothesis has been challenged recently. Bruno Ens, an ornithologist from Netherlands, argued that, most often, divorce in birds corresponds to an unilateral decision made by one pair member to improve its breeding status. The rationale behind the argument is that because the process of pair formation is time limited it sometimes happens that some individuals end up being paired to mates of low quality compared to their own. Such individuals may subsequently improve their breeding status through divorcing their current mate and mating with another individual if a better option becomes available. The individual that initiates the divorce is supposed to benefit from switching mates, whereas its former mate is left as the victim of divorce. Direct empirical evidence in favor of the "better option hypothesis" comes from an experimental study on black-capped chickadees, *Parus atricapillus*. Chickadees are monogamous and territorial during the breeding season, but live in flocks during the winter. A dominance hierarchy for each sex exists within each flock, and males and females of higher social status tend to pair together. The availability of males was experimentally increased through selective mate removal during the nest-building period. Females from neighboring territories then deserted their males for experimentally widowed males of high social status, whereas males of low social status failed to attract females. As in the incompatibility hypothesis, the better option hypothesis considers that poor breeding success is a clue with which individuals can assess the quality of their mate and eventually decide to initiate a divorce. Recently, a review of the empirical evidence in birds has shown that, on average, pairs that fail to reproduce are less likely to maintain their pair bond compared to pairs that bred successfully, although important variation exists between species.

Behavioral and Neuroendocrine Correlates of Monogamy

Monogamy generally involves a process of reciprocal behavioral adjustment between mates. This process is thought to reinforce the pair bond, although pair bond "strength" remains an elusive concept. Ceremonial tail beats in cichlids, greeting ceremonies in seahorses, triumph ceremonies in geese and albatrosses, duetting and allopreening in birds and primates are all thought to play a role in this process. Gibbons (Hylobatidae) are monogamous arboreal apes that live in East Asian rainforests. All species produce loud and long vocalizations. In a large majority of species, mates typically combine their species-specific and often sex-specific songs into duets. Studies conducted in captivity on the siamang, *Hylobates syndactylus*, have shown that the frequency and duration of duet songs was positively correlated with grooming activity and behavioral synchronization, and negatively correlated with distance between mates. This suggests that the production of coordinated duets in siamang pairs is directly related to the intensity of pair bonding.

Besides behavioral characteristics of monogamy, there is some evidence that monogamy also involves some physiological adjustments. In monogamous bird species, males have high levels of testosterone at the onset on the breeding season when they are settling on a territory and start attracting a mate. Once the female has laid eggs, testosterone levels typically drop sharply and levels of prolactin, a hormone involved in parental care, start increasing. The same pattern is not observed in polygynous bird species where males maintain high levels of testosterone throughout the breeding season. Experimental injection of testosterone in males of a monogamous bird species during the breeding season results in males neglecting their brood and resuming courtship activity. More recently, a physiological basis mediating affiliation between pair members has been evidenced in the Australian zebra finch, *Taeniopygia guttata*. In the wild, zebra finches live colonially and maintain long-term pair bonds. Experimental separation of mates in captivity elicits changes in behavior, such as increased song rate and activity in males and increased defecation rates in both sexes. This suggests that pair separation is stressful. Indeed, plasma corticosterone (a hormone indicative of stress) concentrations increased following pair mate separation, even in the presence of individuals of the opposite sex. In addition, corticosterone levels reduced to baseline upon reunion with the pair mate, but not when paired with a new opposite-sex individual.

While rare, monogamy has been observed in several species of mammals, including beavers. This pair is grooming each other.

© Carlyn Galati / Visuals Unlimited.

Whereas knowledge about the endogenous mechanisms of monogamy in birds is just developing, a far deeper understanding of the physiology of pairbonding has been reached in monogamous rodents. Over the last 20 years, studies of the prairie vole, *Michrotus ochrogaster*, have contributed to bridging the gap between ecological and neuroendocrinal approaches to monogamy. In preliminary field studies it was found that male–female pairs of prairie voles are caught in the same trap far more frequently than are pairs of related, polygynous vole species, such as meadow voles, *M. pennsylvanicus*, or montane voles, *M. montanus*. This observation was strongly suggestive of social monogamy. Indeed, subsequent studies revealed that, unlike most rodents, prairie voles engage in long-term pair bonds with both the male and the female in a pair contributing to parental care. Because both monogamous and polygynous voles are easily reared in the laboratory, they proved to be ideal subjects to explore the physiological bases of monogamy. A number of physiological and anatomical studies have implicated two neuropeptides, oxytocin (OT) and vasopressin (AVP) in the mediation of affiliation between pair members, mate guarding and paternal care in voles. Comparative studies between monogamous and polygynous vole species have shown that the neuroanatomical distribution and concentration of OT receptors and AVP receptors show striking and consistent differences, whereas the distribution of other relevant neuropeptide receptors is virtually identical. Recent physiological studies further suggest that OT and AVP may be facilitating affiliation and pair bonding in monogamous voles by modulating particular regions of the brain which are functionally implicated in reward. Although the generality of these findings for other monogamous vertebrate species awaits further investigation, they offer a stimulating perspective to investigating the evolution of monogamy at a mechanistic level.

See also Reproductive Behavior—*Female Multiple Mating*
Reproductive Behavior—*Mate Choice*
Reproductive Behavior—*Sex, Gender, and Sex Roles*

Further Resources

Barlow, G. W. 2000. *The Cichlids Fishes. Nature's Grand Experiment in Evolution.* Cambridge, MA: Perseus Publishing.
Black, J. M. (Ed.) 1996. *Partnerships in Birds. The Study of Monogamy.* Oxford: Oxford University Press.
Bull, C. M. 2000. *Monogamy in lizards.* Behavioural Processes, 51, 7–20.

Cézilly, F., Dubois, F. & Pagel, M. 2000. *Is mate fidelity related to site fidelity? A comparative analysis in Ciconiiforms*. Animal Behaviour, 59, 1143–1152.

Cézilly, F. & Johnson, A. R. 1995. *Re-mating between and within breeding seasons in the Greater Flamingo*. Ibis, 137, 543–546.

Clutton-Brock, T. H. 1989. *Mammalian mating systems*. Proceedings of the Royal Society, 235, 339–372.

Dubois, F. & Cézilly, F. 2000. *Breeding success and mate retention in birds: A meta-analysis*. Behavioral Ecology and Sociobiology, 52, 367–364.

Fuentes, A. 1999. *Re-evaluating primate monogamy*. American Anthropologist, 100, 890–907.

Geissmann, T. & Orgeldinger, M. 2000. *The relationship between duet songs and pair bonds in siamangs* Hylobates syndactylus. Animal Behaviour, 60, 805–809.

Gilette, J. R., Jaeger, R. G. & Peterson, M. G. 2000. *Social monogamy in a territorial salamander*. Animal Behaviour, 59, 1241–1250.

Kleinman, D. 1977. *Monogamy in mammals*. Quarterly Review of Biology, 52, 39–69.

Kvarnemo, C., Moore, G. I., Jones, A. G., Nelson, W. S. & Avise, J. C. 2000. *Monogamous pair bonds and mate switching in the Western Australian seahorse* Hippocampus subelongatus. Journal of Evolutionary Biology, 13, 882–888.

Matsumoto, K. & Yanagisawa, Y. 2001. *Monogamy and sex-role reversal in the pipefish* Corythoichthys haematopterus. Animal Behaviour, 61, 163–170.

Young, L. J., Lim, M. M., Gingrich, B. & Insel, T. R. 2001. *Cellular mechanisms of social attachment*. Hormones and Behavior, 40, 133–138.

Frank Cézilly

■ Reproductive Behavior
Morning Sickness

Morning sickness is the common term for nausea and vomiting during gestation. It actually is a complete misnomer: Symptoms occur throughout waking hours, not just in the morning, and whereas "sickness" implies pathology, healthy women experience the symptoms and bear healthy babies. For these reasons, the medical community uses the acronym *NVP* (short for "nausea and vomiting of pregnancy").

Across the world, about two thirds of pregnant women experience mild nausea to regular vomiting during pregnancy. Symptoms first appear about 5 weeks after the last menstrual period, peak during weeks 6–12, and gradually decline thereafter. Some women experience symptoms severe enough to disrupt daily life, and about 1% of women experience symptoms that require hospitalization (a condition known as hyperemesis gravidarum). However, the vast majority of women have symptoms in the "normal," nondebilitating range.

Although morning sickness is best studied in humans, it may not be unique to us. For example, domestic dogs typically show a sharp drop in food consumption during weeks 3–5 of their 9-week gestation, captive rhesus macaques usually exhibit a decrease in appetite during weeks 3–5 of their 23-week gestation, and captive chimpanzees sometimes experience nausea and appetite irregularities in early pregnancy. Of course, morning sickness may actually be more widespread, but has not been noticed due to difficulties of detecting subtle changes in dietary habits of free-living mammals early in gestation.

The etiology of morning sickness can be analyzed at proximate and ultimate levels. Regarding the former, the underlying physiological mechanisms are well-studied. Reproductive hormones that accompany viable pregnancies set the stage by sensitizing neural pathways that trigger the symptoms. In particular, concentrations of chorionic gonadotropin (hCG) in maternal blood follow a time course that closely parallels NVP symptoms. At the ultimate level, evidence is accumulating that nausea and vomiting protect the mother and her embryo by causing violent and instantaneous expulsion of, and thus temporary aversions to, potentially dangerous foods, especially those that might be contaminated with bacteria and fungi or that contain natural chemicals which could cause birth defects.

This "maternal-and-embryo-protection" hypothesis is supported by five types of evidence. First, morning sickness peaks when embryogenesis is most susceptible to disruption (i.e., during the first trimester). Embryonic tissues are highly sensitive during certain critical periods when cells are rapidly dividing and differentiating into organs. Starting in week 5, the developing central nervous system, heart, and ears become critically sensitive; other organ systems follow soon thereafter. Sensitivities of most embryonic organ systems peak during weeks 6–12 and then decline, except for the central nervous system, which continues to be critically sensitive through week 18. There is a striking correspondence between the most sensitive periods in embryogenesis and the rise and fall in morning sickness symptoms.

Second, morning sickness is associated with positive pregnancy outcomes. Information on the relationship between the symptoms and miscarriages (i.e., spontaneous abortions during the first 20 weeks) is available from nine studies involving 22,305 pregnancies. In every study, women who experienced morning sickness were significantly *less* likely to miscarry than women who experienced no symptoms. There also was an inverse correlation between the severity of NVP symptoms and likelihood the pregnancy would end in a miscarriage. That is, within the normal range of symptoms, women who experienced nausea and vomiting were less likely to miscarry than women who only occasionally felt nauseous.

Third, foods that pregnant women find aversive potentially contain disruptive chemicals and pathogenic microorganisms, but foods that are craved rarely do. This was determined in 20 independent studies of gestational aversions (among 5,432 women) and 21 studies of gestational cravings (among 6,239 women), based on questionnaires administered during pregnancy or soon after parturition. Of nine major food categories, pregnant women most often were aversive to "meat, fish, poultry, and eggs" and "nonalcoholic beverages" (mostly caffeinated). They also found "vegetables" and "alcoholic beverages" to be aversive. Interestingly, the pattern of gestational cravings was virtually a mirror image. Pregnant women most often reported craving "fruit and fruit juice" and "sweets, desserts, and chocolate."

The five food categories for which per capita aversions were significantly greater than cravings were the ones most likely to contain microorganisms (i.e., meat products), plant chemicals (strong-tasting vegetables, coffee, and tea), and alcohol (in sufficient quantities, caffeine and alcohol are teratogens). In contrast, the four food categories for which per capita cravings were significantly greater than aversions were the ones least likely to contain microorganisms or phytochemicals.

Fourth, aversions to foods that potentially contain harmful substances peak in the first trimester of pregnancy when embryogenesis is most sensitive to disruption. Detailed information on dietary preferences of women throughout gestation indicate that per capita aversions to all food categories are highest in the first trimester and decline dramatically thereafter. Aversions were significantly more frequent in the first than the second trimester for six of seven major food categories, and significantly more frequent in the second than the third trimester for three of the seven categories. Women in their first trimester also reported significantly more aversions than nonpregnant "controls" to all seven food categories, particularly to "meat, fish, poultry, and eggs."

Fifth, the occurrence of morning sickness depends on a woman's diet, with symptoms occurring least often among women whose staple foods are unlikely to contain dangerous chemicals and microorganisms. Across the modern world there is considerable variation in the frequency of NVP, from a high of 84% among Japanese women to a low of 35% among women in India, but no one has quantified diets of modern women in relation to morning sickness within or among countries. However, information on both diet and the occurrence of morning sickness among traditional societies exists in a database called the Human Relations Area Files. Interestingly, there are seven traditional societies from Africa, Asia, and North America in which ethnographers reported that morning sickness was never seen. These seven were significantly less likely to have meat as a dietary staple and significantly more likely to have only bland-tasting plants (e.g., corn, beans) as staples than 20 traditional societies in which morning sickness was observed. In most of these 20 societies, meat products were staples, whereas women in societies whose diets consisted primarily of corn were unlikely to encounter foods that are predicted to trigger morning sickness—and they exhibited the lowest incidence of symptoms.

Why is morning sickness not ubiquitous among mammals? There are two possibilities. First, morning sickness may be unnecessary for maternal and embryo protection in mammals that have more effective physiological mechanisms than humans for destroying ingested bacteria and fungi and detoxifying plant secondary compounds. Alternatively, herbivorous mammals might avoid eating the most toxic plants during gestation, and pregnant carnivores might pass up carrion that has been heavily infested with especially virulent or toxic microorganisms.

Further Resources

Fessler, D. M. T. 2002. *Reproductive immunosuppression and diet: An evolutionary perspective on pregnancy sickness and meat.* Current Anthropology, 43, 19–61.

Flaxman, S. M. & Sherman, P. W. 2000. *Morning sickness: A mechanism for protecting mother and embryo.* Quarterly Review of Biology, 75, 113–148.

Forbes, S. 2002. *Pregnancy sickness and embryo quality.* Trends in Ecology and Evolution, 17, 115–120.

Haig, D. 1993. *Genetic conflicts in human pregnancy.* Quarterly Review of Biology, 68, 495–532.

Nesse, R. M. & Williams, G. C. 1994. *Why We Get Sick. The New Science of Darwinian Medicine.* New York: Times Books.

Profet, M. 1992. *Pregnancy sickness as adaptation: A deterrent to maternal ingestion of teratogens.* In: *The Adapted Mind: Evolutionary Psychology and the Generation of Culture* (Ed. by J. H. Barkow, L. Cosmides, & J. Tooby), pp. 327–365. New York: Oxford University Press.

Sherman, P. W. & Flaxman, S. M. 2001. *Protecting ourselves from food.* American Scientist, 89, 142–151.

Sherman, P. W. & Flaxman, S. M. 2002. *Nausea and vomiting of pregnancy in an evolutionary perspective.* American Journal of Obstetrics and Gynecology, 186, S190–S197.

Paul W. Sherman

■ Reproductive Behavior
Sex Change in Fish

Patterns of sexual reproduction vary widely in animals, reflecting the diversity of environments in which they have evolved. Animal behaviorists attempt to identify sexual patterns, trace their distribution and their evolutionary history, and determine possible links with specific environmental conditions. As well, investigators are interested in within-species variation in sexual patterns, mating systems, and social behavior in different environments. Among vertebrates, teleost (bony) fish show the greatest variety in this regard. Scientists are also interested in the effects of fishing on the size of populations of sex-changing species. Exploitation of larger fish on coral reefs is expanding rapidly worldwide, and many of the species being fished change sex.

The sexual pattern most familiar to us is *gonochorism* in which the sexes develop separately and remain separate. Another pattern is *hermaphroditism* in which ovaries and testes occur in one individual, either simultaneously or sequentially. Among animals, many invertebrates are hermaphrodites (e.g., barnacles, most commercially important species of shrimp), but in vertebrates, gonochorism and stable sex-determining mechanisms seem to be the rule. In fact, hermaphroditism is so rare among vertebrates that it used to be considered abnormal. However, work since the 1940s has shown that it is a normal method of reproduction for many species of teleost (bony) fish, although not for cartilaginous fish (sharks, rays, chimeras). Hermaphroditism seems to have evolved at least 10 times in many different groups of teleosts, and the list of fish species that exhibit sex change is still growing.

Hermaphroditism may be either simultaneous or sequential. A simultaneous hermaphrodite produces both ripe sperm and eggs. One part of the gonad (primary sex organ, either testes or ovary in a gonochore) forms the sperm; the other part forms the eggs. These fish can mate with any member of their own species, but very rarely self-fertilize. Some simultaneous hermaphrodites mate first as one sex and then as the other; others function as both sexes at once. Simultaneous hermaphroditism is mostly confined to some bathypelagic (deep-sea) species and to some shallow water marine fish (e.g., some sea basses, hamlets, and groupers, family Serranidae). For simultaneous hermaphrodites, every encounter between mature individuals represents a potential mating opportunity. This is an advantage for bathypelagic species, most of which live at low population densities, but is irrelevant for shallow water species with high population densities. The evolutionary history and possible advantages of simultaneous hermaphroditism in these forms is being investigated.

A sequential hermaphrodite, that is, an individual that undergoes sex change, passes consecutively through both sexes. Sequential hermaphroditism is far more common than simultaneous hermaphroditism and takes two forms. In protandry, an individual develops first into a functional male and then into a functional female (the testes become nonfunctional and are replaced by ovaries). In protogyny, an individual matures first as a female and later becomes a male (the ovaries become nonfunctional and are replaced by testes). Protogyny is more common than protandry and is emphasized in this essay because it has been investigated in more detail.

In some protogynous species in a number of fish families there are actually two types of males, a situation known as *diandry*. Some are primary males (fish that mature as males and remain male throughout their lives), and others are secondary males (males arising from sex-changed females). Put another way, the primary males are gonochores, whereas the secondary males are not. One diandric species, *Thalassoma lucasanum*, is discussed below. Learning why some species are diandric and others are not is another area that interests animal behaviorists.

Sexual patterns are not correlated with specific families of fish. On the contrary, the Sparidae (porgy, scup, sea bream) and Serranidae each contain gonochoristic species as well as sequential and simultaneous hermaphrodites, and the Pomacentridae (damselfishes) include both gonochores and sequential hermaphrodites. On the other hand, some generalizations are possible concerning habitat and sexual pattern. For example, most simultaneous and sequential hermaphrodites inhabit marine rather than fresh waters, and most sequential hermaphrodites inhabit relatively shallow tropical and subtropical waters, especially coral reefs. Correlations have also been made between various environmental factors and mating system types (Petersen and Warner present an overview of this area in their 2003 review of fish reproductive behavior).

For sex change to be a possible alternative to gonochorism, the benefits of changing sex must outweigh the costs. The costs include the energy used for physiological modification of the gonads and the loss of mating opportunities during the time needed to change. Thus energy expenditure must be low, and changeover time should not interfere with mating opportunities. Even assuming that these requirements are met, why should sex change ever be favored over gonochorism? The hypothesis most commonly accepted is known as the size-advantage model, first developed in 1969 by Michael T. Ghiselin and since then expanded and modified by other investigators. It states that natural selection is most likely to favor sex change when an individual's reproductive success, growth rate, or survival probability is closely related to its age or size, *and* when this relationship is significantly different for females and males. For example, if reproductive success is significantly higher for large males than for small ones (and if the difference is not as great between large and small females), there should be selection for protogynous sex change. If not, gonochorism should be favored. This concept can be demonstrated within a single species, the protogynous diandric rainbow wrasse, *Thalassoma lucasanum*. The secondary males are generally large and defend mating sites; primary males are small and less effective at defense. Warner and his colleagues have shown that the ability of large males to control mating sites is significantly lower in large populations than in small ones, because in large populations there is more interference by other males. Thus for *T. lucasanum* the advantage of protogyny should decline with increasing population size, and in fact, most of the males in the large groups are primary males. In small groups, where secondary males can monopolize matings, primary males are at a disadvantage, and in fact, most males in small groups are secondary males.

Work beginning in the 1970s indicated that quite often sex change is socially controlled. This discovery stimulated much research on the behavior and mating systems of sex-changing fishes. In an early experiment, a group of females and males of a protogynous species were kept together in an aquarium for some months, and nothing changed in either the morphology or the behavior of the fish, but when groups of females were isolated from males, the largest female in each group changed to a male after 2 weeks. When this new male was removed, another male gradually developed from the next largest female. This work was repeated in a variety of protogynous species, with similar results. In protandrous anemone fish, too, removal of the female may lead to sex change by her consort male. However, recent research has shown that sex change occurs under a variety of other conditions as well. In some protogynous species, females may change sex even in the presence of a large male. Moreover, in some species the largest females are not necessarily the ones that change sex. The reasons for this are being explored, and the size-advantage model is now being modified to accommodate these exciting results.

Protandry is less common than protogyny, although many species of teleosts are protandrous, including some moray eels (Muraenidae), some sea basses (Serranidae), and the anemone fishes (Pomacentridae). Except for the anemone fishes, many protandric species are pelagic (living in open water), nocturnal, or inhabit waters of limited visibility, hindering detailed investigations. Under what conditions might protandry be favored? Because eggs are larger than sperm, females are usually limited in the number of gametes they can produce, whereas even a small male can fertilize the eggs of females of any size. As a result, a female can maximize her reproductive output (her fecundity) if she is large. However, this alone is not sufficient to favor protandry. Other important factors include the nature of the mating system (spawning pairs, small groups or large groups), where mating is done (in midwater or on the bottom), and whether the eggs are demersal (attached to the substrate) or pelagic (floating). Males cannot monopolize females if the group is large, and pelagic eggs disperse rapidly and cannot be defended. Thus, in species that spawn in large groups in midwater and have pelagic eggs, there is little benefit by being a large male, and if the groups typically contain many large females, protandry should be favored. Most protandrous species observed do spawn in groups in midwater and have pelagic eggs. A marked exception is the anemone fish of shallow tropical and subtropical waters. These are the only protandrous fish whose behavior and social organization have been examined in detail. Anemone fish live in small groups on anemones, which they use for shelter. A group consists of one large female, her smaller consort male, and several still smaller males and juveniles. The female mates only with her consort male; thus mating is monogamous. The pair guards the demersal eggs. If the female dies or is removed, her consort male changes to a female and mates with the next largest male of the group, or if another colony of anemone fish is nearby, he may move over to that colony and attempt to attract that colony's female without changing sex himself. The advantage of protandry would be access by the male to the high fecundity of a large female, but only if the costs of changing sex are less than the costs of moving to a nearby colony.

Regardless of direction, the first sign of sex change is usually a change in behavior, followed by a color change if the sexes are dimorphic (different in appearance), and finally by changeover of the gonads. The behavior may be convincing enough that females will mate with a female exhibiting male behavior, and males will mate with a male who displays female behavior. For example, large female bluehead wrasses (*Thalassoma bifasciatum*) exhibited male behaviors just a few minutes after removal of a large male. They spawned as males within 24 hours, although they did not produce mature sperm until 8 days after sex change was begun. Across species, changeover time varies from 1–2 days to a year, depending on the species and the cues to which it is responding.

Within most species studied, sex change occurs only in one direction, and the change is irreversible. However, one of the biggest surprises of the last decade has been the discovery that individuals of a number of species (mainly gobies, family Gobiidae, but also species in some other fish families) can undergo multiple sex reversals, depending on the sex of the fish they encounter. Previously, this was thought impossible because when a female changes to a male, the ovarian tissue degenerated; and, if reproductive success was higher for a large male than a large female, it would seem unlikely that a large male would ever change back to a female. Petersen and Warner (2003) and Kuwamura and Nakashima (1998) discuss some of the new findings that help to explain this phenomenon. Moreover, other investigators have found in at least one species of fish that undergoes multiple sex reversal, the energetic costs of sex change are similar for protandry and protogyny.

Investigators are also interested in the effects of social interactions on the endocrine system. In some species, increased social status increases the amount of testosterone, but the relationship between hormones, social interactions, and sex determination is unclear. External stimuli may affect the gonads through the hypothalamus and pituitary, but the pathways are not understood. Generally, physiological research on sex change in teleost fish is considerably less advanced than investigations into the social and environmental aspects of sex change.

Why isn't hermaphroditism more common in other vertebrates? It has been suggested that the transition from water to land tended to magnify the differences between females and males. The much harsher terrestrial environment would have selected for increased specialization of males for efficient sperm delivery and of females for the production of larger, more specialized embryos that could cope with the more demanding conditions. Correlated with increasing differences between the sexes in anatomy and physiology, the sex chromosomes become increasingly different as one moves from fish and amphibians to birds and mammals. As specialization proceeded, the costs of changing sex would increase to a point where it would no longer be possible.

Further Resources

Ghiselin, M. T. 1974. The Economy of Nature and the Evolution of Sex. Berkeley: University of California Press.

Kuwamura, T. & Nakashima, Y. 1998. *New aspects of sex change among reef fishes: Recent studies in Japan.* Environmental Biology of Fishes, 52, 125–135.

Petersen, C. W. & Warner, R. R. 2003. *The ecological context of reproductive behaviour.* In: *Coral Reef Fishes: Dynamics and Diversity in a Complex Ecosystem* (Ed. by P. F. Sale), pp. 103–118. New York: Academic Press.

Reinboth, R. 1988. *Physiological problems of teleost ambisexuality.* Environmental Biology of Fishes, 22, 249–259.

Thresher, R. E. 1984. *Patterns in the reproduction of reef fishes.* In: *Reproduction in Reef Fishes*, pp. 343–388. Neptune, NJ: T.F.H. Publications, Inc.

Warner, R. R., 1978. *The evolution of hermaphroditism and unisexuality in aquatic and terrestrial vertebrates.* In: *Contrasts in Behavior* (Ed. by E. S. Reese & F. J. Lighter), pp. 78–101. New York: Wiley Interscience.

Abby L. Schwarz

■ Reproductive Behavior
Sex, Gender, and Sex Roles

Definitions, Darwin's Theory, and Patterns of Parental Care

Biologists define *sex* as meiosis, the cellular processes of chromosomal recombination and reduction division ending in the production of haploid sex cells called eggs and sperm. Others use *sex* to mean copulation. In most organisms, individuals come in two reproductive types, that is, sexed bodies that are defined as *male* or *female* by the relative size of gametes they carry. Females have larger, often immobile gametes called *eggs*; males have smaller, usually swimming gametes called *sperm*. In some organisms, for example, the fruit fly species, *Drosophila hydei*, males have non-swimming gametes that are as large as the females' eggs confounding the definition of sex. When individuals do not fit the classic definitions, the similarities of individuals to males and females in other, closely related species define their sex.

The usual *reproductive roles* of females and males theoretically determine *sex roles*. Sex roles indicate which sex is choosy and which competes over access to the other sex and mates indiscriminately, and ever since Darwin's *The Descent of Man, and Selection in Relation to Sex*, most animal behaviorists have thought of females as choosy and males as competitive and indiscriminate. Geoff Parker, an evolutionary biologist, said that the sexes are what they are because of the size of their gametes. Robert Trivers said that parental investment patterns, including the care males and females give their offspring after zygote formation, determines sex roles. The

Female–Female Sexual Selection

Patricia Adair Gowaty

Charles Darwin defined sexual selection as a kind of selection depending only on the advantage that "certain individuals have over other individuals of the same sex and species, in exclusive relation to reproduction" (in *The Descent of Man, and Selection in Relation to Sex*). Despite the breadth of his definition, many investigators remember his discussion for only two mechanisms, female mate choice and male–male combat, both of which may cause variation among males in *mating success*. Darwin also discussed the possibility that sexual selection would act among females through male mate choice and female–female combat, both also resulting in mating success variation, this time among females instead of males. He also hinted at many more mechanisms of reproductive competition among females, making reading of his 1871 volume a useful source of ideas for current-day research.

Modern evolutionists define the *opportunity for sexual selection* as a function of within-sex variance in mating success. The number of mates a female has is believed to have only a negligible effect on variance in female fitness, whereas male mating success can dramatically affect among-male variance in fitness. Thus the opportunity for sexual selection via mating success is greater among males than among females. Does this mean that reproductive competition (sexual selection) acts more strongly on males than on females? Many suspect that this is the case, but no one knows for sure. There are two reasons for this. First, mating success variation is due to random and adaptive factors, so to measure the true strength of sexual selection, the part of variance in fitness due to chance must be subtracted from the total within-sex variance. If chance explains more of the variance in mating success in one sex than another, reproductive competition may be equally strong in both sexes. Second, the mechanisms of reproductive competition may be different in males than in females. If the strength of sexual selection among males depends upon variance in number of mates, but the strength of selection among females depends upon the variance in offspring quality, the ecological field of reproductive competition for each of the sexes may be different.

What mechanisms of reproductive competition among females may result in variation in offspring quality? Competition over access to resources that limits female reproduction, but that does not limit female survival, could be an important component of variation in offspring quality. Variation among females in their ability to avoid reproduction with particular males, in their abilities to assess, or to respond behaviorally or physiologically to alternative potential or actual mates may result in variation in offspring quality. Variation among females in their abilities to parse among the sperm of different males may also affect the health of their resulting offspring. And, if so, all of these examples would be due to sexual selection. Research on some aspects of female–female sexual selection is just beginning. Because variance differences in components of fitness from female–female reproductive competition remain largely unknown, female–female sexual selection remains a research frontier in animal behavior.

crucial idea here is that if any reproductive attempt fails, the cost to females is larger than the cost to males; thus selection should act to reduce potential losses to individuals, particularly in the sex with the largest investment. Thus, most evolutionary biologists still believe that whenever females invest more in reproduction than males, ancient selection will have favored

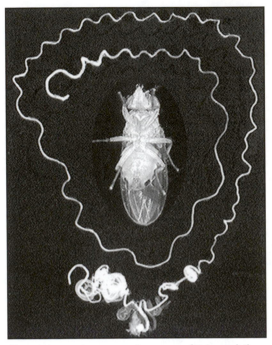

Size symmetric egg and sperm of Drosophila hydei.

Courtesy of Patricia Adair Gowaty.

choosy females, so that now all females are choosy, "coy," and restrained about with whom they mate, and all males are indiscriminate, "ardent," and competitive among themselves over access to females. Other evolutionary biologists object to this conclusion because they say that selection is ongoing and affects the behavior of individuals in real time so that individuals, not sexes, are adaptively flexible in their reproductive behavior.

Another theory to explain sex roles is based on the *potential reproductive rate* (PRR) of females versus males. It is an idea related to parental investment. It says that the operational sex ratio, that is, the number of individuals of each sex available to mate, is determined by reproductive events after mating, including, but not limited to, parental care. Individuals unavailable to spawn immediately after mating are in "reproductive time outs," caring for the kids or recovering from the physiological costs of gamete production or copulation, and so on. The theory says that whichever sex outnumbers the other in the pool of ready-to-mate individuals will be the competitive and indiscriminate sex. The other sex will choose mates from among members of the competitive sex. When the adult sex ratio is even, it predicts males usually will be competitive and indiscriminate,

because females most often have lower PRR (longer parental care) than males. This is an attractive theory to investigators, particularly field animal behaviorists. Instead of measuring sex differences in reproductive time-outs or parental care patterns, they need only count how many males and how many females are searching for mates to predict which sex is the competitive one and which the discriminating one. PRR theory has proved useful for predicting the average behavior of one sex compared to the other, just as the related classic theories based on parental care patterns do, but it does not predict the behavior of individuals. PRR theory also raises questions: Can an individual be choosy and competitive at the same time? Can an individual be indiscriminate, but still unwilling to fight over access to the opposite sex? These questions will only be resolved when investigators focus on individuals, perhaps guided by theories about individuals, rather than sexes.

Nevertheless, many examples are consistent with the logic of the classic sex role theories. In humans and other mammals, mothers gestate and nurse babies, performing parental care activities that fathers cannot. Thus, in mammals mothers invest more in offspring than fathers do, so the parental care and PRR theories predict that female mammals are choosy about mating, but that males are usually competitive and indiscriminate. Reproductive roles sometimes differ from this pattern, so that fathers provision more to offspring than mothers do, as in some bird species in which only males incubate eggs and tend the hatchlings. Biologists (mammals themselves) call these species "sex role reversed," because males seem choosier than females, who compete among themselves over access to males and seem to mate indiscriminately. Despite consistent examples and widespread acceptance of parental care and PRR theories, they remain debatable. Counter examples are abundant, and within many species enormous within-sex, between-individual variation in behavior exists. Because the classic theories do not explain the observed variation, other theories capable of guiding new research are necessary.

Gender: Variation in Social Roles of Reproduction

In many species, same-sex individuals have variable social roles in reproduction. *Gender* refers to within-sex variation in social aspects of sex roles and is defined as the appearance, behavior, and life history of a sexed body. In some fish, for example, adult males exhibit two remarkably different phenotypes, so that during spawning, large territorial males compete with small "sneaker" males to fertilize females' eggs. Large territorial males and small sneaker males are two genders within the male sex. In some species, adult females and males do not reproduce, but forego their own reproduction to help their kin who are breeders. Red-cockaded woodpeckers, *Picoides borealis*, and Florida scrub jays, *Aphelolcoma coerulescens*, for example, sometimes live in groups consisting of a breeding pair and additional adult males, usually sons of the breeding pair or brothers of the male breeder. The additional males are adults, capable of breeding, but instead of breeding, help raise the offspring of the breeding pair. In species with helpers-at-the-nest, breeders and helpers are two genders within a sex. Even within individuals, gender may change with age and/or ecological circumstances. For example, in some human populations like the Hadza of Tanzania, grandmothers often help younger mothers raise their children; thus, a woman might express two genders in her lifetime, first, when she is young, as a breeder, then after menopause, as a helper. Females in other species, including fish, lizards, and many insects, might switch from reproducing sexually to reproducing asexually. Complicating things even further, individuals of some species change sex. Blue-headed wrasses, *Thalassoma bifasciatum*, change from smaller females to larger males. Obviously, sex does not limit an individual's social roles, given their remarkable within-sex diversity in nonhuman animal societies.

Gender diversity also includes homosexuality, which is widespread in nonhuman animals. In animal behavior, homosexuality is defined as same sex behavioral interactions of any kind. Same sex sexual and reproductive behavior catches the attention of animal behaviorists more often than behavior that is not obviously about sex and reproduction. Studies of gulls in which females pair and cooperate in incubation and feeding of nestlings, lesbian sex in monkeys and bonobos, male–male sex in gorillas, homosexual "rape" in Acanthocephalan worms, and the friendly coalitions among males whose members collaborate in force copulations of female dolphins get readers' attention. But studies of male–male winter roosting groups in eastern bluebirds, bachelor herds of elephants and giraffes, and female-bonded social groups in most monkeys, in elephants, lions, and hyaenas are just a few fascinating examples of homosexual animal behavior. Perhaps one of the most interesting conclusions from animal behavior of the last 50 years is that adaptive social behavior often depends upon within-sex social bonds. In a great many well-known species, studied intensively over the last 30–50 years, the core social adaptations are same-sex bonding and affiliative behavior. Investigators have only started to scratch the surface of the extent of and adaptive significance (or not) of homosexuality in animals.

Controversies and Alternative Hypotheses

Sex and gender research, even in animal behavior, is controversial. Some scientists continue to think of sex as an essential, nonchanging, invariant characteristic of an individual. For these people, an individual's sex defines them and limits their social roles to those having to do with heterosexual reproduction. Some think of individuals that differ from the essential ideal for their sex as abnormal, deviant, or mutant. In contrast, animal behaviorists have revealed that gendered social behavior within nonhuman animal societies is as diverse (perhaps more so) and as interesting as in human societies. Some gendered behavior has no apparent adaptive function. Much of it is or is likely to be adaptive, meaning that it enhances the fitness of the individuals expressing it. Or an individual's gendered behavior may benefit its kin or friends in its local population. The conclusion that gender diversity is a normal part of biological diversity is inescapable.

Nevertheless, many theories of animal social behavior still begin with essentialistic or typological ideas about the behavior of males and females, just as the classic theorists of sex roles did. They argued that the sexes are what they are because reproduction, beginning with the size asymmetry of sex cells, usually is costlier for mothers than fathers. The sociobiologists who theorized about sex roles argued that selection favored genes in females for mate choosiness and genes in males for indiscriminate profligacy (i.e., a willingness to mate with any available opposite sex individual) and competitiveness over access to females. These predictions fit with many dramatic observations of the behavior of wild-living organisms: Winner males among elephant seals are often more likely to mate than losers of male–male fights. Female sage grouse actively choose mates from among a set of males performing elaborate displays on leks (mating arenas). Females are often discreet and reserved about copulation, while males seem wildly enthusiastic, willing to (attempt to) copulate with motor bikes (pronghorn antelope) and large flat-bed scales (elephant seals). While the restrained behavior of some females might reflect a lack of ardor, interest, or passion, other explanations are possible. Here are only a few: A female's restraint might be specific to a particular male. It might reflect that mating at a particular time and place may make a female vulnerable to predation or some other threat. Female enthusiasm might take forms difficult for most of us to recognize as ardent. For example, females might indicate their enthusiasm for mating with a particular male by being very still—in some species (e.g., pronghorn) being still is an absolutely necessary female motor response if copulation is to occur. Similarly, there are alternative explanations for most examples of the overwhelming indiscrimination of some males. Are pronghorns that "copulate" with motor bikes and elephant seals that "copulate" with large flat-bed scales doing something else entirely? Could these males be masturbating? The lesson here is that alternative hypotheses must be examined before one concludes that females are invariably restrained and coy, and males "willing to mate with anything."

Importantly, the predictions of parental care theories do not match observations of the behavior of females in a variety of species. Female chimpanzees, for instance, are anything but restrained about mating, sometimes copulating with as many as 10 males within an hour. Experimental studies have revealed behavior inconsistent with parental care theories in female fruit flies in three species, *D. pseudoobscura*, *D. melanogaster*, and *D. hydei*: Females are as active in initiating copulation as males. In *D. pseudoobscura*, a species with no postzygotic parental care, male and female individuals often, but not always, discriminate among potential mates. When they do discriminate (associate for longer with one versus the other "discriminatee" in a testing arena), subsequent breeding experiments have confirmed that subjects assess the fitness that would be conferred on them by mating with one or the other alternative discriminatee. Such fine-scale assessment indicates a degree of choosiness in males not predicted by parental investment theory. The same is true for feral house mice, *Mus musculus*, a species that is typically polygynous, in which males are thought most likely to be indiscriminate. In laboratory experiments, male mice discriminate among females, assessing how offspring health is likely to vary when they mate with one or the other.

Should behavior be called "sex typical," if within a population, individuals vary their behavior flexibly in response to environmental changes or changes in their social environments?

Choosy individuals often change their behavior to indiscriminate and back again, just as indiscriminate individuals often switch their behavior to choosy. Experiments have shown that individuals of the choosy sex in many species flexibly adjust their mating behavior, switching from choosy to random mating as the likelihood of their own death increases. For example, insects including female crickets, *Gryllus integer*, and water striders, *Aquarius remigis*; fish including sand gobies, *Pomatoschistus minutus*, and guppies, *Poecilia reticulata*; and frogs, *Physalaemus pustulosus*, switch from choosy to indiscriminate when predators are around. Another species of fish, female upland bullies, *Gobiomorphus breviceps*, switch from choosy to indiscriminate when their own parasite loads increase. In species in which survival probability presumably declines with age, female house crickets, *Acheta domesticus*, Tanzanian cockroaches, *Nauphoeta cinerea*, and guppies, switch from choosy to random mating. The only species, so far, in which investigators examined if male behavior switched when predators were present showed that pipefish, *Synagnathus typhle*, males also switch from choosy to random mating. Experiments have also shown that when individuals' encounters with others who are ready to mate increase, they are more likely to be choosy than when their encounters are less frequent. It seems that the more opportunities individuals have to mate, the less likely are they to copulate with whomever comes along. For example, female fruit flies, *Drosophila melonagaster*, and guppies are more likely to mate at random when fewer males are around. In tropical African butterflies, *Acraea encedon*, and pill bugs, *Armadillidium vulgare*, males are choosier when more females are around than when fewer females are around. In beaugregory damselfish, *Stegastes leucostictus*, and in threespine sticklebacks, *Gasterosteous aculeatus*, attractive males or males that defend attractive resources are choosier than less attractive males or males defending less attractive resources because their encounter rates with ready-to-mate females is higher. And in mosquitofish, *Gambusia holbrooki*, when males guard females so that female encounters with other ready-to-mate males is decreased, females switch from choosy to indiscriminate. Sex role flexibility will no doubt be documented in many more species as more animal behaviorists observe *individuals* in study populations more carefully. At this point, one is tempted to speculate that sex "roles" seem to be environmentally induced, adaptive switches in the behavior of individuals. Could the rules governing these switches be sex independent, acting on individuals regardless of their sex? This hypothesis, first published by Steve Hubbell and Leslie Johnson in 1987 has not been tested yet.

Investigators mostly have confined study of switches in sex role behavior to females in species with female-biased parental care, and to males in species with male-biased parental care. The best examples of sex role reversal and sex role flexibility are in the fish family Syngathidae, the pipefishes and seahorses. Eggs develop in a pouch on the ventral surface of males, so that males are said to "be pregnant." Thus, the determinants of sex role behavior have been of particular interest in this fish family because these species provide an opportunity to test an important prediction of the selective consequences of high cost parental care explanations for sex roles. If sex roles are reversed when parental care patterns are reversed, this would provide important support for the classic hypotheses. While many continue to cite pipefish as an excellent example in support of the classic hypotheses, the data from studies of pipefish from Sweden paint a different picture. When the encounters with females experimentally decrease, the high-investing males switch to random mating, just as they do when their threat of predation is experimentally increased. Furthermore, female pipefish are often choosy, switching their behavior from choosy to indiscriminate as predation threat increases and as their encounters with males decline. The studies of pipefish by Anders Berglund and colleagues are an excellent example of the utility of experimental studies in species key to theoretical arguments. Pipefish sex roles do not follow the rules predicted by the cost of parental care arguments. Both female and male pipefish adjust their sex role behavior flexibly and adaptively as their ecological circumstances change. Studies of sex role flexibility provide one of the richest opportunities for study by future animal behaviorists. New attention to the origin of sex roles are emerging not just from detailed studies of naturalistic and experimentally induced behavior, but also from quantitative theory that precisely predicts when individuals will switch their sexual and reproductive behavior flexibly and adaptively.

Nature versus Nurture

One of the most enduring controversies in animal behavior swirls around the question: How does life history and environmental variation affect sex role behavior? The nature/nurture controversy centers around two ideas about how phenotypes are determined. One says that sex differences are in our genes, so that once an individual's sex is determined, for instance in mammals by the possession of two X chromosomes or an X and a Y chromosome, there is an inexorable unfolding not only of sexed bodies, but of social behavior as well. The other idea says that sex differences in behavior are due to conditioning and training that young individuals experience during development along with the current environments they experience. The truth may lie somewhere between these two extreme views as argued by other hypotheses for the origins of sex roles.

The *ESS (Evolutionarily Stable Strategy)* sex role theory incorporates elements from the two extreme nature/nurture hypotheses. It says that sex-typical behavior may be due to fixed (universally possessed) genetically influenced *sensitivities* to environmental or social cues that *induce* the expression of flexible behavior *adaptive* for current circumstances. The ESS sex role theory hypothesizes that all individuals, independent of their chromosomal sex, independent of their sex organs, independent of their gamete sizes, and independent of their genders, are sensitive to environmental and life history variables that affect the *time they have to reproduce*. The theory assumes that in all cases, as they encounter potential mates, strategists *assess* the likely reproductive success they would achieve by mating with alternatives. All else being equal, when the differences in fitness conferred are small, indiscriminate mating will more often be favored; and, when the differences in fitness are large, choosy mating will more often be favored. The switch points are determined by the composite value of the fitness conferred by the last two encountered potential mates, and the values of three other variables. All three of these variables determine how much time individuals have: The likelihood that an individual will survive to breed again obviously relates to how much time is left. If all else is equal, the more time that is left, the less cost to being choosy. Number of encounters with ready-to-mate opposite sex individuals affects time left, because encounter rate determines how long an individual will have to search for an appropriate mate. The more encounters, the less time searching takes up, so there is lower penalty for being choosy. The third time variable is the duration of the latency from one copulation to the onset of receptivity to another. All else being equal, long latency decreases total time available, so that relative to other individuals with shorter latencies, each mating subsequent to a long latency will count more toward an individual's lifetime reproductive success, favoring choosy over indiscriminate mating. Each mating that is achieved must make up in fitness all the opportunities lost during the last latency. Here the latency variable is not prospective as it is in the *potential reproductive rate* hypothesis. There is no effect of latency until a strategist reproduces the first time. The differences in models stress that the ESS sex role model is about the effects of the here and now that are important to adaptive, fitness-enhancing reproductive decisions.

The ESS sex role model is consistent with an emerging evolutionary synthesis, exhaustively reviewed by Mary Jane West-Eberhard. Her book, *Developmental Plasticity and Evolution*, shows that phenotypes are often adaptive (enhancing current fitness) and arise because of developmental plasticity, *influenced* by genes and *induced* by environmental variation. The ESS sex role theory says that both genetic influences (on assessment capabilities and sensitivity to factors that affect time left) and environmental inducers are important sources of variation in sex role behavior. It predicts that individuals—regardless of their sex—adaptively express behavior typical for their sex because of social and environmental circumstances typically experienced by individuals of their sex. Rather than resulting from an inexorable unfolding of genes that determine essential characters, genes influence, and environmental factors induce, the sex role behavior of individuals. The ESS sex role model predicts that individuals flexibly switch their behavior in real time, not just evolutionary time, when social and environmental inducers change. It predicts that when the social and ecological environments experienced by the sexes converge, the sex role behavior of males and females will converge. Because the ESS sex role theory has many predictions, some nonintuitive but quantitatively precise and testable, and because it predicts the behavior of individuals, it will be a source of future productive research in basic animal behavior. If observations match its predictions, a new era of understanding of the origins and maintenance of adaptive sex and reproductive behavior of *individuals* will emerge.

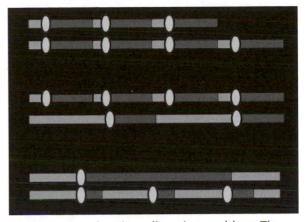

Mating success is a time allocation problem. The top figure represents an idealized reproductive life course for two males. Light grey bars indicate search time; dark grey bars represent post-mating latencies; grey ovals represent encounters and mating. See text for further explanation.
Courtesy of Patricia Adair Gowaty.

See also Reproductive Behavior—*Female Multiple Mating*
Reproductive Behavior—*Mate Choice*
Reproductive Behavior—*Sexual Selection*

Further Resources

Berglund, A. & Rosenqvist, G. 2003. *Sex role reversal in pipefish.* In: *Advances in the Study of Behavior,* Vol. 32, pp. 131–167.

Darwin, C. 1871. *The Descent of Man, and Selection in Relation to Sex.* London: J. Murray.

Gowaty, P. A., Steinechen, R. & Anderson, W. W. 2003. *Indiscriminate females and choosy males: Within- and between-species variation* in Drosophila. Evolution 57 (9), 2037–2045.

Hubbell, S. P. & Johnson, L. K. 1987. *Environmental variance in lifetime mating success, mate choice, and sexual selection.* American Naturalist, 130, 91–112.

Roughgarden, J. 2004. *Evolution's Rainbow: Diversity, Gender, and Sexuality in Nature and People.* Berkeley: University of California Press.

Patricia Adair Gowaty

■ Reproductive Behavior
Sexual Behavior in Fruit Flies—Drosophila

Flies in the genus *Drosophila* represent some of the most important experimental organisms in the world. Commonly called fruit flies, Drosophilids actually belong to a group of animals known as *pomace flies*, which feed on rotting fruits and vegetables. True fruit flies, like the Mediterranean fruit fly (*Ceratitis capitata*), feed on live fruit and are agricultural pests.

In the laboratory Drosophilids make ideal experimental organisms. They are small, easy to maintain, and they produce large numbers of offspring in a short period of time. For almost 100 years, the most famous member of this group, *Drosophila melanogaster*, has been a model organism for the study of genetics. However, in addition to its value for genetic analysis, the behavior of *Drosophila* has also been extensively studied. This chapter will provide an overview of the most well-documented of all *Drosophila* behaviors, sexual behavior.

Historical Beginnings

In 1907, Thomas Hunt Morgan, a professor at Columbia University, instructed a student to place bananas on the windowsill to collect fruit flies. Morgan was entering a growing scientific revolution that would someday be known as the field of genetics. He was searching for an animal model to test Gregor Mendel's theories on inheritance. His efforts with the bananas yielded a colony of *Drosophila melanogaster*, a species that would someday be found in genetics labs throughout the world.

Alfred Sturtevant, a young scientist in Morgan's lab, realized that *Drosophila* were ideal subjects for behavioral analysis. Sturtevant was specifically interested in male courtship behaviors, which he first described in 1915. Like other behavioral scientists, Sturtevant knew that some of the most elaborate and remarkable behaviors seen in the natural world are those that animals perform while courting. For some species, the only true social interaction between individuals occurs during sex, and courtship displays are often the cues that bring the animals together. Courtship signals are used to convey information about an animal's species, condition, fitness, and intent. Successful courtship leads to successful mating, which leads to production of the next generation and perpetuation of the species.

Sexual Behavior in *Drosophila*

Like the vast majority of animal species, reproductive behavior in *Drosophila* follows a pattern known as *female choice*. Within this system males often appear to be indiscriminate, ready to copulate with any willing female. Females, on the other hand, appear to be sexually restrained and discriminating. Females control mating by choosing males based on their resources, appearance, and courtship behaviors. Female choice probably evolved in response to differences between the sexes in terms of *parental investment*, or the costs associated with reproduction. For females the costs are typically much higher, involving large metabolic input for the production of eggs, which the female may care for after fertilization. By contrast, sperm production is far less costly. A sperm cell is really nothing more than packaged DNA with a tail to propel it to the egg. Such differences in parental investment may have led to the evolution of distinct mating strategies between the sexes. Males can afford to provide sperm to any receptive female, increasing their fitness by mating as often as possible. For females, reproductive success may depend on their ability to discriminate between males, choosing to mate only with males who can prove their fitness.

Male Courtship Behaviors

When a mature *Drosophila melanogaster* male encounters a sexually attractive fly (typically a female of the same species, but not always), he initiates a series of stereotyped courtship behaviors (see the drawing). The male approaches the female and assumes a position in which he is facing her abdomen (*orientation–A*). If the female moves away, which is common, the male chases her (*following*) and will touch her with one of his front legs (*tapping–B*). While maintaining his orientation toward the female, the male will then perform the courtship song by extending one wing and vibrating it (*courtship song–C*). The male will also attempt to contact the female's genitalia with his mouth parts (*licking–D*), and curl his abdomen in an attempt to mate (*attempted copulation–E*). If the male's courtship has sufficiently increased the female's receptivity she will mate (*copulation–F*).

The types of behaviors performed by *Drosophila* males during courtship are known as *fixed action patterns*, or *FAPs*. FAPs are innate (genetically programmed) behaviors that are performed the same way each time. While the sequence of the behaviors may vary, the performance of each individual behavior does not change. Performance of each behavior is thought to be like playing a neurological tape that guides the movements of the male. A key element to FAPs is that they can be performed without prior experience or learning. This can be easily demonstrated in *Drosophila* through the use of isolation experiments. In these experiments, *Drosophila* males, isolated from other flies for their entire lives, perform all of the courtship behaviors correctly the first time they encounter a female.

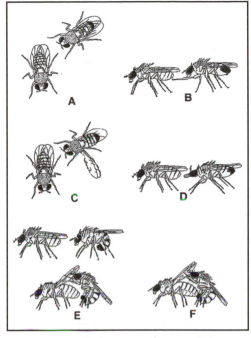

The courtship behaviors of Drosophila melanogaster.
Courtesy of Scott P. McRobert.

Of all the male courtship behaviors, the courtship song has received the most scientific study. The song is complex, species-specific, and serves the critical function of increasing female receptivity to copulation. Males whose wings have been surgically removed will perform courtship, but are far less successful in convincing females to mate. However, when a tape recording of male courtship song is played through a loudspeaker, females will mate with wingless males as if the males were normal.

Female Behavior

The role of *Drosophila* females in sexual behavior is more subtle than that of the male. Females stimulate males sexually through the production of courtship-stimulating, or aphrodisiac, *pheromones*. Pheromones are chemicals produced by one animal that cause a change in the behavior of another animal. For *Drosophila melanogaster*, the primary aphrodisiac pheromone is a 27-carbon compound known as (Z, Z)-7, 11-heptacosadiene. Visual cues also appear to be important (blind males, and males in the dark, can court and mate, but their behaviors are not completely normal).

The key behaviors performed by females during courtship are related to their willingness to mate. In many *Drosophila* species, unreceptive females perform rejection behaviors such as kicks and wing flicks. Receptive females often signal their willingness to mate by slowing down, assuming a position that exposes their genitals to the male and opening vaginal plates that would otherwise physically prevent copulation.

Mated Females

During copulation, *Drosophila* males deliver a cocktail of compounds in addition to sperm. One of these compounds, known as *sex peptide*, has dramatic effects on the female. Within hours of copulation, sex peptide causes the female to stop production of the aphrodisiac pheromone, thus rendering her less attractive to males. In addition, sex peptide acts on the female's central nervous system, causing her to be less receptive to male courtship. Other peptides in the ejaculate also have similar effects on the reproductive behavior of the female.

The evolution of the sex peptides appears to be a way for males to protect their sexual investment. Females who mate with two males in succession will preferentially utilize the sperm of the second male. Therefore, it benefits a male who has succeeded in copulation to provide the female with compounds that reduce the chances that she will mate again in the near future. The effects of the sex peptides last about a week, during which time the female will lay hundreds of eggs. At this point the effects of the sex peptides will have worn off, and the female will be sexually attractive and sexually receptive once again. However, a side effect of the sex peptides is that they actually shorten the life of the female fly.

The Study of Sexual Behavior in *Drosophila*

One of the advantages of *Drosophila* is that their sexual behavior is relatively easy to study in the laboratory. Unlike many organisms, whose behavior in captivity becomes abnormal, *Drosophila* are willing to demonstrate their sexual behavior anywhere, at any time. Typically, a study of sexual behavior begins with the collection of immature flies (recently emerged from their pupal cases), which are separated by sex and housed in vials until the experiment begins. To observe courtship, a mature male and female (typically 3–5 days old) are placed into a small plastic mating chamber and observed under a dissecting microscope. To assess a male's courtship vigor, or the sexual attractiveness of the female, the *courtship index* (the percentage of an observation period during which the male is performing any of the courtship behaviors), is recorded. Other measurements typically taken include the *copulation latency* (the amount of time until copulation takes place), *copulation duration* (the amount of time the flies spend copulating), and the *copulation frequency* (the percentage of pairs of flies that mate during a given period of time). Analyses of these measures have provided a tremendous amount of information on the mating behavior of different *Drosophila* species as well as the effects of mutations on sexual behavior.

The Genetics and Neurobiology of *Drosophila* Courtship

To study the effects of genes on behavior, mutations have been created through the use of *mutagens* such as x-rays or DNA-altering chemicals. Not all of these mutations are specifically sexual behavior mutations, but have effects on some aspect of the fly, which, in turn, affects sexual behavior. For instance, the first mutation discovered in Thomas Hunt Morgan's laboratory caused the eyes of flies to be white rather than the normal red. This mutation (known as *white*), affects vision and all behaviors affected by vision. With respect to sexual behavior, white-eyed males take longer to initiate courtship, take longer to copulate, and sometimes even face away from the female when singing. Thus, while this gene is not a courtship gene per se, it has been used to document the importance of visual cues during courtship.

Similar to the *white* mutation, mutations known as *Olfactory–C* and *parasmellblind* affect the ability of a fly to detect chemical cues. Males carrying these mutations don't respond normally to the female aphrodisiac pheromone, resulting in less intense courtship and a lowered rate of success. As with *white*, these genes have provided information on the importance of cues (in this case chemical) on the performance of normal courtship behavior.

Wing mutations can affect performance of the male courtship song. For instance, a mutation known as *raised* causes flies to hold their wings upright. Mutants cannot fly, and mutant males cannot sing. Since the courtship song is critical to increasing female receptivity, *raised* males are less likely to copulate than normal males. Mutations in a gene known as *period*, or *per*, also affect courtship song, but in a more complex way. The *per* gene affects all rhythmic behaviors within the fly, including activity cycles (waking and sleeping) and the timing of emergence from the pupal case. It turns out that the courtship song is also rhythmic, with intervals between bursts of song that fluctuate sinusoidally. Males affected by a *per* mutation produce a song lacking the correct rhythm. Some *per* alleles speed up the song, some slow it down, and some cause males to sing with no rhythm whatsoever.

Mutations also exist that are relatively specific to sexual behavior. The most famous of these is *fruitless*, which affects a male's ability to choose sexual partners. Males that are homozygous for *fruitless* do not attempt to mate with females. They do, however, court normal, sexually mature males, and elicit courtship from other males by synthesizing sexually attractive pheromones. In addition, *fruitless* males perform an abnormal version of the courtship song. While all aspects of the *fruitless* are not completely understood, it appears to represent a gene necessary for normal sexual behavior in males.

Finally, although there are hundreds of other studies and other genes demonstrating genetic effects on sexual behavior, it is worth noting a fascinating study on the neurobiology of male courtship in *Drosophila*. This study, performed by Jeffrey C. Hall of Brandeis University, was directed at mapping the regions of the central nervous system (CNS) involved in each of the male courtship behaviors. To do this, Hall

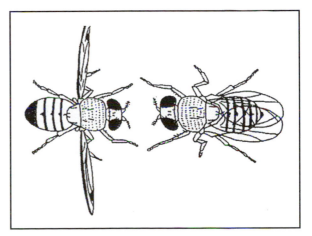

Drosophila male facing a female.
Courtesy of Scott P. McRobert.

created *gynandromorphs*, a type of mosaic which contains both male and female tissue in the same body. Each gynandromorph is distinct, with a random mixture of male and female cells. Hall looked for flies that were essentially female, with a few male cells in the head and, hopefully, in the brain. After analyzing the courtship behavior of each gynandromorph, the fly was frozen and its nervous system painstakingly sectioned and studied. Using genetic markers to identify male and female tissue, Hall was able to create a neurological map of the CNS. The early courtship behaviors, *orientation, following,* and *wing extension* require male cells on one side, in the back of the brain. *Licking* requires male cells in another region, further back. And performance of the *courtship song* and attempted *copulation* requires male cells in regions of the thoracic ganglion (a CNS region in the thorax, similar in some respects to the spinal cord).

Conclusions

Although best known for their use as model organisms for the study of genetics and development, *Drosophila* have also made great contributions to the field of animal behavior. They demonstrate complex sexual behaviors, which can be studied easily in the laboratory. And by combining behavioral studies with genetic techniques, scientists have utilized *Drosophila* to gain a greater understanding of the interactions between genetics and behavior.

Further Resources

Greenspan, R. J. 1995. *Understanding the genetic construction of behavior.* Scientific American, April 1995, 72–78.
Hall, J. C. 1994. *The mating of a fly.* Science, 264, 1702–1714.
Sturtevant, A. H. 1915. *Experiments on sex recognition and the problem of sexual selection in* Drosophila. Journal of Animal Behavior 5, 351–366.
Tompkins, L. 1984. *Genetic analysis of sex appeal in Drosophila.* Behavior Genetics, 14, 411–440.
Tompkins, L. 1998. *The development of male- and female-specific sexual behavior in* Drosophila melanogaster. In: *Genome Analysis in Eukaryotes: Developmental and Evolutionary Aspects* (Ed. by R. N. Chatterjee & L. Sanchez), pp. 120–148. New Delhi, India: Narosa Publishing House.
Weiner, J. 1999. *Time, Love, Memory: A Great Biologist and his Quest for the Origins of Behavior.* New York: Alfred A. Knoft, Inc. Publishers.

Scott P. McRobert

■ Reproductive Behavior
Sexual Cannibalism

One of the most bizarre sexual behaviors is sexual cannibalism: the killing and eating of one sex by the other during courtship, copulation, or after copulation (an alternative, perhaps less vivid, term for this behavior would be "intraspecific intersexual predation"). Sexual cannibalism has been reported in over 90 species—mainly the arthropods plus a few mollusks. Observations before the 1980s were usually anecdotes or studies in captivity. Subsequent research has documented sexual cannibalism in nature. Many species that exhibit sexual cannibalism are predatory, with the females being larger than the males. Thus, most occurrences involve the females attacking and eating the males. Three groups of predators have become infamous for their *femme fatales:* spiders, praying mantids, and scorpions.

In spiders (Class Arachnida, Order Araneae), sexual cannibalism occurs in at least 50 species. Females may lunge at males during the courtship phase as the males are approaching. Such an attack precludes sperm transfer by the male. Alternatively, females may violently lunge at the males during or after sperm transfer. In most species, males attempt to avoid attacks by females. Multiple observations on three species in nature (*Agelenopsis aperta, Argiope aemula, Latrodectus hasselti*) give a wide range for the frequency of the cannibalism of the male, from 10% to 82% of all sexual encounters. An exceptional behavior occurs in *Latrodectus hasselti*, the same genus as the North American "black widow," and *Araneus pallidus*. In these species, the male somersaults into the female's mouthparts during sperm transfer. These seem to be rare cases of male complicity in cannibalism.

In mantids (Class Insecta, Order Mantodea), sexual cannibalism has been recorded for over 20 species. Despite what popular literature might suggest, the insemination of the female does not require the decapitation of the male. Repeated observations exist for four mantid species in the field (*Iris oratoria, Mantis religiosa, Stagmomantis limbata, Tenodera aridifolia sinensis*), where the cannibalism of the male accounts for 17–31% of sexual encounters. As in spiders, females may attack and consume males during the courtship and approach phase. Cannibalism and copulation can co-occur in one of three ways. First, the female may attack the male before he mounts her. As the female chews on the male, his headless body may break free of her hold. Endogenous neural activity in the male's thoracic and abdominal ganglia enables him to crawl along her abdomen, find her genital opening with his penal hooks, and copulate. Second, the male may mount the female, and she then may reach back and begin feeding. Third, the female may hunt down the male after he has dismounted from her.

In scorpions (Class Arachnida, Order Scorpiones), sexual cannibalism has been reported in at least 11 species. The frequency of sexual cannibalism in nature is unclear, since observations of scorpions are largely anecdotal or confined to captivity. Scorpions have elaborate mating behavior, marked by the male grasping the female's pincers and leading her to a site where he deposits his spermatophore. When cannibalism occurs, it is usually done by the female, who may attack the male after she accepts the spermatophore. Sexual encounters are characterized by quick, violent escape behaviors by the smaller sex, most likely to avoid being attacked by the other sex.

Sexual cannibalism is an extreme, rather literal, manifestation of reproductive conflict between the sexes. Why female-on-male cannibalism occurs has received much attention. An early explanation centered on suicidal behavior by the males, wherein the males somehow facilitated their own cannibalism. Despite this romantic (or macabre) image of a male "giving it all for love," suicidal behavior by males is rare. The most overt examples are the male spiders that somersault into their mates' mouthparts. In the vast majority of other species, males appear to take steps to avoid being cannibalized, especially before sperm transfer occurs. Males may retreat or flee from aggressive females, and typically strike back as females are attacking them.

It is clear that the risk of cannibalism poses a dilemma for males. To reproduce, they must make contact with a predator—the female—who has the advantage of larger size in many species. Researchers propose that the risk of cannibalism has shaped male mating behavior over evolutionary time. Courtship displays and the orientation of the male's approach may be ways of reducing the chances of an attack by the female. Because reproduction involves the very real risk of death, one can expect a male to be selective about his mates, either to avoid approaching an aggressive female or to ensure that his mating investment brings the highest return. Some male spiders prefer to approach females that are already eating prey. Recent work

More than 20 species of mantids engage in sexual cannibalism.
Courtesy of Mike Maxwell.

on spiders and mantids suggests that males prefer virgin females or females with many eggs. Both kinds of females may be less likely to attack males. Furthermore, males may fertilize a high proportion of a virgin's eggs, or may fertilize a high absolute number of eggs by mating with a fecund female.

Yet, cannibalism can offer certain reproductive benefits to the males. This realization fuelled early arguments for male suicide as an explanation for the occurrence of sexual cannibalism. An immediate payoff is that the eaten male inadvertently converts his body into eggs, thereby contributing to an extra batch of offspring. This notion is appealing for those species in which the male is reasonably large, as in many mantids and scorpions. Research demonstrates that the consumption of a male mantid can increase female reproductive output, although this effect seems to depend upon the female's nutritional status, or even the species-specific peculiarities of food digestion and egg development. In many spiders, the male is much smaller than the female, and seems to represent a slight *hors d'oeuvre*, not a main course. In addition, a female spider's egg load appears to be fixed at adult emergence in some species. Presumably through the combined effects of small male size and female reproductive physiology, the consumption of a male has not been found to increase female reproductive output significantly in several spider species.

Depending on the species in question, the consumption of the male may or may not boost his partner's reproductive output. To make matters worse for the male, his fatal mate may copulate with other males, as has been observed in mantids and spiders in nature. This prospect of sperm competition throws doubt on whether any eggs manufactured from the eaten male's body will actually be fertilized by him. So, is cannibalism a total loss for males? Perhaps not. In some spiders and mantids, an interesting consequence of cannibalism is that copulation is prolonged. The act of cannibalism prolongs sperm transfer in spiders, which appears to result in higher fertilization success for the eaten male. Headless male mantids can remain *in copula* for over 20 hours in nature, possibly acting as semiliving "chastity belts." Thus, although cannibalism in the absence of sperm transfer is clearly not beneficial for males, the possibility of providing material for more eggs or enhancing fertilization success may weaken selection against cannibalism co-occurring with insemination.

Explanations for the evolution of sexual cannibalism must include female interests. A minimalist, nonadaptive hypothesis posits that adult females that cannibalize males simply fail to recognize them as receptive mates. This "mistaken identity" hypothesis seems well-suited for cannibalism occurring in the absence of copulation, but it is difficult to reconcile with the behavioral biology of many species. Males of visually acute species, such as mantids and jumping spiders, use courtship displays, but may be cannibalized nevertheless. Strong arguments against mistaken identity are the spiders *Portia labiata* and *P. schultzi*. In these species, females perform displays only toward males of the same species, and they respond to displays given by males. Yet, these females may attack males during courtship or sperm transfer.

A likely explanation for the occurrence of sexual cannibalism points to female nutritional requirements. A female may simply attack a male when she is food-deprived or lacking a certain nutrient or vitamin. Regardless of the male's size, his body is composed of the compounds needed for producing healthy eggs and offspring. Several studies on mantids and spiders indicate that starved or food-limited females are more likely to cannibalize males. Although this "female hunger" or "economic" hypothesis appears to go some way toward explaining the evolution of sexual cannibalism, there are other factors that influence whether a female attacks a male of the same species. Importantly, the male is not simply potential prey; he is also a potential mate. A hungry female must allow at least one male to inseminate her. Furthermore, she will likely mate with several males to ensure that she receives enough viable sperm or to ensure genetic variability or even superiority among her offspring. Indeed, recent work on the spider *Argiope keyserlingi* demonstrates that virgin females are less aggressive and cannibalistic toward males than are mated females. Thus, a hungry female must exercise some discretion about whom and when in the mating sequence to attack, so that she does not jeopardize sperm receipt.

From the above, one can expect a female's nutritional and mated status to jointly influence her decision about attacking a given male. Various hypotheses point to other potential factors. For example, a female may cannibalize a male as an expression of mate choice. In the ultimate form of mate rejection, the female may simply kill and eat the male because she views him as a substandard mate. Recent research on spiders suggests that a female can control the duration of sperm transfer, and hence the fertilization success of the male, by initiating cannibalism. Another idea, proposed for spiders, is that cannibalism by the adult individual is a "development by-product" of selection for rapacious feeding behavior as a juvenile. By this hypothesis, cannibalism of males by adult females does not necessarily provide a reproductive benefit to either party. A third hypothesis points to competition between females. Once a female becomes sperm-satisfied, she may cannibalize males to remove them from the mating pool, so as to keep her neighboring rival females sperm-limited. If her rivals lay fewer eggs, then her offspring will have fewer immediate competitors upon hatching.

Male-on-female cannibalism also occurs in nature. Although male arachnids occasionally consume females in a sexual context, this behavior has been examined most rigorously in crustaceans. In the amphipods *Gammarus* spp., males are larger than females. A male clings to a female for several days before she molts, after which the pair mates. When males are experimentally starved, roughly one third cannibalizes the clutched females. Male paddle crabs (*Ovalipes catharus*) similarly cling to females before molting and mating. Once the female molts, the male may cannibalize her without copulation, or even after copulation. The latter seems maladaptive, and may be an artifact of captivity.

See also Predatory Behavior—*"Sexual Cannibalism": Is It Really "Sexual" or Is It Just Predatory Behavior?*

Further Resources

Andrade, M. C. B. 1996. *Sexual selection for male sacrifice in the Australian redback spider.* Science, 271, 70–72.

Arnqvist, G. & Henriksson, S. 1997. *Sexual cannibalism in the fishing spider and a model for the evolution of sexual cannibalism based on genetic constraints.* Evolutionary Ecology, 11, 255–273.

Elgar, M. A. 1992. *Sexual cannibalism in spiders and other invertebrates.* In: *Cannibalism: Ecology and Evolution among Diverse Taxa* (Ed. by M. A. Elgar & B. J. Crespi), pp. 128–155. Oxford: Oxford University Press.

Gould, S. J. 1984. *Only his wings remained.* Natural History, 93, 10–18.

Maxwell, M. R. 1999. *Mating behavior.* In: *The Praying Mantids* (Ed. by F. R. Prete, H. Wells, P. H. Wells & L. E. Hurd), pp. 69–89. Baltimore: Johns Hopkins University Press.

Polis, G. A. & W. D. Sissom. 1990. *Life history.* In: *The Biology of Scorpions* (Ed. by G. A. Polis), pp. 161–223. Stanford: Stanford University Press.

Roeder, K. D. 1935. *An experimental analysis of the sexual behavior of the praying mantis* (Mantis religiosa, L.). Biological Bulletin, 69, 203–220.

Schneider, J. M. & M. A. Elgar. 2001. *Sexual cannibalism and sperm competition in the golden orb-web spider* Nephila plumipes (Araneoidea): Female and male perspectives. Behavioral Ecology, 12, 547–552.

Mike Maxwell

■ Reproductive Behavior
Sexual Selection

Charles Darwin's theory of evolution by natural selection explains why plants and animals have complex physical and behavioral characters (= adaptations) that increase an individual's chances of surviving. Only individuals that survive get to reproduce. Those that die do not. Why then do some animals perform behaviors that make them more visible to predators, or behaviors that consume energy that could presumably be better invested into growth, disease prevention, or looking after their offspring? Surely these traits should have been removed by natural selection? For example, why do some male bowerbirds in New Guinea spend hours collecting feathers to adorn their territory? And why do peahens spend many days watching and comparing displaying peacocks before deciding with whom they will mate? More generally, why do animals ever indulge in activities that will reduce their lifespan? One answer can be found in the study of sexual selection. It explains why males perform energetic courtship displays, fight each other during the mating season and possess extravagant physical characters which females often lack, like gaudy colors, elongated feathers, and horns. Sexual selection also accounts for female choosiness when selecting mating partners.

Elaborate behaviors and characters that confer no apparent survival advantage posed a special problem for Darwin. His solution was to develop the theory of evolution by sexual selection. In fact, natural selection does not depend on survival alone. Ultimately, the fittest animals are those with traits that maximize their genetic contribution to future generations by producing the most descendants. Surviving for many years and reproducing each year is one obvious way to achieve this goal. But another option arises if there is a trade-off between lifespan and mating success. Sometimes individuals are prepared to accept a shorter lifespan in return for increased reproductive success while they are alive. Sexual selection explains how animals can develop traits that increase their reproductive success per breeding bout.

Darwin's main insight was to recognize that there is a critical source of conflict between the sexes. It takes two—a male and female—to create offspring. If there were always as many males as females and breeding occurred synchronously in the population, there would be no opportunity for either sex to increase its reproductive success. This is, however, rarely true. In most cases there are more males than females available to mate at any given moment, so males must compete for fertile females. So any behavior that increases a male's success in competition with other males for mating opportunities could be beneficial, even if there are costs in terms of an earlier death.

To use Darwin's words, sexual selection selects for traits because of "the advantage, which certain individuals have over other individuals of the same sex and species, in exclusive relation to reproduction." However, whether a sexually selected behavior is maintained ultimately requires that the increase in immediate reproductive success it confers does not outweigh any subsequent reduction in lifespan. Net natural selection always favors characters that increase the total number of descendants produced.

Darwin noted that male–male sexual competition takes two main forms. First, there is direct competition between males during physical contests. Fighting selects for weapons like antlers, tusks, horns, spurs and greater body size (which is why males are often bigger than females in species like elephant seals where males directly compete for females). Males with these characters tend to win fights over females and are therefore more likely to mate and sire offspring. Of course, as we all know from military history, the development of weapons often leads to "arms races." It is the strength of your weaponry relative to your rivals that determines victory or failure. This means that larger and larger weapons are selected for so that, over evolutionary time, sexually selected traits become increasingly costly to produce or maintain. Think, for example, of the increased energy a moose must expend to support the weight of a bigger set of antlers. Again, however, there is an upper limit to elaboration. This is reached when the mating advantage conferred to the bearer is balanced by a reduction in lifespan. In reality it is likely that males will differ in their ability to bear the immediate costs of these sexual characters (and, by extension, the negative effect of the character on lifespan will vary among males). In consequence, some males will still be able to afford to produce slightly larger weapons than others. Most biologists accept that direct competition has led to sexual selection for male traits that increase fighting success, even at the cost of reducing male survivorship.

Darwin also noted a second form of sexual competition which he described as "the ability to differentially attract, seduce, overpower or manipulate the opposite sex." If sexual selection through female choice is a valid explanation for costly male displays, then females must prefer to mate with males with more energetic courtship or elaborate sexual traits. In general, a female obviously benefits from being choosy if this directly increases her lifetime fecundity (number of offspring produced). For example, female damselfish that preferentially mate with males that court more intensively benefit because these males are better parents who are less likely to eat their offspring. The theoretical problems start, however, when we want to understand how a male trait like courtship rate or body coloration can signal this type of benefit reliably. As we all know, advertisements often fail to deliver the products that they promise. If male signals continually deceive females, then females would be better off if they mated at random and saved the time and energy costs of being choosy. An ongoing area of research therefore asks why preferred male traits are, on average at least, honest signals of so-called direct benefits that increase female lifetime fecundity.

A second possibility is that sexually selected male traits that females find more attractive provide information about a male's genetic quality. These behavioral or physical characters may indicate to a female that a male will pass on genes to his offspring that will make them more successful. Specifically, genes that make his sons more attractive thereby increase their mating success and/or genes that make his offspring live longer by increasing their ability to resist disease or stay in good physical condition. The idea that females choose males with "good genes" has proved controversial though, and it is currently the most active area of research in sexual selection.

In sum, males often perform elaborate displays that are not essential for successful reproduction; and females are often more choosy than is necessary simply to ensure they mate with a member of the same species. These so-called secondary sexual characters do not differ between the sexes for purely functional reproductive reasons. Instead, like human advertisements or military weapons, they only exist because there is competition for access to limited resources. In animals, this limited resource is the opportunity to mate with members of the opposite sex. It leads to sexual selection for characters that make individuals (usually males) more attractive to the opposite sex or characters that increase success during direct physical contests for mating opportunities.

See also Reproductive Behavior—*Female–Female Sexual
 Selection*
 Reproductive Behavior—*Female Multiple Mating*
 Reproductive Behavior—*Mate Choice*
 Reproductive Behavior—*Mate Desertion*
 Reproductive Behavior—*Sex, Gender, and Sex Roles*

Further Resources

Andersson, M. B. 1994. *Sexual Selection*. Princeton, NJ: Princeton University Press.

Birkhead, T. R. 2000. *Promiscuity: An Evolutionary History of Sperm Competition*. Cambridge, MA: Harvard University Press.

Cronin, H. 1991. *The Ant and the Peacock*. Cambridge: Cambridge University Press.

Diamond J. 1998. *Why is Sex Fun? The Evolution of Human Sexuality*. New York: Basic Books.

Gould, J. L. & Gould, C. G. 1989. *Sexual Selection (Scientific American Library, No 29)*. New York: Freeman & Company.

Judson, O. 2002. *Dr. Tatiana's Sex Advice to All Creation*. New York: Owl Books.

Ridley, M. 1995. *The Red Queen: Sex and the Evolution of Human Nature*. New York: Penguin.

Zuk, M. 2002. *Sexual Selection: What We Can and Can't Learn about Sex from Animals*. Berkeley: University of California Press.

Michael Jennions

■ Reproductive Behavior
Shredding Behavior of Rodents

The shredding behavior of rodents, especially of the genus *Peromyscus* is usually overlooked or casually mentioned in biological behavior. *Shredding* is defined as the tendency of rodents to pull apart or tear materials (especially paper) by using their sharp front teeth (incisors).

When a group of locally-trapped laboratory-confined mice of the subspecies *Peromyscus maniculatus osgoodi* (white-footed or deer mouse) was presented with standard lavatory paper towels in their cages, shredding varied greatly among individuals. It appeared that these variations could be measured with a fair degree of accuracy, because in a 24-hour period, when giving one paper towel to one mouse, there was usually a piece of the towel left unshredded. One could estimate, therefore, the amount of towel that was left, or conversely, the amount of towel that was shredded. For example, if a quarter of the towel was still left, one could simply say that three quarters of the towel had been shredded. This made possible an evaluation of the behavior in *quantitative* terms, rather than the usual qualitative observations. Since shredding by individual mice under normal conditions could be compared with shredding under experimental conditions, the results could then be combined to indicate the general tendencies of a large group of mice.

The principal research was an investigation of two important environmental conditions, *temperature* and *light*, that in preliminary studies seemed to have the most profound effects on the shredding behavior of these mice. The relationship of this behavior to *nest building* was also examined.

Standard cages were used in this study, one mouse being confined to one cage for a 24-hour period during each test. Each cage was 7 in wide by 7 in high by 9.5 in deep (17.8 cm × 17.8 cm × 24 cm) and fitted into a battery (rack) that held a total of individual cages, 30 on each

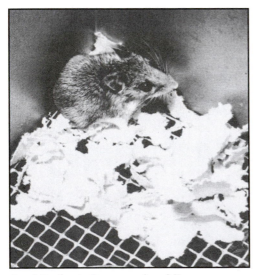

A white footed mouse shredding paper.
Courtesy of Oakleigh Thorne II.

side, manufactured by Bussy Products of Chicago. They had 0.25 in (.64 cm) mesh hardware cloth floors and fronts with sides and backs of sheet metal. Standard equipment for each cage was one empty tuna fish can, open at the top, for a nesting container, one drinking fountain with water, and one 8 oz glass food jar. The mice were fed ample quantities of whole-grain barley and Purina dog chow checkers.

The fact that shredding behavior in these mice is intimately connected with nest building becomes immediately obvious to the experimenter working with these animals. The quality of the nest built, therefore, was recorded in order for it to be compared to the quantity of shredding. Nests were rated on a scale from none to poor to fair to good, and steps in between, and were translated into a number scale from 0–9. These subjective qualitative grades were compared to the quantitative shredding data. A temperature and light control chamber, or zootron, was designed and constructed with a plywood floor and a 2 in by 2 in (5 cm × 5 cm) frame covered with Celotex. The outside was covered with rolls of insulation at least 1 in thick. The unit was large enough (6 ft high, 6 ft wide, and 3 ft deep or 1.8 m × 1.8 m × .9 m) to accommodate the 60-cage battery. Two doors on the front allowed the battery of cages to be rolled out for data gathering, maintenance, and setting up for the next 24-hour test. Fans inside the zootron kept the air circulating, and air temperature could be modified by an attached air conditioner coil or electric heater, depending on which temperature condition was desired. Four 48-inch fluorescent lamps (two on the floor and two on the ceiling) provided light. When these were off, the zootron was completely dark inside. An electronic control panel plus thermoregulators inside the zootron insured the desired temperature and light conditions for each experiment. The three temperature settings were: *high* of 33°C (91.4°F) *medium* of 21°C (69.8°F), and *low* of about 8°C (46.4°F).

Sixty mice were tested over a period of 27 days, the shredding and nest building performance of each mouse being recorded daily. Besides the conditions of light and dark, or low, medium, and high temperature, a random sequence permitted analysis of the effects of small and large decreases or increases in temperature, the effects of going from light to dark, or dark to light, and the possible effect of one condition on a subsequent condition (or carry-over). The entire sequence of experiments was started after the mice had previously been given one whole paper towel to shred daily for several days at *medium* 21°C room temperature in normal light. After gathering the numerical data for each mouse, numerical *group totals* for shredding and nest building were determined (see the table on p. 942). These results are also shown graphically in the two histograms on p. 943.

Effects of Temperature

The results clearly show that, compared to shredding at medium temperature, the behavior was intensified considerably at low temperature and practically eliminated at high temperature. This variance was statistically significant.

At low temperature, there was a definite improvement in the nest building quality of the mice as a group compared with that at medium temperature. At high temperature there was little or no nest building. This variance was also statistically significant. The improvement in nest quality involved both the utilization of more shreds and the construction of a nest that provided better insulation against heat loss from the body surface of the mouse. In spite of the general improvement in nest quality shown by most mice, there were a few that built no nest at low temperature, yet they shredded all or most of their paper toweling.

There was indication of a slight amount of what might be termed "carry-over" effect; that is, when the temperature condition of the previous night was extreme (either high or

Numerical Group Totals for Shredding and Nest Building

(Intensive experiment with temperature and light)

Test	Day	Condition (Temperature–Light)	Total Shredding	Total Nest Building
A	1	Medium–Dark	4082	221
B	2	Medium–Dark	3967	223
C	3	High–Dark	564	0
D	4	Low–Dark	5597	345
E	5	High–Dark	669	7
F	6	Medium–Dark	3304	252
G	7	Medium High–Dark	2323	153
H	8	Medium–Dark	3731	274
I	9	Medium–Light	2040	163
J	10	Low–Light	4651	380
K	11	High–Light	158	9
L	12	Low–Light	4337	337
M	13	Medium–Light	2220	184
N	14	Medium–Light	2181	184
O	15	Medium–Dark	3803	276
P	16	Low–Dark	4689	404
Q	17	High–Dark	470	19
R	18	Low–Dark	5066	397
S	19	Medium–Dark	2543	269
T	20	Medium–Dark	3087	278
U	21	Medium–Light	1268	128
V	22	High–Light	54	0
W	23	Low–Light*	3784	352
X	24	High–Light	164	13
Y	25	Medium–Light	1941	196
Z	26	Medium–Light	173	240
AA	27	Medium–Dark	3439	328

Notes:

This table is represented graphically in the figure on p. 943.

*Low for the first half, then gradual increase to High

Shredding: Each mouse could make a high score of 100 each day if 100% of its paper towel was shredded. Sixty mice, therefore, could make a possible score of 6,000 during a one-day test.

Nest Building: Each mouse could make a high score of 9 each day if a good nest was built. Sixty mice, therefore, could make a possible score of 540 during a one-day test.

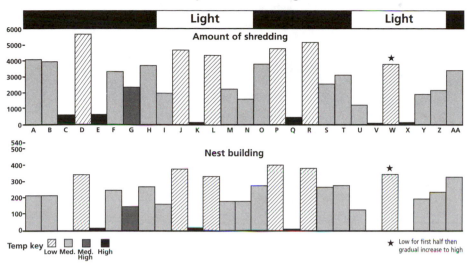

A graph of the data from the table on p. 942.
Courtesy of Oakleigh Thorne II.

low), it appeared to have some influence on the shredding during the subsequent night. These differences were checked statistically and showed one comparison was not significant, two were in favor of "carry-over," and one was significant in opposition to showing a difference. Results, therefore, were unclear and more study is needed. An interesting result from some preliminary experiments showed that amount of shredding increased when mice were deprived of shredding material for a period of time before being tested.

Effect of Light

Light caused a general decrease in the amount of shredding, which was especially noticeable at medium temperature. Elimination of light caused a marked increase in shredding. Again, this was statistically significant. It is interesting to note that when the light was on, the quantity shredded by the mice at medium temperature was always about half or less of the quantity shredded at low temperature. When the light was off, however, the amount shredded at medium temperature by the group usually exceeded half of the amount shredded at low, and was once as much as 81% of the low value. The combined effect of high temperature and light together resulted in the least amount of shredding (where the light was on and the temperature high).

Miscellaneous Observations

Shredding is performed with the incisors (sharp front teeth), the material being held by the front feet. The resulting shreds from any particular mouse were usually very distinctive and consistent. Although there were often many sizes of shreds present in the cage or nest after a given amount of shredding time, there was usually a predominance of one kind, so that

Various examples of shredding by mice.
Courtesy of Oakleigh Thorne II.

individual mice could usually be classified as coarse shredders, medium shredders, fine shredders, and the like. Some shreds were almost square and others long and narrow, depending on the individual mouse. In general, the longer the period of time allowed for shredding, the greater was the extent of shredding. Shreds are usually finer in a nest that has resulted from a week of shredding than one resulting from only a day of shredding. This is typical of most mice and indicates that they work over the shreds several times until a certain degree or limit of "fineness" is reached, depending on the individual. The placement of nests by mice was fairly constant, either inside the nest can, or outside it in a particular corner or section of the cage, with a preference for one of the rear corners.

The degree to which shredding is innate and the advent of the behavior were both tested on a small scale when a few young mice were raised in a cage with no shredding material present from the time of their birth to the time that they were weaned (as soon as their eyes were beginning to open, at about 14 days of age). When they were placed in individual cages with a whole paper towel supplied to each, the young mice began to shred slightly at about 16–20 days of age, and eventually became normal shredders and nest builders within 1–2 months. The evidence from these few individuals tested would indicate that the tendency to shred, at least in *Peromyscus*, is a species-predictable behavior, which implies a hereditary basis.

In order to get a hint of how learning might be involved in shredding, two separate litters of white-footed mice were raised to maturity with their female parents, both of which were "very fine" shredders. They were given an ample supply of paper towels at all times. At 1 month of age for one liter, and at 47 days of age for the other, the young mice were removed and tested individually. At first, most of them shredded finely or very finely, as did their female parents; but after a time, their shredding patterns began to vary more and more; that is, there was a distinct divergence in the behavior toward more individualistic patterns of shredding.

There was no indication that such factors as age, sex, stage of the estrous cycle, or previous environment had any important effect on the shredding behavior. It was observed several times, however, that just before or after parturition (birth of young) there was an increased amount of shredding and nest building in cages containing mated pairs of adult mice.

Shredding as Thermoregulation

Evidence from this study would indicate that the shredding behavior has survival value at low temperature, since it is so closely correlated with nest building, and since the amount of shredding is increased to such a marked degree at low temperature. The author considers that shredding, therefore, is primarily a thermoregulatory behavior, through its relationship to nest building. There seem to be other aspects of this behavior, however, that have psychological implications. (eg., the increased shredding after being deprived of materials to shred for a period of time, or the apparent "carry-over" effect of the extreme temperature conditions of the previous day). Since the present study was ecological, and not psychological, these factors need to be studied further by some scientist with these interests in mind.

It is possible that none of the three experimental temperatures (8°, 21°, and 33°C) used in the principle work with temperature and light constituted any appreciable stress situation for the mice, because experiments done 50 years ago by Stinson and Fisher (1953) indicated that even though *Peromyscus* preferred temperatures of 20°–30°C, they did not completely avoid temperatures as low as 6°C. In the wild, they would certainly encounter such temperatures quite commonly. These researchers found that individual mice had preference ranges that were narrower than the group preference range, and that these fell at various points between 20° and 30°C. These variations could account for the great variation in shredding found at 21°C. (medium temperature) in the present study. Because individual temperature preferences probably lie on both sides of 21°C, one would expect a wide variety of shredding reactions to this temperature, which was indeed the case. At high (33°C) temperature, however, most or all of the mice would be under a nonpreferred heat condition and would not be motivated to shred. At low (8°C) temperature, on the other hand, most or all of the mice would be under nonpreferred cold condition, and thus would be highly motivated to shred. The present study certainly indicated this to be true. If these assumptions are correct, they would account for the narrower individual ranges of shredding variation found at the two extremes of temperature (33° and 8°C) compared with the wide range found at medium temperature (21°C). The high (33°C) temperature was probably the more disturbing of the two extremes to the mice, for it was only this condition that so strikingly reduced the shredding and nest-building behaviors.

Since shredding and nest building were extremely pronounced at low (8°C) temperature, here is an example, therefore, of two possible survival mechanisms that are stimulated, at least under laboratory conditions, well in advance of the real stress situation that would arise from further lowering of the temperature. If these two mechanisms (shredding and nest building) operate in the wild, mice would have increased drive to build protective nests (other than brood nests) in early autumn, well in advance of the colder winter season. This certainly would have survival value, and it is reasonable to assume selection could have operated in such a manner to genetically fix shredding and nest building by favoring those mice that exhibited these behaviors. For example, if a mouse in the wild was confronted with a sudden temperature decrease and only coarse materials (such as leaves or bark) were available for use in nest building, shredding could then become essential for survival. Perhaps the size of the material would act as a cue for shredding.

Further study of certain aspects of the shredding behavior might be of economic importance, for it is quite clear that high temperature (heat) and/or bright light have repellent qualities with respect to shredding, and either one or both of these might be effectively used (as a repellent) under certain conditions to prevent shredding damage by mice. It is felt that with a better understanding of the shredding behavior, economic problems with which it is involved can be more intelligently approached and solved.

It would be interesting to experiment with the effects of various hormone injections on the amount of shredding by mice. Previous studies indicated that the hypothalamus was most likely the level of the central nervous system responsible for energy control, that is, the regulation of body temperature and activity. The author guesses that the shredding behavior of *Peromyscus* might be under the control of the hypothalamus acting on the anterior pituitary, which in turn would act on the thyroid. This implies that thyroxin is the controlling substance in the shredding behavior, but other hormones (such as steroids) might be involved.

The presence of light must reduce shredding and nest building through reduced activity, since *Peromyscus* is much less active in light (daytime) than in the dark (night), but reduction of nest building due to light, although clear at medium temperature, was not very apparent at either low or high temperatures. In fact, there was better nest building during the first low temperature *light* period than during the first low temperature *dark* period. This emphasizes the importance of high and low temperatures as *overriding* environmental factors. At *medium* temperature, on the other hand, *light* becomes more of a controlling influence on both shredding and nest building. This indicates that experiments should be carried out at *medium* temperature when one is testing *other* kinds of environmental conditions, otherwise quantitative differences in shredding caused by anything other than temperature are likely to be obscured.

One of the great advantages of the shredding behavior, as mentioned earlier, is that it can be used as a *quantitative* indicator of many behavioral changes in various experimental procedures where mice are kept in small cages. Since each mouse is fairly consistent in its manner of shredding under *normal* conditions, this can be compared with its performance under *experimental* conditions, and data from individuals can be combined to give a *numerical* measure of a behavioral change in a large group of mice. A thorough genetic study of shredding might be undertaken, not only because it is a behavior that can be measured quantitatively, but also because variations in shred size are produced that distinguish one individual mouse from another. It is possible that this individuality in the manner of shredding could be examined for its inherited and learned components.

Summary

Shredding behavior of *Peromyscus maniculatus osgoodi*, the white-footed or deer mouse, was studied in the laboratory using paper towels as the standard material to be shredded. The drive to shred was strong in both sexes when mice were confined in small cages, and increased when animals were deprived of shredding material for a time.

Shredding is considered to be a *thermoregulatory* behavior through its relation to nest building.

High (33°C) temperature reduced shredding and virtually eliminated nest building, while low (8°C) temperature greatly enhanced them.

Presence of light reduced amount of shredding, especially at medium (21°C) temperature, and usually lessened the quality of nest building, except at low temperature.

Carry-over effect of one extreme temperature upon another was indicated in the shredding and nest building, but was not statistically significant.

Shredding can be used as a *quantitative* indicator of *behavioral changes* in various experimental procedures with caged mice.

Oakleigh Thorne II

■ Reproductive Behavior
Sperm Competition

Organisms are behaviorally adapted to pass their genes on to the following generations. How do they achieve this? In 1871, Charles Darwin advanced the idea that males may compete among them for having sexual access to females, whereas females would choose which males would father their offspring. Darwin called this provocative idea *sexual selection*, and although researchers in the last century gathered evidence supporting it, they also realized that females of practically all animal species copulated more than once previous to fertilization. Given this, researchers rethought that sexual selection, via male–male competition or female choice, was unlikely to finish when a male secured a copulation but continued within the female body. One researcher that provided the grounds for this postcopulatory sexual selection idea was Geoffrey Parker, whose observations in the 70s motivated a new line of research which is now known as sperm competition.

The idea of sperm competition stemmed from detailed observations of reproductive behavior in dung flies carried out by Parker. Males of this species wait for females in cow dung, a resource that females use for laying their eggs. Males aggressively defend these sites against other males and "exchange" the use of these places for copulation with females. Parker observed that during copulation some males were either able to take over the copulating male or disrupt the couple after copulation, and then they would copulate with the female. Given that insect females are able to store sperm in specialized receptacles, the idea of sperm of different males still competing within the female genitalia emerged. This hypothesis had some support from examining the paternity of a female that had copulated with different males, which indicated that offspring were fathered by different males. Parker realized that sperm competition was far from unique in this species because females of practically all insects mated more than once and with different males. Furthermore, sperm competition was also likely to occur in other animals since females of most species also mate more than once

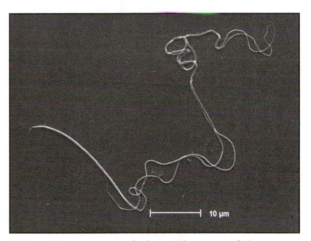

A bizarre sperm morphology. The case of the golden egg bug (the sperm head is the wider extreme appearing below on the left)
Photo by Francisco García-González.

previous to fertilization, and sperm are able to survive for considerable periods of time inside the female. Thus, although the idea that sperm competition originally emerged using insect models, now it is commonly accepted that sperm competition represents a widespread phenomenon in animal reproduction. Behavioral ecologists and evolutionary biologists now recognize that male–male competition and female choice—processes originally ascribed to occurring previous to copulation—can continue after copulation.

Sperm competition can be defined as the competition that emerges when the sperm of more than two males compete for fertilization of a certain female's set of ova. Sperm competition has served as an evolutionary engine for both male and female adaptations—for males to increase fertilization success and for females to have control over mate choice. Here, two sexual instances at which intense selection will occur (previous to, and during and after copulation) will be detailed, showing the kinds of traits that have been shaped by this competition in both sexes.

Before Copulation

Substantial evidence has outlined how sperm competition has fueled the evolution of male traits that prevent a female from engaging in further matings. Of those traits occurring before copulation, the more explicit are those in which males undertake behaviors such as precopulatory mate guarding, preventing other males from sexually approaching a female. In some birds, males remain close to their female mates until they have laid all their eggs. In a more extreme example of this guarding companionship, water strider males jump on to the female's body and, even when she shakes her body vigorously to make him fall off, some males successfully hold on for some time.

Presumably, the function of at least some part of this precopulatory pairing is not related to sperm transfer, but is used to prolong the time until the sexual activity of other males has decreased. By doing this, the mating male will reduce the chances of other competitors to mate with his female during that day. Sperm competition may have served as the evolutionary basis for male structures to physically secure female mates. For example, damselfly and dragonfly males are provided with clasper-like structures at the end of the abdomen which grasp the females' prothorax. Several studies have shown that in using such grasping behavior, males prevent takeovers by other males. In some of these species, males transfer their sperm when the reproductive activity of other males is nearly over (in some species the grasping behavior may last as long as 7 hours).

During and After Copulation

This is the sexual instance that has been most thoroughly investigated. Three subsets will be depicted emphasizing the great variety of traits involved in sperm competition.

1. Morphological Adaptations

One of the most extraordinary male adaptations to impede fertilization by other males is that related to male genital morphology. In 1979, Jonathan Waage illustrated one such example with damselflies. He showed how, during copulation, the male penis reached those places where females store sperm from previous copulations and remove it. Such an extraordinary process was achieved by the specialized morphology with which the penis tip was endowed: pointing spines able to trap the stored sperm. After this, the copulating male transferred his own sperm to the female. Since this pioneering and groundbreaking study, other researchers have unraveled how this sperm manipulation ability is also used by other insects, including beetles and crickets. This research line has also shown that insect genitalia have evolved not only spines, but other morphologies for the same aim; some examples are a scoop-like, or an enlarged and barbed penis, among others.

But genitalia have diversified to include a variety of other functions. A function which seems widespread in several animal species is stimulation of the female sensory system to induce the female to favor the sperm of the copulating male or to reduce the chances of using the previously stored sperm. Females are vastly irrigated with sensory terminations within their genitalia. This complex sensory system has evolved for reasons related to increasing efficiency during fertilization. In insects, for example, sensory terminations in the vagina provide information about when an egg is ready to be fertilized and, therefore, when the sperm should be released from the sperm storage organs. It has been shown that males of various insect species (eg., crickets) stimulate these female sensory devices via their penis movements during copulation, achieving a variety of female responses. Examples of these responses are: inducing the expulsion of other males' sperm to the outside, and inducing the transport of the copulating male's sperm to sites where its use for fertilization will be more likely, among others. In these species, the penis morphology has evolved particular forms to perform the stimulation. Male genitalia may sometimes take on more bizarre functions.

Sperm transfer in leeches, octopuses, bed bugs, and flatworms, among other animals, is of a hypodermic nature. The bed bug penis has a syringe-like morphology that pierces the female body in different regions, introducing the spermatozoa, which route themselves to the ovaries via the *haemocoel* (the insect circulatory system). In some species, females have even evolved particular sites to receive the injected sperm. Interestingly, the female immune system kills most spermatozoa when these are trying to reach the ovaries. Not surprisingly, this form of transfer has been called *traumatic insemination.*

One radical form of adaptation for putting up with sperm competition pressures involves variable morphologies of sperm. These different forms include variation in sperm length (e.g., up to 58 mm or 2.3 in long in some *Drosophila* flies), form (up to five different sperm types in a wasp), and the presence or absence of genetic material (sperm with and without genetic material in butterflies, mollusks, among others). Related to this last type of variation, in butterflies, for example, two sperm forms are present: long, fertile, nucleated sperm, and short, unfertile, anucleated sperm, called eupyrene and apyrene sperm respectively. Recent studies in butterflies have started to elucidate the roles of these variants. The apyrene sperm seems to delay female's sexual receptivity (so further matings with other males are prevented). On the other hand, when there is the chance of sperm competition (other competitor males in the area), it has been observed that males transfer more eupyrene sperm. According to this observation, this sperm seems to function to surpass, in number, the spermatozoa of other males.

2. Physiological Adaptations

Usually, sperm is embedded in a seminal fluid. Interestingly, this fluid is comprised of an incredibly enriched array of chemical substances. One possible function for this chemical cocktail is to reduce a female's sexual receptivity. Studies in fruit flies have shown that female receptivity can be reduced in as much as 9 days following copulation when only the fluid, but not the sperm, was provided. Seminal substances act on the female's sensory system, inhibiting her sexual propensity. Additionally, these substances promote egg production and oviposition. Given that these induced activities are energetically costly, females have their survivorship reduced as a consequence of mating. Substances in seminal fluids can further prevent a female's sexual encounters in a different way. In butterflies, crickets, mice, and marsupials, among others, after sperm transfer, males deliver a glue-like substance within the vagina which acts as a plug that physically prevents future copulations. Another notable adaptation, which is also widespread, is the presence of spermatophores. These structures are frequently chitinous in composition and bear sperm within them. In some species with internal fertilization (e.g., crickets and katydids), the spermatophore is passed on to the female along with another structure called spermatophylax. While the former has a fertilization aim, the latter conveys a nutritional function—it is composed of proteins that the female absorbs and uses for egg production. It has been shown that the longer the female takes in digesting the spermathophylax, the greater number sperm of the copulating male are moved to the sperm storage organs.

A set of adaptations which have been elucidated recently are those of sperm numbers passed on during copulation. Using mathematical models, G. A. Parker predicted that a male may adjust his sperm numbers according to the risk of sperm competition he is facing. For example, if the number of male competitors were small, males would not transfer as many spermatozoa as they would if the number of competitors increased. Studies in a variety of taxa (e.g., insects, fish) have uncovered that males behave according to Parker's predictions. Furthermore, he also predicted that if the number of competitors is too high, copulating males will not transfer as many sperm because the chances of surpassing other males will be reduced if all other males increase their sperm numbers, too. Tests of these ideas have also proven Parker is right. Supporting this idea, other researchers have found that, across a variety of taxa, animals have evolved larger sperm-producing structures in species depending on the levels of sperm competition risks. For example, in primate species whose females mate with different males during the estrus period, testicles are usually larger than in species whose females mate with the same male.

Another set of adaptations are "antiaphrodisiac" hormones that males deliver to females to inhibit their sexual activity. Females of a number of butterflies "advertise" their sexual receptivity by emitting male-attracting hormones. Presumably, the male hormone has evolved to counteract the female's sexual calling and, therefore, reduce the risk of sperm competition. A subtly related example also takes place in mice. Males that have taken over a female harem produce hormones which promote abortion in already pregnant females. The newcomer stands out in fertilization success by reducing the success of the last harem-dominating male.

3. Behavioral Adaptations

Reducing the sexual chances of other males may take more overt forms. Extending the time spent in copulation has been widely demonstrated in arthropods and may underpin five possible functions.

1. Sperm Manipulation and Transfer. In insects such as dragonflies and damselflies, males may increase the time incurred in copulation in order to displace the sperm the female has stored from previous males.

2. Mate Guarding. Males may prolong copulatory time to reduce the chances of other males of copulating with the same female. In this way, the male may become a sort of living plug which is the case of, for example, dung flies and locusts.

3. Multiple Insemination. Males may use the extended time to transfer more sperm (via single or repeated ejaculations) and/or seminal substances (such as hormones) to the female. In some beetle and fly species, how much sperm is delivered depends on the risk of sperm competition: If other potential competitors are detected, more sperm is passed.

4. Male Courtship. Females may demand some sort of courtship which may only be accomplished if males stay for relatively long periods. In some beetles, for example, males perform internal genitalic-mediated stimulatory movements. Females bias paternity in favor of some males according to how long the male continues producing the stimuli.

5. Sexual Conflict. Extended associations during copulation may reflect a conflict of interests: Males may demand that females use their sperm, wheres females may want to do otherwise. In the bean weevil, the male genitalia, after sperm transfer, evert spine-like structures which get "anchored" in the vaginal tissue causing serious damage. The more times a female copulates and the longer she remains in copulation, the more reduced is her chance for survival. Not surprisingly females repeatedly kick their mates to prevent the wounding. Apparently a male gains from damaging the female by reducing the chances his female partner will mate again (to prevent her

Genital contact in a beetle. One of the groups that has been most intensively studied is the insects.
Photo by Adolfo Cordero Rivera.

from further wounding) and by inducing her to lay eggs as soon as possible (since the female may "perceive" her chance of survival has been diminished and would "rush" to leave offspring).

An Important Note: Female Roles in Sperm Competition

Up until now it seems as if females were playing a passive role in the evolution of male sperm competition. Nevertheless, females have been ancillary in the evolution of such male traits even when these traits seem too aggressive. First at all, the fact that it is within females where reproduction takes place, provides females with enormous opportunities to direct the evolution of male adaptations. Think, for example, of insect males that have access to and remove the sperm from the female's sperm storage organs. Such access required some sort of female cooperation, otherwise females would have evolved inaccessible structures as in other insects in which females are endowed with extremely long vaginal ducts. More obvious explanations of female roles are, for example, those cases in which the female clearly exerts control over the fate of sperm. Wild fowl females eject the sperm obtained from copulations of subordinate males. This control also takes place in other animals—mammals and insects—for which males have very little influence. It has been proposed that females may incite the competition between males through the form of sperm competition. For example, females of different bird species actively solicit copulations from other males when their mate is not present. Such control of reproductive decisions and propensity to favor sperm competition leads to the question of what females gain from all this. A female may derive two kinds of benefits: direct (nongenetic, advantages for her fecundity or survival) and indirect (genetic benefits for her offspring) benefits. One example of direct benefits is the nuptial gift received previous to or during copulation. In scorpionflies, for example, males donate a prey to the predator female during copulation. Since females take more time to eat large compared to small gifts, it has been found that the larger the gift, the more sperm is transferred during copulation. Furthermore, it has been shown that this "free" meal helps the female to produce more eggs or to survive for longer. Indirect benefits include the provision of genetic information that augments the survival chances of offspring or their sexual capabilities during mate competition (e.g., being a good male in securing female mates). In the particular case of a female that promotes sperm competition, the benefit she would obtain is that the characteristics of those successful sperm may be passed on to the male offspring. This "sexy son" hypothesis applies equally well to extremely aggressive male traits, for example the wounding property of the bean weevil penis. If the female gains in having male offspring with similarly successful characteristics, then the costs for females are not as high because their sons will be successful, too.

Of course, males would not remain passive to these female filters, and possibly some of the male aggressive traits may have become too costly for females who receive reduced benefits from them. This idea can be the case for some animals such as the water striders, in which females face considerable costs when mating with aggressive males. Males in this species have evolved clasper-like devices to secure females during mating whereas females have counterevolved anticlasper structures. Female adaptations like this are rare in other animals and, in fact, females of different species have evolved traits that, far from deterring males from mating, seem to filter them. Examples of this are incisions in which clasper devices are inserted, nervous terminations that males can stimulate, sperm being stored in places that males can reach, and particular body places that males can genitalically pierce, among others. These adaptations apparently serve to filter males via female cooperation. In water striders, on the contrary, it seems that mating became too costly for females and that is why they have evolved structures to avoid matings. What is true, nevertheless, is that females, far from being passive players, can be ascribed as architects of sperm competition, as Patricia Gowaty, a behavioral ecologist, has called them.

The field of sperm competition is still growing up, and important gaps still remain to be clarified. Some questions to be uncovered are: What benefits (direct and/or indirect) do females gain from male sperm competition traits? How extended are those cases of species whose males have controlled female reproductive decisions? Given the rapid evolutionary nature of male–female interactions in sperm competition traits, what is the effect in relation to speciation? And there are other questions. Surely, equally exciting answers, as those we have already found, will be discovered shortly.

Further Resources

Birkhead, T. R. & Møller, A. P. (Eds.). 1998. *Sperm Competition and Sexual Selection*. London: Academic Press.

Darwin, C. 1871. *The Descent of Man and Selection in Relation to Sex*. London: John Murray.

Eberhard, W. G. 1996. *Female Control: Sexual Selection by Cryptic Female Choice*. Princeton, NJ: Princeton University Press.

Parker, G. A. 1970. *Sperm competition and its evolutionary consequences in the insects*. Biological Reviews, 45, 525–567.

Simmons, L. W. 2001. *Sperm Competition and its Evolutionary Consequences in the Insects*. Princeton, NJ: Princeton University Press.

Waage, J. K. 1979. *Dual function of the damselfly penis: Sperm removal and transfer*. Science, 203, 916–918.

Alex Córdoba-Aguilar

■ Robotics
Animal Robots

Imagine your dog "talking" to an animal robot! Scientists in Paris and Budapest are doing just that. Luc Steels and Frederick Kaplan at Sony Computer Science Laboratory in Paris pursue research in artificial intelligent using a four-legged robot called "AIBO" (Artificially Intelligent roBOt). This robot can sense the environment, walk toward and away from things it sees, crouch, bark, wiggle its ears, and wag its tail. Hungarians Eniko Kubinyi and Adam Miklosi have field-tested AIBO with real dogs. The dogs apparently view the robot as lifelike. Puppies approach and try to play with AIBO. Adult dogs curl their lips, growl, and even attack AIBO if it approaches them while they are feasting on meat.

The scientific field of animal robotics is just emerging. The design and anatomy of animals has previously been used as inspiration for building robots that move in efficient ways (called "biomimetic robots," which mimic biology). Now animallike robots are being created to interact with live animals with the express goal of better understanding the behavior of the live animals themselves. Since the study of animal behavior is called *ethology*, we could call this new field *ethorobotics*. This innovative method has the potential to answer questions about behavior that have been difficult to answer with previous methods for studying animal behavior.

Suppose that you want to understand the meaning of a particular animal display. For example, male lizards can expand the colorful skin on their throats, called dewlaps, to produce vivid visual displays. What would you do if you wanted to find out what the dewlap display means?

The most direct way to understand behavior is simply to observe it carefully. Much of ethology has been based upon observation. You might go outside on a warm day, settle yourself quietly by a large banyan tree, for example, and watch until you see a small brown anole lizard darting up one of the thick roots that arc away from the base of the tree. The lizard may stop suddenly and stand motionless. Then, very slowly, he starts to extend his dewlap. You see it first as a thin slice of orange by his neck, and then the movement accelerates and the shape of the lizard is suddenly transformed, with the large round curve of the dewlap fully extended before you. You can describe the movement; you can document when and where it occurs. But how do you figure out what the display *means*?

A behavior "means" something if it causes other animals to react in a predictable way. With the observational method, we can watch for other, neighboring, lizards, and document their responses to the display. By observing whether the other lizards approach or run away, for example, we have an idea of whether the display was friendly or aggressive. But how do we know for certain what caused the neighboring animals' behavior? What if it was a hawk swooping down below the branches of the tree that caused the other lizard to run, rather than the dramatic display of the first lizard?

Ethologists have found ways to ask the animals what displays mean. One method is to use playbacks. We can play an audio recording of a bird song back to live birds in the field, for example, and see what they do. We can show videos of lizards extending their dewlaps and flexing their legs in "pushup" displays, and then see how live lizards respond. Playback is a nice method because it removes the variability of the real life observational situation by allowing us to control the environment. We can conduct repeated playback trials, varying only the display that we are interested in and keeping everything else constant. If an animal responds to one trial and not to another, we will know exactly what caused its change in behavior. But can animals see video screens? We have convincing evidence that they can. Video playbacks have now been successful with monkeys, birds, lizards, and fish, for example, and more animals are being tested. But is a flat, 2-dimensional video or LCD screen as salient (meaningful) as a real, 3-D figure? And what if the animal responds; how can we manipulate the video image to respond back, to create a "conversation"?

Imagine if we could create realistic 3-D robotic animals models that could interact in real time with live animals. We would create an endless variety of displays, limited only by the hardware and software design of the robot. The robot could produce displays in new combinations as it interacts with the animal subject. We could then use the record of the behaviors produced by the robot and the responses made by the live subject to determine the impact of each behavior on the recipient. In other words, we could find out the *meaning* of each behavior. This ideal scenario is still the stuff of dreams, since the first animal robots that have been made are not yet sophisticated enough to have a highly variable repertoire of behavior or to be able to react in real time to live subjects. But the dream is quickly becoming reality.

At least three other research teams, in addition to the AIBO group, have now built animal robots that have communicated successfully with live animals. These robots are programmed to do a few simple behaviors activated by direct motors or remote control.

Although they only perform a few behaviors, the behaviors were carefully chosen to be key to successful communication with the animal. One of the first "animal-bots" was a bee, built by Danish and German researchers in the early 1990s. Axel Michelson and colleagues built a bee that wiggles, dances, and hums like a real bee. The model didn't look exactly like a bee, but that didn't matter because bees communicate in the dark of their hive, so they couldn't see the robot anyway. Bees perceive each other with senses of taste, touch, sound, and vibration. The scientists were able to use the model bee to communicate to real bees about the location of food items in their environment. The robot bee "told" the live bees where the food was, and the live bees found it!

Recently, a research group built a robotic bird in Australia. Gail Patricelli and her colleagues built a beautiful model of a female bower bird which can crouch, fluff its wings, and turn its head like a real bird. In this species, the male builds an elaborate "bower" for the female. This is a nest, constructed of sticks and twigs, that resembles a throne with an arch overhead. The females travel around and try out the bowers until they find one they like. Gale and her collaborators built such a convincing female robotic bird that they fooled the males, who courted the robot vigorously, hoping to win her as a mate!

A rendering from video frame illustrating aggressive behavior of a male dart-poison frog toward the electromechanical model frog (left) placed in his territory.

© 2003 National Academy of Sciences, U.S.A.

We can also use robots to test questions about how combinations of signals affect meaning. This is important because natural behavior includes many components. When humans laugh, for example, we produce a sound (an acoustic component), our mouth moves and the skin around our eyes may wrinkle, and your whole body may even shake (multiple visual components). In some cases, we may also touch a friend on the arm to emphasize how helpless our laughter has left us (a tactile component). Why do we include so many components? Are the components redundant with one another or do they provide extra information? We don't know.

Animal displays are also made of multiple components. The iguana, for example, a relative of the smaller anole lizard, hisses while it does its visual display. We can investigate the impact of both visual and auditory channels on behavior by building tiny speakers into the robot, so that the robot can produce sounds as well as visual displays. The sense of smell is also of paramount importance to many animals. Perhaps the field of chemical ethology will advance to the point that we can manufacture specific pheromones (animal scents) for each emotion to add to the milieu of signals being produced by the robot. By observing the responses of live subjects to each component of the display separately, as well as to the entire suite of signals together, we could better understand what the signals mean and why animal displays are so complex.

Peter Narins and his colleagues in Vienna have built a small robotic frog to test questions about which sensory channels are necessary for effective communication in the dart-poison frog. The robot, dubbed the "robo rana," has one main moving part: its throat pouch, which inflates into a round bubble. Inflation of the throat pouch is used to create croaking sounds in real frogs. Robo rana is mounted on a circular pedestal that can rotate to point the model in different directions. It also has an imbedded speaker used to produce prerecorded croaks. When the robot expanded its throat and croaked, live frog in the area approached and even attacked this apparent intruder. Both the motion of the throat pouch and the vocal signal were necessary to elicit the response from the live frogs.

The success of these new animal robots at communicating with real live animals is exciting, and makes me wonder. . . how long until my beloved yellow lab, Tundra, has her very own Robo-Fido to play and commiserate with while I am away at the office?

See also Robotics—*Artificial Pets*

Further Resources

Kubinyi, E., Miklósi, A. Kaplan, F., Gácsi, M., Topál, J.,Csányi, V. 2004. *Social behavior of dogs encountering AIBO, an animal-like robot in a neutral and in a feeding situation.* Behavioural Processes, 65, 231–239.

Michelsen, A., Andersen, B. B., Storm, J., Kirchner, W. H. & Lindauer, M. 1992. *How honeybees perceive communication dances, studied by means of a mechanical model.* Behavioral Ecology and Sociobiology, 30, 143–150.

Narins, P. M., Hodl, W. & Grabul, D. S. 2003. *Bimodal signal requisite for agonistic behavior in a dart-poison frog,* Epipedobates femoralis. Proceedings of the National Academy of Science, 100, 577–580.

Partan, S. and Marler, P. 1999. *Communication goes multimodal.* Science, 283, 1272–1273.

Patricelli, G. L., Uy, J. A., Walsh, G. & Borgia, G. 2002. *Sexual selection: Male displays adjusted to female's response.* Nature, 415, 279–280.

Sarah Partan

■ Robotics
Artificial Pets

Artificial pets are robotic toys with petlike behavior. Unlike traditional robots, which are made to be intelligent tools that serve their owner, artificial pets are autonomous creatures that elicit *attachment* from their owner.

Artificial pets provide a fascinating arena for examining the relationship between behavior and attachment. Physically, many are unremarkable: One of the most successful artificial pets to date, the Tamagotchi, was a small plastic key-chain egg with simple animations on a low-resolution screen. Yet people became extremely attached to them, giving high priority to caring for them and mourning them when they "died." It is the behavior of the artificial pet that fosters this attachment.

Although artificial pets have been developed in a variety of forms, from the simple key-chain pets to complex robots such as AIBO, an artificial dog, they share several key behaviors: They appear to act autonomously, they are dependent on their owner for nurturance, they require frequent interaction, and they develop in response to their owner's actions.

Autonomy: An artificial pet acts—or, more precisely, appears to act—autonomously. This means that its actions seem to be internally motivated, it appears to have its own goals, feelings, and desires. It does not necessarily accept the commands of a human and instead makes its own demands on the person. When machines work exactly as we expect them to and do what we request of them, we think of them as simply machines. It is when they do not work as expected that they appear to have a will of their own, and we ascribe intelligence to them.

A child in Ohio plays with her Tamagotchi.
© *Thierry Mamberti / Corbis Sygma.*

Dependence: Most artificial pets start as "infants," which elicits nurturing and affection: We instinctively take care of the young. Throughout their lifespan, the pets are designed to require their owner's help in order to thrive and survive. If the owner does not "feed" or "entertain" them they become ill or even die. The pet's dependence makes the owner feel responsible for it.

Interaction: Feeding, cleaning, and playing with the pet all involve interacting with it—and the pet becomes integrated into the owner's daily routine. Having spent a considerable amount of time and energy on the pet, the owner becomes invested in its well-being.

Development: Artificial pets are designed to develop in response to the owner's treatment of them. A pet that is well cared for will be healthier and more tractable. The owner is thus encouraged to take pride in their pet's well-being.

Artificial pets are a good example of how we use *metaphorical thinking* to conceptualize behavior. If we think of them as games, the time spent playing with them is entertainment and somewhat self-indulgent; if we think of them as animals, time spent playing with them is caretaking, an act of responsibility and altruism. It is obsessive to leave a meeting or dinner because a game requires attention, but it is reasonable to do so if a pet is in need. The metaphor we use to think about them changes how we understand the interface, act toward the object, and judge the behavior of others towards similar objects. Simply calling an interactive program a pet is not sufficient for people to think of it as one.

Not every responsive toy that is marketed as a pet manages to achieve that metaphorical status, and comparing those that are perceived as pets with those that are not can help us understand some of our beliefs about behavior, autonomy and agency. For example, screen-based pets did not achieve anywhere near the popularity that key-chain pets did. They had two signifiant differences. First, the key-chain pets were embodied—the pet is the whole physical package, not just the image on the screen. Seeing the pet as an object, rather than a program, helped lend it credence as a creature.

Second, the key-chain pets could not be turned off, whereas the screen ones could. A pet that could be turned off lacked urgency, and the ability to switch it on or off broke the illusion of aliveness.

There is considerable controversy about the social implication of artificial pets. Do they teach children to nurture—or are they wasting empathy on artificial, commercial objects, empathy that should instead be turned toward real beings? Some argue that they are useful for teaching responsible behavior in a safe setting. Others say that they erode responsibility, by providing people with pseudoanimals that can be neglected to the point of simulated death without moral repercussions. Some proponents of artificial pets hope they can be used as companions for the elderly, providing the emotional support that a real pet provides, without the need for care that an animal has.

The evolution of artificial pets will occur in an environment of market forces and technological development. Today, the trend appears to be towards more complex and "intelligent" pets, robotic creatures that are designed to develop a long term relationship with their owner. It will be interesting to see what qualities of real animals are replicated in these artificial beings, and what unique features are developed for them; it is an evolution that will provide a new perspective on the coevolution of humans and domesticated animals.

See also Robotics—*Animal Robots*

Further Resources

Kaplan, F. 2000. *Free creatures: The role of uselessness in the design of artificial pets.* In: *Proceedings of the 1st Edutainment Robotics Workshop* (Ed. by T. Christaller, G. Indiveri, & A., Poigne), September GMD-AiS.
Pesce, M. 2000. *Toy Stories.* New York: New York Academy of Sciences.

Web Sites

Samp, J. 1997. *Critical thoughts about Tamagotchi.* http://demo.neopets.com/presskit/articles/recent/seattletimes.html
http://www.virtualpet.com/vp/research/research.htm
http://www.neopets.com

Judith Donath

■ |Self-Medication

Animal self-medication, or *zoopharmacognosy*, is a relatively new field in the discipline of animal behavior. Nonetheless, evidence of health maintenance or self-medicative behaviors are widespread throughout the animal kingdom. The study of such behaviors adds a new dimension to our understanding of the evolutionary axiom—"survival of the fittest." Undoubtedly, being born with the "right genes" exerts significant long-term physiological control of an individual's health and lifetime reproductive success. However, in the short-term, all individuals can benefit from behavioral flexibility to tweak and maintain the system on a day-to-day basis in response to both internal and external challenges imposed upon their health. Parasites are considered to have exerted great influence on animal behavior in this respect. From insects to chimpanzees, evidence suggests that, indeed, animals have behavioral strategies for limiting the affect of parasite disease on their health and reproductive fitness.

The woolly bear caterpillar (tiger moth), alters its diet depending upon whether it is parasitized by tachinid fly larvae or not. Parasitized caterpillars under controlled outdoor conditions were found to prefer poisonous hemlock to lupine, two of their most important food sources, when given a choice. While this alkaloid "cocktail" does not kill the parasites (if it did it would probably kill the caterpillar too!), these caterpillars were more likely to survive the infection than those that were not given the choice to feed on hemlock. Other less well-studied examples suggest that often some very puzzling food choices play an important role in suppressing parasite infection in much larger animals like the camel, rhinoceros, wolf, and wild bison.

Chimpanzees in the Mahale Mountains of Western Tanzania, showing patent signs of illness such as diarrhea, lack of appetite, weakness, and parasite infection, ingest the African medicinal plant, "bitter leaf" (*Vernonia amygdalina*), a tree in the Composite family. This has been described as a means for controlling infections of a strongyle nematode, the nodular worm (*Oesophagostomum stephanostomum*). Chimpanzees show visible recovery from their symptoms and a significant drop in the nematodes excreted egg count within 24 hours after the ingestion of a limited amount of the strongly bitter pith. Across Africa, people traditionally use bitter leaf as treatment for a number of intestinal and blood parasites and related gastrointestinal upsets. In the laboratory, wide antiparasitic and anticancer activities have also been demonstrated from the 13 newly discovered compounds in this plant. While dietary protein has been found in the pith, the extremely low observed frequency of use compared to other protein rich foods available in their diet, and the small amounts ingested at any one time, do not point to any particular importance as a food source to Mahale chimpanzees. This appears more likely to be a plant used mainly for its medicinal value by sick animals that subsequently recover from that illness.

Another response to parasite infection is a behavior known as leaf swallowing. First reported in Gombe and Mahale chimpanzees, this peculiar ingestive behavior is now known to occur in at least 24 different social groups of chimpanzees, bonobos, and lowland gorillas across Africa. All of the 34 different plant species, whose leaves are folded in the mouth and swallowed whole, share the common property of being rough and bristly. Such leaves contain high amounts of lignin, silica, and other compounds that make the leaf difficult to

Linda, a female chimpanzee of Mahale M group swallowing a leaf of Aspilia mossambicensis.

Courtesy of Michael A. Huffman.

digest. A plant antiherbivore defense mechanism, apes exploit and enhance these properties by not chewing the leaves before swallowing them on an empty stomach, often first thing in the early morning. As a result, the 10–100 leaves swallowed in a single session often cause diarrhea within 6 hours, expelling the undigested leaves and purging the ape of adult strongyle nematodes at a ratio of two leaves per one adult worm. The expulsion of tapeworm fragments has also been associated with leaf swallowing by chimpanzees in the Kibale and Budongo forests of Uganda. Leaf swallowing has also been seen in Canadian snow geese and grizzly bears, and is associated with the expulsion of tapeworms prior to winter migration and hibernation. This is proposed to rid the animals of nutrient sapping invaders at critical times when all their reserves must be fully utilized.

Topical self-medication of items to repel insects and ectoparasites is also a widespread phenomenon in the animal kingdom. One version of this behavior involves the application of the antipredator defenses of acid secreting ants and toxic millipedes into their fur and feathers. Anting behavior, as it is called in birds, is practiced in over 200 species, and the formic acid secretions have been shown to repel or kill lice and feather mites effectively. In mammals, this behavior is known as fur rubbing and is thought to repel mosquitoes and other pesky flying insects as to well as to relieve itchiness caused by bites or fungal infection. Two primate species, the black lemur in Madagascar and the wedge-capped capuchin of northern South America, will capture and sometimes bite and kill millipedes, rubbing their defensively secreted, smelly, toxic compounds into their fur. Other mammals rub plant substances into their fur. The Kodiak and brown bear of North America rub the chewed roots of a plant commonly known in the western states of North America as "Indian Parsley" or "Bear Medicine" (*Ligusticum porteri*) into their fur. The plant has demonstrated antiviral and antibacterial properties. The white-nosed coati of Panama take resin oozing out of slashed bark of the tree locally known in Panama As "carana" (*Trattinnickia aspera*) and vigorously rub the resin into their fur. Capuchin monkeys and spider monkeys will rub crushed and chewed citrus fruits or the stems or leaves of a number of different pungent smelling plant species into their fur. Capuchins exhibit fur-rubbing behavior most frequently during the rainy season, when the need is greatest for repelling insects like mosquitoes and bot flies that can transmit health-threatening parasites and irritate the skin.

Health is closely linked to reproduction of individuals and thus survival of the species. A recent study on the sifaka, another prosimian of Madagascar, provides an interesting example of the effects of a "medicinal" diet against parasites and its direct implications for reproduction. Too much tannin in the diet will prevent the absorption of protein and therefore should normally be avoided, particularly by pregnant or lactating females whose energy and protein requirements increase 2–10 times greater than normal. At this time all available protein first goes to the growth of the fetus, then to body maintenance, and lastly to immune function. This is why pregnant females are more prone to parasite infections than others. However, it was found that pregnant or lactating female sifakas in the Kirindi forest significantly increase their intake of tannins from the group's normal diet, as well as seek out high tannin-containing plants not eaten by other group members. In light of the potential harm, why do pregnant sifakas increase tannin intake? Veterinary and pharmacological research into the beneficial uses of tannins in appropriate amounts points to prophylactic antiparasitic action, increase in weight, and the stimulation of milk secretion as a probable explanation for this seemingly maladaptive craving.

As all of the above cases show, animals have learned to use the very defenses that have evolved as strategies to prevent plants and insects from being preyed upon. The saying, "one man's poison is another man's medicine," seems to hold true for the rest of the animal kingdom.

See also Cognition—*Imitation*

Further Resources

Carrai, V., Borgognini-Tarli, S. M., Huffman, M. A., & Bardi, M. 2003. *Increase in tannin consumption by sifaka* (Propithecus verreauxi verreauxi) *females during the birth season: A case for self-medication in prosimians?* Primates, 44, 61–66

Engel, C. 2002. *Wild Health*. Boston: Houghton Mifflin Co.

Huffman, M. A. 2002. *Self-medicative behavior in the African great apes: An evolutionary perspective into the origins of human traditional medicine*. BioScience, 51, 651–656.

Huffman, M. A. 1997. *Current evidence for self-medication in primates: A multidisciplinary perspective.* Yearbook of Physical Anthropology, 40, 171–200.

Moore, J. 2002. *Parasites and the Behavior of Animals*. New York: Oxford University Press.

Zimmer, C. 2000. *Parasite Rex*. New York: The Free Press.

Michael A. Huffman

■|Sharks

The largest predatory fish on Earth, great white sharks (*Carcharodon carcharias*), known to all oceans of the world, are most often found along the coastlines of California, southern Australia, the Mediterranean, and South Africa. Named for their white undersides and grey backs, great whites can grow over 20 ft (6 m) long and weigh over 4,500 lb (2,040 kg). Subject to much popular myth and misconception, great whites, also known as white sharks, are elusive creatures. Very little is known about their social interactivity, mating, and migratory behaviors. Once characterized as lone man-eaters, recent research has shown that great whites are highly organized social animals, efficient hunters with little interest in consuming humans.

White sharks are *endothermic*, generating heat from their own metabolism. They prefer swimming either within 15 ft (4.5 m) of the surface or at depths of 900–1,500 ft (274–457 m),

A great white shark cruises off the coast of Australia.
Courtesy of Corbis.

A great white shark jumps high into the air, catching a giant petrol bird off the coast of South Africa.
© Brandon Cole / Visuals Unlimited.

but also dive to over 2,000 ft (610 m) deep. Individual white sharks are identified by dorsal fin irregularities, scars, and discoloration. Actual coloration varies from light grey to almost black topsides. Their jaws contain multiple rows of teeth. Although fearsome in appearance, modern observation suggests that white sharks also use their mouths for holding and sensing in much the same way as dogs. Nonfeeding sharks have been seen to mouth objects nonaggressively. Healthy adult females often bear scarring from other sharks on the side and back of the head near the gills, which some researchers suggest is due to being held during courtship or mating activities. These activities have never been observed in the wild, and white sharks have never been successfully kept in captivity.

Females carry developing eggs internally, bearing live young. Slow to mature and slow to reproduce, white sharks are now a protected species in the United States. According to fossil records, various species of *carcharodon* (sharp- or jagged-toothed) sharks have been around for about 57 to 65 million years. It is unknown how long individuals live. Scientists measure shark age by counting calcified bands in vertebrae of captured specimens. The largest captured sharks, over 22 ft (7 m), were about 30 years old.

Young sharks feed primarily on fish and other *pelagic* (open sea) organisms. Juvenile teeth are elongated and narrowed, well-suited to grabbing and holding slippery prey. At around 10 or 12 years of age, the pelagic teeth are replaced by broad, flat teeth more ideally suited to flaying and stripping flesh. Adult sharks feed primarily on sea mammals, especially pinnipeds, which include seals, sea lions, and sea elephants, but have been known to consume many types of prey from tuna to dolphin. The only known fatal attack upon white sharks other than by humans was by a female orca, or killer whale, seen to be protecting her calf in the waters near San Francisco Bay in 1999. There were no more shark sightings in the area until the following year.

Contrary to popular myth, human victims of white shark attacks have rarely died from injuries. Attacks on humans are most often reported to be by adolescent white sharks ranging from 12–16 ft (3.5–5 m) in length, at the age when converting from a fish to a mammalian diet. Victims, most often swimmers, kayakers, or surfers, are also usually located in shoreline waters frequented by the white shark's primary food source: seals and sea lions. Dietary confusion may be the leading cause of these shark attacks. Survivors almost universally report a single bite; however, even one bite can cause massive bleeding and trauma; therefore, swimmers and surfers should avoid areas where great whites have been spotted, and where their favorite food source, seals and sea lions, are known to congregate.

More active in daylight hours and moderately clear waters, great white sharks are visual hunters, preferring rocky headlands where deep water lies close to shore. Lurking at the bottom or near outcroppings until a prey silhouette is seen on the surface, white sharks strike one tremendous blow, usually fatal, often aimed at the head or soft underbody. The sharks then retreat to the bottom, resting and waiting for their prey to die, and return later to devour the carcass, taking as much as 400–500 lb (180–225 kg) of flesh per bite. Their pinniped victims can weigh as much as 2,000 lbs (900 kg). After the initial strike, feeding white sharks swim leisurely and are relatively docile.

Sharks—Mistaken Identity?

Erich K. Ritter

Many people believe that white sharks, *Carcharodon carcharias*, bite surfers because of their resemblance to seals when viewed from below, and thus "mistaken identity" is assumed to be the cause of these accidents. However, when this theory is examined more closely, "mistaken identity" seems unlikely. Questioning such assumptions is a method used to gain a clearer understanding of an animal's possible motivation.

Below are five areas in which the "mistaken identity" theory is questioned: wound pattern, coevolution of prey and predator, search-image, incidents involving non-seal-eating sharks, and statistical incidence. On close examination, they suggest that exploratory behavior, rather than mistaken identity, is the more likely motivation for many of these incidents.

Bite Pattern

During a predatory bout on a seal, the initial bite is usually forceful enough to impair the motion of the mammal and prevent its escape. Bite patterns on surfers, however, are predominantly superficial. Superficial bites occur when a shark mouths an object to examine it; the bites are insufficient to handicap a seal. Some wounds on surfers may appear quite severe and are usually caused by the surfer as he or she reacts to the initial bite (called "secondary wounds").

Sharks don't have hands; they explore items with their mouths, much like human infants. Exploratory bites are the most common types of shark bites, and when a predator examines an unfamiliar object, it usually does so with caution.

Coevolution of Predator and Prey and Search Image

That surfers may trigger an "unfamiliar object" approach and/or reaction among white sharks is also indicated by the shark's evolution. White sharks and seals have been sharing the sea for more than 10 million years. It is very likely that the sharks have inherited strategies to hunt seals, and are aware of all the possible shapes, motions, and positions of seals in the water (i.e., search images). Surfers move and are positioned (sitting, prone on surfboard, standing) very differently from seals, and thus it is unlikely that a surfer triggers a search image for a white shark.

Incidents by Non-seal-eating Sharks

Another argument supporting exploratory behavior is that blacktip sharks, *Carcharhinus limbatus*, which are responsible for more bites on surfers than white sharks worldwide, do not feed on seals and do not even live in areas where seals occur. So even if the "mistaken identity" theory had merit, it could not explain why blacktip sharks make similar "mistakes."

Statistical Incidence

From a statistical viewpoint, despite the number of sharks swimming in proximity to surfers, bites by sharks on surfers are extremely rare. If sharks routinely confused surfers with prey items, the incidence of shark bites on surfers would be

(continued)

Sharks—Mistaken Identity? (continued)

far higher. The low incidence of bites suggests that "mistaken identity" on the part of the shark is unlikely.

Conclusion

As with many a behavioral question, it can never be really known what a specific animal's thoughts or motivations are for a particular behavior. And so, taking the animal's point of view may be the best method to gain an understanding of its motivation. If you are a shark and there is something intriguing, floating or moving at the surface—but unknown and potentially dangerous—what will you do? Examine it—cautiously!

Further Resources

Collier, R. S., Marks, M. & Warner, R. W. 1996. *White shark attacks on inanimate objects along the Pacific coast of North America.* In: *Great White Sharks: The Biology of* Carcharodon carcharias (Ed. by A. P. Klimley & D. G. Ainley), pp. 217–222. San Diego: Academic Press.

Klimley, A. P., Pyle, P. & Anderson, S. D. 1996. *The behavior of white sharks and their pinniped prey during predatory attacks.* In: *Great White Sharks: The Biology of* Carcharodon carcharias (Ed. By A. P. Klimley & D. G. Ainley), pp. 175–191. San Diego: Academic Press.

Strong, W. R. 1996. *Shape discrimination and visual predatory tactics in white sharks.* In: *Great White Sharks: The Biology of* Carcharodon carcharias (Ed. By A. P. Klimley & D. G. Ainley), pp. 229–240. San Diego: Academic Press

---

Great white sharks have been observed in every major ocean of the world. Their populations are most densely located in five coastal areas—California, Japan, southern Australia, South Africa, and the Mediterranean. Genetic markers indicate that great whites in the Sea of Japan and along the California coastlines may interbreed. The same markers indicate that these populations do not overlap or interbreed with south Pacific white sharks found along the coast of Australia. Similarly, these sharks do not appear to interbreed with Atlantic or Mediterranean white sharks.

Northern California white sharks have been tracked to the east central Pacific, and one as far west as Hawaii, using radio transmitters. Their actual migratory pattern and behaviors during the rest of the year are largely unknown. They may migrate as far as Japan and the coast of Alaska before returning to northern California in late summer. In an area west of San Francisco, California, dubbed "the red triangle," males return annually, females every other year, congregating from August through November. Breeding grounds for seals, sea lions, and sea elephants around the Farralone Islands and other isolated shorelines in the area provide a primary food source. Researchers have identified and cataloged the same sharks returning year after year. White sharks appear to kill individually, but feed in groups. The same sharks are often seen feeding together. An analysis of the number of seal kills, and calculating the number of calories in the vast amount of seal flesh consumed per shark during the autumn season, indicates they may not feed for the duration of the year, not unusual among large marine creatures such as whales or sea elephants. In the "red triangle," known for more great white attacks on humans than any other place in the world, there have been about 25 attacks on humans in the last 50 years. Of those, only two attacks were fatal.

Once considered lone killers, researchers have found that great whites are highly social animals, feeding in groups of 6–10 with one considerably larger female close by, typically over 21 ft (6.5 m) in length and weighing about 4,800 lb (2,200 kg). Fossilized teeth indicate much larger sharks have existed and may still exist.

Further Resources

Cousteau, J-M. 1995. *Cousteau's Great White Shark*. New York: Harry N. Abrams.

Ellis, R. 1995. *Great White Shark*. Stanford, CA: Stanford University Press.

Klimley, A. & Ainley, D. 1998. *Great White Sharks: The Biology of* Carcharodon carcharias. San Diego: Academic Press.

Matthiessen, P. 1997. *Blue Meridian: The Search for the Great White Shark*. New York: Penguin.

Web Sites:

PSRF—Pelagic Shark Research Foundation www.pelagic.org
   Dr. A. Peter Klimley and Dr. Burney Le Boeuf, UCD Bodega Bay Marine Lab, are the world's foremost authorities on the ethology and behavior of white sharks and northern elephant seals.

PRBO—Point Reyes Bird Observatory http://prbo.org
   Peter Pyle, bird and shark wildlife observer, has been studying the feeding behavior of great white sharks at Farralone Islands Research Observatory for over 22 years. He spoke at the Sausalito Bay Model in March 2002.

http://www.zoo.co.uk/~z9015043/gws conserv.html
   This web site contains a detailed history and description of the great white shark.

http://www.prbo.org/marine/sharkref.htm
   This is a detailed page of further references pertaining to the great white shark.

Ethan Hay

■ |Siblicide

Siblicide can be defined simply as any process of sibling aggression that causes or contributes to death in very young animals (some striking parallels also occur in plants). This phenomenon was noted anecdotally in old natural history accounts dating back to Aristotle, but began to receive serious scientific attention only in the last 25 years. Previously, it was dismissed as a bizarre novelty practiced by a few great eagles and owls that nest in remote places and are discouragingly hard to study. But in recent years, its biological significance has come into focus, and various forms of the behavior have been discovered in birds that nest in dense colonies or nest boxes, logistical features that facilitate careful exploration. As a result, many egrets, pelicans, boobies, and even a few gulls are now known to lay more eggs than the parents are willing or able to support, a pattern that automatically leads to a lethal version of "musical chairs" wherein one or two nestlings are bullied and end up dying, either directly from injuries or indirectly as the result of socially enforced starvation.

The processes of siblicide are important because they reveal some unsavory, but subtle aspects of natural selection. Because natural selection is fundamentally a reproductive race, one generally expects parents able to produce the greatest number of offspring to leave the most descendants (bearing copies of their genes). But the quality of the offspring produced can be as important as their numbers, such that fewer high-success progeny may pay higher long-term dividends to parental fitness. Understanding that food and other key resources are often insufficient to support all brood mates well thus sheds light on the simplest riddle posed by siblicide—why young animals often kill their closest kin in the nest. A dependent offspring facing lethal food shortages must often choose between its own survival and that of its brother or sister: Without any need to make sophisticated calculations or understand why, the habit of self-promotion is usually favored by the dispassionate ledger of natural selection.

A deeper puzzle concerns why the parents create so many offspring in the first place. If cattle egret parents were to lay just two eggs, for example, there would be no need for heated sibling rivalry, bloody pecking, and lingering death for the chick hatching from the third egg: Parents can almost always provide well for two. Yet they go on to lay an additional egg or two, which sets the whole dynamic in motion. Why? At present, three incentives have been identified for parents to produce extra eggs, thus precipitating the need for subsequent bloodshed. First, the food base may chance to be richer than usual, such that the designated victim need not be terminated. In some species, abundant food actually suppresses the sibling aggression; in others, it appears that the fighting continues, but the individual at the bottom of the dominance hierarchy receives enough food to pull through. Second, the laying of an egg does not guarantee the subsequent hatching of a viable chick, so parents may lay extras that can serve as back-ups, as "insurance," against the premature demise of one of the core brood members. Third, in some species, even a doomed chick may provide temporary benefits for its stronger siblings, perhaps as a blanket against nocturnal chilling or even as a meal during food shortages. If any or all of these incentives outweigh the costs associated with building and maintaining an extra offspring, parents do well to aim high initially and then let sibling aggression downsize the family, as needed.

Such consideration of parental strategy clears up another nagging mystery. Before siblicide was observed closely, it was assumed that parents would try to interfere with the destructive fighting behavior among their nestlings. After all, the reasoning ran, parents are in the business of reproduction, so aggression that destroys a large fraction of the output is likely counter to parental interests and should be opposed: Siblicide may profit the bullies, but it presumably reduces parental fitness. But when the process was actually observed in detail, it turned out that parents do nothing of the sort. Indeed, in the half dozen or so species that have been studied well, parents act as if quite indifferent to the whole drama. This indifference makes no sense if parental interests are being harmed, but it fits neatly with the view that parents are engaged in a larger gambit (initial overproduction followed by downsizing).

As research into siblicide continues, many variations on the theme are emerging. In addition to the birds that have received the most study, there are strong indications of such manifestations as prenatal sib-cannibalism in live-bearing sharks and salamanders, precocial weaponry (teeth) for use against brood mates in pig and hyena litters, wholesale slaughter of brothers by special warrior sisters in parasitoid wasps, and water-soluble sib-killing chemicals deployed by seeds in tree pods. A general picture is emerging that predicts fatal sibling competition whenever concurrent offspring are crowded into a closed space ("nursery") where they rely on finite supplies of key resources. From the starting point of parental overproduction, siblicide evolves repeatedly to restore equilibrium between supply and demand.

See also Caregiving—*Parents Desert Newborns after Break-in!*
        Infanticide

Further Resources

Mock, D. W., Drummond, H. & Stinson, C. H. 1990. *Avian siblicide*. American Scientist, 78,438–449.
Mock, D. W. & Parker, G. A. 1997. *The Evolution of Sibling Rivalry*. Oxford: Oxford University Press.
Mock, D. W. 2004. *More than Kin and Less than Kind: The Evolution of Family Strife*. Cambridge, MA: Harvard Press.

Douglas Mock

Social Evolution
Flowering Plants and the Origins of Primates

Primates (prosimians, monkeys, apes, and humans) are characterized by a number of morphological traits. These include: five fingers and toes, and grasping hands and feet; nails instead of claws; forward facing eyes and stereoscopic vision; and reduction of the sense of smell. In the early 1900s, these morphological characteristics were thought to be adaptations for living among the branches of trees. This theory was later elaborated upon and became referred to as the "arboreal theory" of primate evolution.

Recently, Duke primate biologist Matt Carmill rejected the arboreal theory of primate evolution and argued that if many arboreal mammals exist that do not share the anatomy of primates, we must conclude that there is something wrong with the arboreal theory of primate evolution. In fact, of the 14 orders of terrestrial mammals, 9 have arboreal forms. Many of these animals have highly successful adaptations to life in the trees and do not have the characteristic traits developed by the primates.

For example, tree squirrels fill a specific set of niches in the forest canopy, but unlike primates, tree squirrels lack orbital convergence and bear claws on their digits instead of nails. Another example is the arboreal sloth who has large, nearly immobile hook-like claws instead of prehensile hands. Arboreal kangaroos (yes, there really are such things) have no notable primate-like traits.

In an elegant use of the comparative method, Cartmill compared the function of specific traits shared by primates and other animals in an attempt to determine the precise niche that early primates might have filled. He found that grasping hands and feet are universal among animals that habitually forage in the small, terminal branches of shrubs and trees. He also concluded that forward facing, convergent eye orbits are largely restricted to predators. They are particularly marked in such animals as owls, hawks, and cats—animals which depend on vision for the detection of prey. Thus, Cartmill argued that the earliest primate adaptations involved visually oriented predation on insects in the lower canopy and undergrowth of tropical forests.

However, since Cartmill first presented his theory, the diet of a number of prosimians has been studied in detail. Most small nocturnal primates (those most closely resembling the earliest primates) feed mainly on crawling insects, many of which are captured on the ground. Very few species of primates are known to have a diet including a greater proportion of insects than plants, and the dietary pattern found among the vast majority of primates (over 85%) is omnivory (eating both plants and animals). Furthermore, the general anatomy of the digestive tract of primates reflects adaptations for an omnivorous diet.

In fact, the reduction of the sphere of vision imposed by very low light intensities renders nocturnal vision insufficient for a highly insectivorous diet. Finally, the only mammals that possess a complex visual system similar to that of primates are fruit bats (Megachiroptera). It seems likely, therefore, that visual predation per se is not a sufficient explanation for the visual adaptations of primates.

We might look to the evolutionary history of primates to delve deeper into this question. The Paleocene–Eocene boundary, about 55 million years ago, was a period of rapid change that involved coincidental adaptive shifts in a number of plant and animal groups, including the primates. It is in the context of the interrelationships between these groups that we might find an alternative hypothesis for the origin of primates.

We might hypothesize that, like most living primates, the earliest primates were *omnivores*, feeding on small-sized objects found in the terminal branches of trees. Thus the novel adaptive shift involved two aspects: 1) becoming well adapted to feed in the small branch milieu; and 2) including a high proportion of plant material in the diet. The most important difference is that in this model the ecological resource providing a new basis for exploitation is identified—the fruit and flowers of the angiosperms or flowering plants, and the insects attracted to these products.

Angiosperms are the grouping of plants that produce flowers to enhance their prospects of fertilization, and fruit to enhance their ability to disperse their seeds. Both systems require the cooperation of animals. Flowers provide a reward of sweet nectar to animals who, in visiting the flower, will inadvertently rub off some pollen and carry it to the next flower. Fruits are a sweet, nutritious offering to animals that ingest the seeds and then disperse them. Angiosperms have become very successful because of this elaborate "coevolutionary" system.

During the Paleocene epoch, about 55–65 million years ago, there was a major radiation of angiosperms. This radiation coincided with the origin and diversification of many mammals that began to use the trees and to exploit this new niche. A number of these mammals had dental morphology that suggests a switch to plant foods. The new feeding niche included the small branch milieu of the newly radiating angiosperms, which offered an array of previously unexploited resources, for example., flowers, fruits, floral and leaf buds, gums, nectars, and also the insects that feed upon these items. During this early period, the mammals beginning to use these new resources did not look like modern primates; most of them had claws and did not have forward facing eyes.

However, these coevolutionary interactions between flowering plants and animals appears to have reached a threshold at the Paleocene–Eocene boundary. In fact, the Paleocene–Eocene boundary was the time of disappearance of many ancient species of angiosperms and of the appearance of a diverse array of modern groups of these plants. Modern evergreen tropical rain forests also first appeared and became widespread during the Eocene. It is at this same time that the earliest modern-looking primates appeared and diversified. This also coincides with the first record of fruit-eating birds and fruit bats.

Thus, the establishment of biological interactions between angiosperms and their pollinators and dispersers is reflected in the rapid appearance of modern families and genera in the Eocene epoch. The evolution of modern primates parallels that of other herbivorous mammals, of plant-eating birds, and of modern angiosperms; it also appears that many of these organisms were linked in a tight coevolutionary relationship. At present, frugivorous birds, bats, and primates are the most important seed dispersers in tropical rain forests, and flowering plants depend upon them for their continued existence.

The evolution of modern primates, therefore, as well as that of fruit bats and fruit-eating birds, may be directly related to the evolution of improved means of exploiting flowering plants. Furthermore, the particular pattern of Paleocene extinctions of earlier arboreal mammals may be related to the rapid evolution and radiation of the primates, bats, and plant-feeding birds.

Some clawed arboreal mammals persisted into the Eocene epoch, thereby overlapping in time with primates. In contrast to these clawed mammals, the new primates possessed a divergent toe and thumb and flattened nails to produce effective grasping organs. It is generally agreed that these adaptations would have allowed Eocene prosimians far greater access to fruits and flowers, as well as plant-visiting insects, making them much more efficient at locomoting and foraging in the small terminal branches of bushes and trees than were clawed mammals.

However, how do we know if this hypothesis is correct, or if the earliest primates were visually oriented predators. Is there a way to test the alternatives? In fact, Matt Cartmill has suggested a way that this could be done. If we knew something about the order in which the various primate peculiarities were acquired, we may be able to answer this question. For example, if the first primates had grasping feet and blunt teeth adapted for eating fruit, but retained small, divergent orbits, the primate–angiosperm coevolution hypothesis would be supported. Now we might just have this evidence.

In 2002, a remarkably well-preserved, late Paleocene (56-million-year-old) skeleton of a plesiadapoid (*Carpolestes simpsoni*) was discovered in Wyoming. This fossil differed from previous Paleocene mammals in being very complete, including tiny hand and foot bones, along with an intact skull. This fossil allows us to test the proposed theories concerning the evolution of modern primates.

Carpolestes lacks primate visual specializations such as convergent orbits. However, it has a divergent and opposable big toe, with a nail, that was adapted for strong grasping of small diameter supports similar to that of the

A reconstruction of the early primate-like mammal, the Carpolestes.
Courtesy of Robert W. Sussman

primates. It also has correlated grasping specializations in the hand. In fact, the hand and foot bones of *Carpolestes* bear a remarkable resemblance to those of modern primates. Furthermore, the morphology of the molar teeth are consistent with a fruit-eating diet.

This fossil find is consistent with the hypothesis that early primates evolved grasping first and convergent orbits later, and is inconsistent with the visual predation hypothesis. On the other hand, the sequence of acquisitions is that predicted by the angiosperm exploitation theory.

Thus, it appears that modern primates derived from an ancestor similar to the late Paleocene fossil, *Carpolestes simpsoni*. Furthermore, the combination of morphological traits shared by primates indicate that the adaptive shift accompanying the appearance of the primates was the occupation of new locomotor and feeding niches made available by the co-evolving angiosperm plants of tropical rain forests. Although bats and birds can reach the terminal branches of large rain forest trees by flying to them, primates need their grasping appendages to obtain the same advantage. In fact, apart from a few primate-like marsupials, primates are the only major taxonomic groups of nonflying vertebrates to regularly exploit the terminal branch niche of the tropical forest.

See also Social Evolution—*Social Evolution in Bonnet Macaques*

Further Resources

Bloch, J. I. & Boyer, D. M. 2002. *Grasping primate origins*. Science, 298,1606–1610.

Cartmill, M. 1992. *New views on primate origins*. Evolutionary Anthropology, 1,105–111.

Clark, W. E. LeGros. 1963. *The Antecedents of Man*. New York: Harper and Row.

Rasmussen, D. T. 1990. *Primate origins: lessons from a neotropical marsupial*. American Journal of Physical Anthropology, 22,263–277.

Sussman, R. W. 1991. *Primate origins and the evolution of angiosperms*. American Journal of Primatology, 23, 209–223.

Sussman, R. W. 2003. *Primate Ecology and Social Structure. Vol 1: Lorises, Lemurs and Tarsiers*. Boston: Pearson.

Robert W. Sussman

■ Social Evolution
Optimization and Evolutionary Game Theory

Evolution by Natural Selection and Its Implications

It is strikingly apparent, when we take the time to really watch animals, just how complicated and different they are, both in the way they look and in what they do, compared to each other and to us. Take the fact that most honeybees are female and sterile and will commit suicide readily for the sake of their colony, for instance. They also "tell" their hive mates about where the juiciest, nectar-full flowers are by waggling their bodies and dancing in circles. Compare that with the closely related leaf cutter ant who also works exclusively for the sake of her hive and family, but instead of gathering nectar she harvests leaf cuttings and composts them deep underground to fertilize communal gardens of nutritious fungi. Such awesome variety undermines the notion that animals are merely robots, slaves to a narrow range of reflexes and stimulus–response associations as championed by many early observers of animal behavior, particularly the behaviorists such as Skinner, Thorndike, and Pavlov. Indeed, as with biological diversity of any sort, animal behavior, with its species specificity and bewildering intricacy, is likely to have been shaped in the furnace of natural selection in a world where the raw materials for life are limited and so the rate at which organisms multiply outstrips the capacity of their environment to support them all, those organisms with the best tricks for coping with life are likely to win out and persist. If these good tricks are inherited by future generations, then animals will be born that tend to match their environment more and more with each generation. This environmental matching is called *adaptation* and gives life the appearance of being designed since the lineages that persist are likely to possess features and do things that work very well in the worlds in which they live. For instance, many fish possess the familiar "herring bone" arrangement of swimming muscles, which happens to allow the muscles to contract at a rate that maximizes the power produced. Any fish with a slightly different muscle arrangement will not be as powerful a swimmer and will tend not to find food as efficiently or will suffer a higher risk of being caught by a predator. Such individuals tend to leave relatively few offspring that will inherit the inefficient muscle arrangement and leave fewer progeny themselves, and so on. In the end, fish with "herring bone" muscles will predominate.

Now, the accumulation of design via natural selection is unlikely to proceed unchecked. This is because the good tricks that prevail in environments are the best around, not necessarily the best that are ever possible. Nature only "selects" from alternatives that are around at the same time, which will depend heavily on what designs and good tricks worked in the past and have been inherited. Limits to adaptation as a result of this dependence of biological systems on history are often called *constraints*. These are not the only checks, however, on the ability of organisms to adjust to meet the demands of a changeable world. Of course, the structures and behaviors that organisms develop and maintain will also be limited, fundamentally, by what is possible in a natural universe—the laws of physics and chemistry. For instance, opting to fly limits the maximum body size

Acknowledgement: Gilbert Roberts (University of Newcastle, UK) is responsible for the concept of the "social stake" and recognizing that it can be used to generalize the factors that allow cooperation to be an ESS in the *Prisoner's Dilemma*. My territoriality game is based on his reformulation of the Prisoner's Dilemma.

that individuals can attain. This is because, above a certain body mass the muscles required to lift the body off the ground would weigh so much, since bigger muscles are needed to generate more power, that they could not also be lifted. Such functional limitations generate what are called "trade-offs," which, if specified—along with the constraints on the system, can be used to predict from first principles what designs are likely to predominate in a particular system. Returning to the issue of body size and flight muscles, for instance, the kind of trade-off that can be used to predict the lifestyles of different sized flyers is derived from the fact that larger animals are better protected from predators and able to travel farther, but they will also be limited by their weight in their ability to maneuver and take off and land. For this reason, it is perhaps not surprising that the largest flying birds live in open habitats without many obstructions and with convenient take off and landing spots like cliffs or large bodies of water.

The Logical Consequences of Adaptation

If we adopt the premise that an animal has evolved by natural selection, there are two main types of prediction that can be made and tested against observations. This is a necessary part of any attempt to gather objective knowledge—to do science—in a complex world given that we, the observers, are likely to be biased, if only by the limits of our senses.

Differences between Species Will Reflect the Kinds of Worlds They Live in

If an animal's lineage has become progressively better matched to the environments its ancestors were born into, lived, and died in, comparing different species systematically, along with details of their environments and how they are related to each other, allows us to identify which features and behavior are adaptations to which features of the environment. For example, in the 1960s John Crook studied 90 species of weaver birds—small finches that live throughout Africa. He noted that species living in forests tended to eat insects solitarily, defend territories, and build cryptic nests in monogamous pairs. Species living in the savannah, on the other hand, tended to eat seeds in flocks and nest in large conspicuous colonies. From this, he reasoned that food supplies had selected for the different social systems he observed. On the one hand, insects are relatively hard to find and catch and so defending an exclusive patch of forest is essential, as is having two parents to find food for young. On the other hand, seeds in the savannah are very abundant when they are found and easy to eat. This means that looking for food in groups is essential since more eyes help to locate the bonanzas, which can be easily shared. Group living includes nesting because it's hard to hide nests in the open savannah, and so big communal nests are safest. This "comparative approach" is used extensively to understand animal behavior from an evolutionary perspective.

Organisms Should Possess "Good Tricks" for Solving Their Day-to-Day Problems

Another consequence of evolution by natural selection is that animals are expected to possess features and do things that function well in their environments. This insight allows us to analyze animal behavior (or indeed any feature of an organism) in the way that an engineer might study a strange piece of machinery—we can ask what the behavior is for, and if we do so in a careful, rigorous manner, we can understand it thoroughly and profoundly.

Furthermore, the notion that animals should possess "good tricks" to help them in their daily lives also allows us to specify what we mean by function—that which allows animals to maximize their lineages' ability to persist over evolutionary time in the face of competition with other similar lineages. Or, in other words, animals should tend to behave in ways that maximize their Darwinian fitness when they behave adaptively. Therefore, one way to study animal behavior systematically and scientifically is by constructing rigorous, testable hypotheses deduced from that which Theodosius Dobzhansky—one of the founding fathers of modern biology—asserted famously, everything in biology "makes sense": evolution by natural selection. We do this by using mathematical tools borrowed from economics: optimality theory—traditionally called "utility theory" in economics—and game theory.

Optimality Theory: Specifying Good Tricks in Simple Environments

Imagine a bird, say a European starling, walking across your lawn, all the while searching for food—insects and worms just below the soil's surface (and occasionally in the grass). When it finds an edible insect or worm, it must decide whether or not to consume it. Now this might at first glance appear to be a trivial issue. Surely the starling should just eat anything edible it comes across. However, if we analyze the problem facing the bird from a design perspective, by breaking it into parts and determining their relationships, the issue becomes less clear-cut and yields some surprising insights such as that the well-designed starling should either always, or never, eat some classes of food regardless of how common they are in the environment!

To begin to analyze behavior from a design perspective, we must decide carefully what type of options an animal has available to solve the problem facing it. For the foraging starling faced with a potential prey, the "decision assumption" is straightforward: The bird can either *eat* a food item or abandon it to *search* for better fare. Then, we must also decide by what criterion the animal's behavioral options are to be assessed. So, what should be important to a starling that is looking for food, from an adaptive perspective? Strictly speaking what is really important to any adapted animal is its Darwinian fitness—behavior that enhances fitness the most is likely to persist long enough to be observed. However, when we attempt to rank foraging options, for instance, specifying their consequences for the persistence of an animal's lineage over evolutionary time would be unworkably complex and involved. Instead, we typically choose a "currency," which we assume relates the immediate consequences of an animal's actions to its fitness. For a foraging starling (and many other foragers), more energy is assumed to be better since it allows metabolic requirements to be met, and energy can then be spent on important nonforaging activities such as fighting, fleeing, and reproducing. Yet time is also important because animals are likely to be pressed to meet their daily feeding requirements and are usually less able to perform their other vital activities while they are foraging. Therefore, we can assume that a well-designed, foraging starling will behave so as to maximize the amount of energy it obtains as quickly as possible, or its long-term net rate of energy gain, R, from foraging. Specifically:

$$R = \frac{E_g - E_l}{T_f}, \tag{1}$$

where E_g is the energy gained from foraging, E_l the energy lost doing so, and T_f is the time spent foraging.

Finally, we need to identify the factors that limit and define the relationship between an animal's actions (its decision variables) and the payoffs it receives in terms of the currency we have identified. For the foraging starling, the constraints it must contend with when deciding whether to eat a prey item or continue searching are likely to include the following. On the one hand, a starling searches for topsoil invertebrates by probing into the soil with a closed bill and opening it while it is buried to create a small hole. While its bill is open, its eyes swivel to focus down the sides the gaping bill and scan the hole for movement. Because of this searching technique, it is not possible for a bird to search while it is consuming food, which always occupies some of its foraging time, called "handling time." In addition, the search technique employed usually means that prey are located one at a time and independently of one another. This means that the availability of prey in the environment can be described by a single quantity, the rate at which items are encountered (λ) from which the average time spent searching per prey item (s) can be calculated:

$$s = \frac{1}{\lambda}. \tag{2}$$

Furthermore, to keep things simple and clear, we are going to assume that the starling will only encounter two prey types: "good" and "poor," distinguished by their energetic "profitability"—the net amount of energy each item yields (e_g or e_p) divided by its handling time (h_g or h_p). By definition then:

$$\frac{e_g}{h_g} > \frac{e_p}{h_p}. \tag{3}$$

Facing the foraging problem outlined above, a starling finding a good prey should always consume it in order to maximize its net rate of energy gain since it will only reduce the payoff specified in (1) if it ignores the prey and searches for another; If it ignores a good prey, then the best it can hope for is to locate another good prey that will take s_g time to find. Specifically, since:

$$\frac{e_g}{h_g} > \frac{e_g}{h_g + s_g},$$

whenever good prey is found the best thing for the starling to do is to eat it.

Alternatively, however, when the bird comes across a poor prey item, our optimality analysis reveals that the best thing for a rate-maximizing starling to do is less clear cut. It must choose between a less-than-best payoff from eating the poor item now, or abandoning it to obtain the best possible food s_g time in the future. In other words, the starling should eat a poor prey if, and only if, a poor prey now is more profitable than searching elsewhere for a good prey, or if:

$$\frac{e_p}{h_p} \geq \frac{e_g}{h_g + s_g}. \tag{4}$$

An important insight from this simple analysis is that whether or not to eat a less-than-maximally profitable prey type has nothing to do with how common or rare it is, it has only to do with the availability of better prey in the environment. In (4), only how often maximally profitable prey are encountered (s_g) matters in the decision as to whether to accept all prey encountered or specialize on the best that are out there. This runs counter to what one might expect intuitively: Surely if a type of prey is very common should you not consume it, regardless of whether or not it is the best possible type out there? This is where analyzing the problem formally shows its value, since it exposes the kind of mistakes we can make if we only use "common sense" to understand animal behavior. This very simple model also illustrates a common principle when analyzing animal behavior from a design perspective: "lost opportunity." Generally speaking, decisions about exploiting resources can be assessed by comparing the potential gains from exploitation with the potential loss of the opportunity to do better. Here, whether or not to consume a poor prey depends on the cost of losing the opportunity to consume a good prey in the meantime.

Okay, so this is all very well in theory, but how does it help us to understand what real starlings actually do when they are foraging on lawns? Well, we can use simple models like the one outlined above to make predictions, which we can then test against observation. For instance, we can use the inequality in (4) to specify precisely how rare good prey need to be to make poor prey worth consuming. Rearranging (4), we can specify a critical search time for good prey S_{crit} as:

$$S_{crit} = \frac{e_p \cdot h_g}{e_p - h_g} \tag{5}$$

If good prey are encountered at shorter intervals than S_{crit} then it never pays to eat a poor prey, and so they should be ignored, and the bird should specialize only on good prey. Otherwise, it should consume every prey it finds. Constructing precise, testable predictions like this allows us to utilize Darwin's insight to organize how we collect and interpret information about the biological world. Indeed, we can use such logic to ask a range of general questions about animal behavior. How good are organisms at their jobs? What are animals designed to do? What mechanisms control their behavior? For instance, using the foraging model outlined above—called the Optimal Diet Model—we can go out and measure the energetic content of different prey and the time it takes a predator to consume each item. We could then observe the foraging behavior of starlings on lawns with different densities of topsoil invertebrates and assess whether those with relatively low densities of the better prey have starlings that forage less selectively on them. On the other hand, we could control when birds encounter different quality prey in a laboratory—by presenting food on a conveyer belt, for instance—and observe whether varying the intervals between encounters with good prey around the predicted threshold causes birds to be selective or not, as predicted.

In fact, both such approaches have been carried out on a range of different species of foragers. Broadly speaking the pattern predicted by the theory has been observed: Animals appear to rank their prey according to how profitable they are, and varying encounter rates with more profitable prey changes predator selectivity. However, predators fail consistently to either always, or never, include poorer prey in their diets—in other words, there is little evidence for an actual *threshold* encounter rate as predicted by (5), either in the laboratory or in the field. Instead, the mammals, birds, fish, crustaceans, and insects observed tended to increase their preference for more profitable food as it became more common, but not in a very consistent and/or strict manner. The reasons for this seem to fall into two camps. On the one hand, it appears that individuals of the same species vary in how they handle and search for food and/or how much they value energy—in other words, there is a significant amount of variability in some of the fundamental parameters of the model between different animals and in different contexts. In addition, an important assumption underlying the reasoning—that animals are perfectly informed about their foraging options—is often untrue. If foragers are uncertain about whether or not a prey item is actually good or not, then it makes sense for animals to occasionally consume food that we might know to be of low profitability, even if they are trying to specialize on profitable food.

Nevertheless, this failure of the model (and others) to predict observed behavior has inspired researchers to consider other fundamental factors influencing foraging behavior and enriched the study of animal behavior as a whole. In particular, by starting simply and explicitly, it has become very clear that an animal's history can have a profound influence on the adaptive value of its decisions—the best thing a forager can do often depends on its body condition, which determines the value of feeding and depends on prior foraging success, among other things. Moreover, it is often the case that the payoffs associated with decisions depend critically on what others decide to do, which will be influenced by a focal individual's behavior. Such problems are technically called "games," the solution of which has been made famous by the Nobel-Prize winning economist John Nash.

Evolutionary Game Theory: What to Do When the World Bites Back

An outstanding problem for post-Darwinian biology is the existence of animal behavior that results in an individual sacrificing its own fitness for that of another. Think back to the honeybee or leaf cutter ant who is willing to sacrifice her life defending her colony or who abstains from reproducing so that she can work to help her mother to reproduce. This kind of behavior is called *altruism*, defined as any action that results in a net benefit to a recipient at a net cost to the actor. Surely, thinking adaptively, isn't such behavior unlikely to exist? In a world full of altruists, any selfish individual would have a field day and leave many more descendents than average, who will exploit the remaining altruists in turn, and so on. From this perspective, we should never expect to see altruistic behavior, or only rarely, and it should not persist for very long. In fact, we see many instances of such behavior, from the relatively cheap, short-term altruism of grooming in many species of primates, to the lifelong, extreme altruism of sterility in the social insects. To analyze the issues involved and understand how such behavior can persist in a Darwinian world, it is necessary to extend the kind of design analysis outlined above to consider problems for which individual fitness depends on what others do. This involves the use of evolutionary game theory, which was introduced to evolutionary biology from economics by John Maynard Smith—an engineer turned biologist—in the late 1970s.

An Example of Evolutionary Game Theory

Imagine a pair of animals, say a male and female chickadee defending their territory, confronting an intruder who has ignored their warning calls and is attempting to take over their patch. Each member of the couple must decide whether to *attack* the intruder or to *hang back* and hope the other drives the trespasser off. Imagine that they have to decide in an instant and have no way of responding to what the other decides to do during that particular confrontation. Individuals suffer an immediate cost to attacking c, while they enjoy a benefit b from holding onto the territory. If both members of the pair attack they will always drive the intruder away, whereas if only one attacks, it will drive the intruder off half the time and lose the territory the rest of the time—all individuals, including intruders, are evenly matched for their fighting abilities. If neither fights then they lose their territory and gain no returns in the immediate future. To analyze the problem in more detail, each member of the pair can be treated as identical. This means that we can arbitrarily choose an individual and ask what is the best thing it should do, and that will tell us what kind of behavior will be selected for since it will also be the best thing for its partner to do. Therefore, if the focal individual decides to attack the intruder and its partner also attacks, it will suffer the cost of attacking, but it will also always hold onto the territory resulting in a net payoff of $b - c$. However, if it attacks, and its partner hangs back, then it will pay the cost while only holding onto the territory half the time and losing it otherwise: $b/2 + 0/2 - c$. Alternatively, it can decide to hang back, and if its partner attacks, it will avoid the cost and have a 50–50 chance of holding onto or losing the territory: $b/2 + 0/2$, whereas if its partner also hangs back, then they always lose the territory at no cost. The first table summarizes an individual's payoffs as a function of what it decides to do and what its partner does in this territoriality game.

Payoffs for Basic Territory Defense Game

Payoffs to a focal individual's actions (rows) as a function of its partner's actions (columns) for the basic territory defense game described in the text.

Focal Individual	Partner	Attack	Hang back
Attack		$b - c$	$\dfrac{b}{2} - c$
Hang back		$\dfrac{b}{2}$	0

Now, imagine that attacking an intruder is rather costly, but the territory is rich enough to compensate for the cost if access to it can be guaranteed in the immediate future. Specifically:

$$\frac{b}{2} < c < b. \tag{6}$$

In this situation, the best the focal individual can do is to hang back. This is because the payoff to attacking, averaged over every possible thing that its partner could do, is only:

$$\frac{b - c + \dfrac{b}{2} - c}{2} = \tfrac{3}{4}b - c.$$

While by hanging back, on the other hand, on average it would gain:

$$\frac{\dfrac{b}{2} + 0}{2} = \tfrac{1}{4}b,$$

which, since $c > b/2$ from (6), is always greater than $3/4\, b - c$. Furthermore, in a population of attackers, if an individual decides to hang back, it will obtain a higher payoff than a randomly chosen attacking individual in the population (who will be paired with another attacker); in other words, $b/2 > b - c$ when (6) is true. This suggests that mutants that hang back are likely to spread in populations in which everyone else attacks. On the other hand, it can be shown similarly that when hanging back becomes common, it can never be invaded by individuals that attack when $c > b/2$. This result has a number of interesting implications.

The bottom line is that it suggests that neither individual could do better by doing anything other than hanging back when an intruder challenges for their territory, and when hanging back is common, it never pays an individual to attack. Therefore, we say that hanging back is likely to be selected for and be stable over evolutionary time—it is an *evolutionarily stable strategy* (ESS) in our model. At first glance, however, this is somewhat surprising. Even though the value of defending a territory outweighs the cost of doing so, because you are unsure of help from your partner, it pays to never actively defend it against intruders that are no stronger or fiercer than you are. In addition, because of this insecurity, everyone ends up losing out on the benefits of defending territories together; remember $b - c > 0$ since $b > c$ in (6). This is an example of the so-called tragedy of the commons, whereby the common good is often sacrificed because of the potential to cheat.

Finally, analyzing the problem facing a pair of territorial chickadees from a design perspective as we have done here may provide insights into the year-round, joint territoriality by mated pairs seen in species like the Carolina chickadee. On the one hand, our analysis suggests that we might expect the kind of territory defense behavior we see to be relatively cheap since "attacking" would pay and be an ESS (evolutionarily stable strategy) if $c \leq b/2$ in our game. This may help explain why much of the territory defense we observe in nature involves displays and vocalizations, which are relatively cheap. Nevertheless, however, it is not unheard of for pairs of small birds like chickadees to actively attack intruders on their territories, which is territorial behavior that carries a high risk of injury or death, and so—as we have shown here—cannot be explained by considering only the immediate costs and value of territory defense.

Indeed, one key issue that has not been dealt with in our analysis so far is the fact that pairs of chickadees will rarely face only a single intrusion on their territories while they are together. This likelihood of repeated intrusions means that individuals will have a stake in their partner's well-being as well as their own—against any particular intruder, keeping your partner in good shape is valuable as it will maximize the opportunities for cooperative territoriality in the future, or for back-up should you be unable to defend the territory for any reason. We can investigate the implications of this issue in our analysis by taking account of this "social stake" (s), which, for any interaction between individuals, can be thought of as the rate at which the actor's fitness (W_A) changes as a function of changes in the recipient's fitness (W_R). Specifically:

$$s = \frac{\partial W_A}{\partial W_R}. \tag{7}$$

Adding this factor to our territoriality game changes the payoffs to a focal individual's actions since the payoffs to its partner, discounted by the social stake it has in the partner, also start to count. Now, if both animals attack, the focal individual gains $b - c$ as before, but since its partner also gains $b - c$, the total payoff to the focal individual becomes $b - c + s(b - c)$. Likewise if it attacks but its partner hangs back it gains $b/2 + s(b/2) - c$ in total since it gains $(b/2) - c$ while its partner gains $b/2$. The payoffs to hanging back can be specified in a similar way and are shown in the table on p. 978.

This time when we calculate at the average payoff to attacking and compare it to the average payoff to hanging back with the costly territory defense described in (6), which is too complicated to detail here, attacking turns out to always be the best thing to do, and, when it is common, will resist invasion by mutants that hang back when:

$$s \cdot b > c. \tag{8}$$

Payoffs to "Hanging Back"

Payoffs to a focal individual's actions (rows) as a function of its partner's actions (columns) for the territory defense game with a "social stake" in its partner's fitness.

Focal individual ／ Partner	Attack	Hang back
Attack	$b - c + s\,(b - c)$	$\dfrac{b}{2} + s\dfrac{b}{2} - c$
Hang back	$\dfrac{b}{2} + s\left(\dfrac{b}{2} - c\right)$	0

Thus, relatively costly cooperative territory defense can be selected for when individuals have a stake in their partner's well-being, sufficient to satisfy the conditions set out in (8). Such a stake can arise for reasons other than the potential for future interactions with each other as discussed here. In territorial chickadees, another potential social stake is that an individual may gain from its mate's ability to find and deliver food to its young during the breeding season, which will depend on its well-being.

Furthermore, the evolutionary game outlined above is actually a version of the classic model used to analyze the evolution of cooperation and altruism in a selfish world: the *Prisoner's Dilemma*. Indeed, after much analysis, evolutionary biologists and economists have identified a range of factors that can lead to cooperation as an ESS and the tragedy of the commons being avoided, many of which can be thought of as social stakes. For instance, if being in a group is beneficial for protection from predators or keeping warm, then individuals will have a social stake in the well-being of other group members. In fact, being related to each other in itself generates a social stake among individuals, which is proportional to the degree to which they are related. This is because an individual's fitness is determined by its lineage's persistence over evolutionary time, which is also influenced by its relatives—so the success of a relative is equivalent in fitness terms to your success discounted by how much you are related to each other. Therefore, "relatedness" (r), which is the likelihood of sharing by descent a randomly chosen trait relative to the likelihood of sharing the trait with an average member of the species, can be substituted for s above. By doing this (8) becomes Hamilton's Rule for the evolution of altruism:

$$r \cdot b > c.$$

named after William Hamilton III, a famous Oxford evolutionary biologist. Indeed, Hamilton's Rule has been used to understand the extreme altruism demonstrated by social insects like bees and ants. It turns out that, because of the way that sex is determined in bees and ants, workers are more closely related to their sisters ($r = 0.75$ in many species) than they would be to their own offspring ($r = 0.5$, typical of sexually reproducing organisms). This means that it actually pays female workers to sacrifice reproducing themselves to ensure that their mother produces as many sisters as possible! It is by shedding light on such strange examples of the things that animals do that the formal analysis of design, as embodied by optimality and evolutionary game theory, demonstrates its value to the study of animal behavior.

See also Aggressive Behavior—*Dominance: Female Chums
or Male Hired Guns*
Aggressive Behavior—*Ritualized Fighting*
Cooperation

Further Resources

Dawkins, R. 1989. *The Selfish Gene,* 2nd Edn. Oxford: Oxford University Press.
Dugatkin L. A. & Reeve H. K. 1998. *Game Theory and Animal Behavior.* New York: Oxford University Press.
Krebs, J. R. & Davies, N. B. 1993. *An Introduction to Behavioural Ecology*, 3rd Edn. Oxford: Blackwell Science.
Stephens D. W. & Krebs J. R. 1986. *Foraging Theory*. Princeton, NJ: Princeton University Press.

Sasha R. X. Dall

■ Social Evolution
Social Evolution in Bonnet Macaques

Primates represent an extremely fascinating group of mammals, particularly due to the extremely variable structure of their social organizations. Basically, primate social structures fall into four main categories, although variations do occur within them:

1. *Solitary/semisolitary*, in which individuals range and forage alone within separate home ranges. In many nocturnal species, two or more individuals may group to sleep together, although the composition of such sleeping groups may often fluctuate. Examples include bushbabies, lorises, pottos, mouse lemurs, and orangutans.

2. *Unimale groups*, in which one adult male, several females, and their dependent offspring live together. Takeovers occur periodically when stranger males challenge and oust the reigning male. Examples include gelada baboons, hanuman langurs, and guenons.

3. *Multimale groups*, in which several males and females of different age categories live together. Subclasses within this include *typical multimale–multifemale troops*, seen in rhesus macaques and pig-tailed macaques; *fission–fusion communities*, seen in the common chimpanzee, in which males form the core of the group while females freely move between groups; and *multilevel bands* with small unimale harems coalescing into larger foraging clans and still larger resting troops, seen in the hamadryas baboon.

4. *Pair living*, in which an adult male and a female live together with their dependent offspring. Examples include tarsiers, the fork-marked lemur and gibbons.

In addition, *unifemale groups*, characterized by a single adult female and several males, are known in some tamarin populations, while *all-male "bachelor" bands* tend to occur where the norm is a unimale society, as, for example, in hanuman langurs.

Ecological and social pressures have together been invoked to explain the formation of primate social structures and the pattern of behavioral interactions between individuals within and across such groups. Primate social evolution is usually hypothesized from comparative studies across different primate taxa; the actual evolution of new stable groups with different demographic structures and individual behavioral strategies, driven by changing ecological and social forces within an existing population, has only rarely been observed. Long-term studies are, therefore, essential if a direct understanding of the selective forces driving demographic and behavioral evolution in natural primate populations is to be achieved.

The bonnet macaque (*Macaca radiata*), a cercopithecine primate found only in peninsular India, usually lives in large troops of 8 to 60 individuals; such multimale troops typically contain several adult males and females, as well as juveniles and infants of both sexes. In the Bandipur National Park, southern India, however, there seems to have evolved, in recent years, small stable unimale troops, each containing a single adult male, a few adult females, and their dependent young; such troops now constitute 52% of the macaque groups in this area.

Certain features of these unimale troops are remarkably different from those seen in the multimale troops. These are:

1. a significant bias toward female infants (1:4), which appears to be due to a genuine skew in birth sex ratio;
2. reduced number of male subadults, juveniles, and infants;
3. emigration of adult, subadult and, juvenile females; and
4. atypical behavior of the adult resident male.

Interestingly, the first three characteristics appear to have evolved in response to the fourth—the reproductive monopolization practiced by the resident adult male in these groups.

Unlike adult males in multimale groups, the adult males in unimale groups are strikingly intolerant of other males. Behavioral manifestations of this intolerance include herding of the group females, the display of severe aggression toward subadult males within the troop, active group defense with severe injuries inflicted on potential competitors, and the successful prevention of male immigration into the troops. This behavior of the resident male may have led to increased male emigration, resulting in the depletion of subadult males in these troops. Since male emigration at a relatively young age could potentially entail heavy costs, females in unimale troops may also have been selected to bear and raise daughters, who would remain in the troop and contribute to the matriline, rather than sons, who would almost invariably be driven away. The observed

A bonnet macaque family huddles together in a tree at the Nagarahole National Park, India.
© *Kevin Schafer / Corbis.*

female emigration could be a behavioral response to the potential lack of mate choice that exists in these groups. Dual sex dispersal of this nature is rare among cercopithecines and virtually unknown among macaques.

An ecological factor that may have significantly affected the bonnet macaque population within the Bandipur National Park in the recent past is its increasing tourist traffic. Tourists typically feed the macaques with nutritionally rich human foods, leading to intense competition within and between troops. The clumped nature of these meager foods, coupled with the patchiness of natural food resources in the summer months, appears to severely restrict group size; such an ecological regime can support possibly only small groups of macaques. Since bonnet macaque females typically form stable core groups in a particular area, this may have led to the evolution of unusually small troops of closely related females in Bandipur, troops that can be easily monopolized reproductively by a single dominant adult male.

The coexistence of these two distinct kinds of societies within the same population of bonnet macaques raises several interesting issues relating to the ecology of these organizations and the life-history strategies adopted by individuals living in them. More generally, however, it points to the extent to which a primate species can exhibit flexible behavioral patterns, ecological adaptability, and a labile social organization, all of which may enable evolution along new and often unpredictable routes.

See also Cognition—*Tactical Deception in Wild Bonnet Macaques*
Social Organization—*Social Knowledge in Wild Bonnet Macaques*

Further Resources

Kappeler, P. M. 2000. *Primate Males: Causes and Consequences of Variation in Group Composition.* Cambridge: Cambridge University Press.

Sinha, A. 2001. *The Monkey in the Town's Commons. A Natural History of the Indian Bonnet Macaque.* Bangalore: National Institute of Advanced Studies.

Smuts, B., Cheney, D. L., Seyfarth, R. M., Wrangham, R. W. and Struhsaker, T. T. (Eds.). 1987. *Primate Societies.* Chicago: University of Chicago Press.

van Schaik, C. P. 1989. *The ecology of social relationships amongst female primates.* In: *Comparative Socioecology: The Behavioural Ecology of Humans and Other Mammals* (Ed. by V. Standen & R. A. Foley), pp. 195–218. Oxford: Blackwell.

van Schaik, C. P. 1996. *Social evolution in primates: The role of ecological factors and male behaviour.* Proceedings of the British Academy, 88, 9–31.

Anindya Sinha & Sindhu Radhakrishna

■|Social Facilitation

Social facilitation is defined as an increase in the frequency or intensity of a behavior caused by the presence of others performing the same behavior. Commonly this phenomenon is referred to as "contagion," because a particular activity spreads from one individual to another. Contagious yawning is an example of social facilitation that is familiar to everyone. Behavior could also spread from one individual to another via observational learning, but social facilitation is limited to activities already present in an individual's behavioral repertoire. Often social facilitation has been confused with increases in a particular activity in the mere presence of other individuals of the same species. However, in such situations, the individuals involved may not be performing the same behavior, and thus mere presence effects are distinct from social facilitation.

Social facilitation can be an important aspect of living in groups, since the behavior of one individual can induce the same behavior in other members of the group. Behavior of group members can, therefore, become synchronized at a much finer scale than synchrony due to, for example, time of day or weather. Such fine-scale synchrony allows groups to act as a cohesive unit. In some cases, this can have important adaptive benefits, such as when all group members flee from or mob a predator that may have been detected by only one individual. Social facilitation of feeding may allow generalist species to expand their diets. For example, capuchin monkeys are more likely to eat novel foods if a group member is seen feeding. However, the adaptive value of contagion in behaviors such as yawning, preening, and dust bathing is unclear. Evidence from both humans and ostriches suggests that yawning may be a form of visual communication that signals transitions from wakefulness to sleepiness and vice versa, and may help groups to synchronize their activity patterns. (Yes, ostriches yawn, as do most vertebrates.)

Social facilitation of courtship and copulation can lead to reproductive synchrony in species that breed in colonies. In the 1930s, Fraser Darling proposed that reproductive synchrony could

Social facilitation of yawning may result from yawning acting as its own releasing stimulus via innate visual feature detectors. Not only humans yawn.
Courtesy of Corbis.

lead to increased survival of offspring because so many young would be present simultaneously that predators would be unable to take them all. There are many examples of socially facilitated reproductive behavior in colonial seabirds. However, contagious behavior may not spread throughout an entire colony. Reproductive synchrony would then be present only within small subcolonies, the colony as a whole would be asynchronous, and predator satiation or swamping may not occur. While several studies have shown advantages to breeding near the peak period of reproduction in a colony, others have shown advantages to breeding early. Even if it is advantageous, reproductive synchrony can result from factors other than social facilitation, such as short breeding seasons, synchronous arrival at the nesting grounds, and social stimulation by colony noise or by mere presence effects.

The proximate causes of social facilitation are not well understood. Partly this is a result of confusion over definitions. Confusion may also result from different mechanisms acting in different species or in different situations. For example, social facilitation of nest building may result from an individual's attention being directed toward nesting material by the presence of another individual building a nest and carrying nesting material. However, an increased pecking rate by chicks in the presence of others pecking at a high rate requires a different mechanism since the stimulus was not previously being ignored. Social facilitation of yawning, on the other hand, may result from yawning acting as its own releasing stimulus via innate visual feature detectors. Where causal explanations have been proposed, information on physiological mechanisms are still lacking.

Many of the suggestions made by Clayton in his 1978 review of social facilitation are still valid today. More research on the mechanisms, development, and adaptive value of social facilitation is clearly needed. In addition, few studies have examined the details of the social interactions involved. Are certain individuals more likely to be the facilitators or the responders, and if so, why? For example, contagious preening in captive mynahs tends to spread from dominant to subordinate individuals. Why? What are the effects of group size and spatial scale on the spread of contagious behavior? Most studies of social facilitation have simply demonstrated that it occurs. There are many examples of social facilitation, but few well-studied examples.

Further Resources

Clayton, D. A. 1978. *Socially facilitated behavior*. Quarterly Review of Biology, 53, 373–392.

Gochfeld, M. 1980. *Mechanisms and adaptive value of reproductive synchrony in colonial seabirds*. In: *Behavior of Marine Animals Vol. 4: Marine Birds* (Ed. by J. Burger, B. L. Olla & H. E. Winn), pp. 207–270. New York: Plenum Press.

Palestis, B. G. & Burger, J. 1998. *Evidence for social facilitation of preening in the common tern*. Animal Behaviour, 56, 1107–1111.

Provine, R. R. 1996. *Contagious yawning and laughter: Significance for sensory feature detection, motor pattern generation, imitation, and the evolution of social behavior*. In: *Social Learning in Animals: The Roots of Culture* (Ed. By C. M. Heyes & B. G. Galef, Jr.), pp. 179–208. New York: Academic Press.

Visalberghi, E. & Addessi, E. 2000. *Seeing group members eating a familiar food enhances acceptance of novel foods in capuchin monkeys*. Animal Behaviour, 60, 69–76.

Brian Palestis

■ Social Organization
Alliances

On the morning of June 22, 1941, troops from Hitler's Third Reich stormed into Russian territory, breaking a nonaggression treaty in an invasion that cost the lives of 20 million Russians. But Hitler miscalculated. His invasion of Russia not only failed to take Moscow, but it launched an alliance between Russia and the Allied powers that led ultimately to the defeat of Nazi Germany. History turns on the fate of alliances between nations, villages, and individuals; Julius Caesar was well on his way to becoming the Emperor of all of Rome until he fell victim to the alliance of Brutus and Cassius. Some believe that alliance formation is fundamental not only to human history, but to the evolution of our large brains as well. In this view, a Machiavellian world of complex, shifting alliances is negotiated successfully only by the brainiest of individuals.

The origins of complex human alliances, including warfare, may be older than humanity itself. Our closest relatives, common chimpanzees, engage in alliances worthy of a daytime soap opera. In a community of chimpanzees, the top-ranking alpha male can monopolize fertile females, but his is a precarious position. Out of his view, the beta male may be grooming the third-ranking male, forming an alliance to topple the alpha. Shades of Brutus and Cassius? Strategy must be met with counterstrategy. Discovering the grooming session, the alpha may attack the third-ranking male, punishing his transgression, only to groom him later when the beta is not around. Aggression among rival males in a community may be severe, but even if the alpha could, it may not be in his interest to kill the beta. They need each other. Large groups of males from neighboring communities make regular stealthy patrols into their territory, hunting for solitary chimpanzees. Together, the alpha and beta can fend off an attack, but if either is caught alone, he will be killed.

Human and chimpanzee societies are highly unusual among mammals because the most prominent bonds are among males. Males remain in the social group of their birth. Kinship probably plays an important role in contests between chimpanzee communities and many smaller human groups. However, within chimp communities and in human interactions at all levels, relatedness might be only one factor of many incorporated into an individual's alliance strategy.

Most primates are like most mammals; females stay home, and males emigrate. Many species of Old World monkeys share a pattern of alliance formation among females. Members of a matriline have adjacent ranks, resulting in a rank among matrilines. A younger smaller female from a high-ranking matriline can chase a larger female away from a foraging area because she knows that other members of her matriline will support her. Occasionally lower-ranking matrilines will join forces in mob attacks against one individual, then another from a higher-ranking matriline, causing a long-lasting change in matriline rankings. At the level of the social group, all of the females are an alliance. Encountering another group at a preferred foraging area, they will unite to chase off the interlopers.

Unrelated males in female-bonded primate groups may also form alliances, often of a temporary nature that some prefer to call coalitions. A male savannah baboon will solicit help from another male before he attacks a higher-ranking male with a female consort. If their joint attack is successful, either male may end up with the female. From his perspective, the soliciting male has some chance of acquiring the female if he attacks with a partner, but no chance at all if he makes a solo charge.

Both male and female lions ally with same-sex relatives, but in different groups. Females remain in their home pride, defending their territory and their kills against other prides. By synchronizing births, females increase their chance of producing a large cohort of male offspring that can emigrate together as an alliance. A male enjoys greater reproductive success in a large group that is more effective than small groups at taking over and defending prides. Sometimes small alliances of two or three males include nonrelatives—evidently it is better to be in an alliance with a nonrelative than to be alone.

A dominant male Camargue horse may take on a subordinate ally in his efforts to defend a harem against intruders. The subordinate takes on the lion's share of the fighting and sires only a quarter of the foals, but does better than similar subordinate males that adopt the alternative solitary "sneaky" mating strategy. Even though the alliance partners are no more closely related than any other stallions in the herd, their relationship might last a lifetime. In some species alliances between dominant and subordinate or "satellite" individuals are more ephemeral. Pied wagtails defend territories along a riverbank where the birds feed on insects washed ashore. If there is abundant food and many intruders, it pays the territory holder to share with a satellite that helps with territory defense. The satellite is evicted when the food supply dwindles to the point where it does not pay the dominant to share, but will be tolerated again on more productive days.

The alliances of lions, horses, and pied wagtails are all between individuals of the same group against "nonmembers." Alliances of this sort are absolutely everywhere, including any insect, bird, or mammal that forms groups for the purpose of resource defense. Unlike alliances within groups, these between-group conflicts do not provoke ideas of "Machiavellian" intelligence. Relationships between groups are relatively simple because they are always hostile. That isn't true for allies within groups—allies like our beta and third-ranking male chimpanzee. They may be friends or foes, depending on a multitude of factors that each must consider if they are to be successful.

Outside of primates, such complex within-group alliances are rare. A remarkable example is found in an Australian society of bottlenose dolphins. Within a huge social network of hundreds of dolphins, males form "alliances of alliances." Pairs and trios of males cooperate to sequester receptive females for days or weeks at a time. Teams of alliances cooperate to take females from other groups or to defend against such attacks. Alliances that cooperate on one day may be rivals on another. Imagine how much more complex strategic alliance decisions might become if males must consider the impact of their decisions on both levels of alliance. To find a similar example of multilevel male alliances within a large social network, we must return to Rome, the World War II battlefields, or any human society, where nested alliances are a ubiquitous feature of social organization.

See also Antipredatory Behavior—*Sentinel Behavior*
Antipredatory Behavior—*Vigilance*
Cooperation
Dolphins—*Dolphin Behavior and Communication*
Friendship in Animals
Social Organization—*Herd Dynamics: Why
    Aggregations Vary in Size and Complexity*
Social Organization—*Monkey Families and Human
    Families*
Social Organization—*Reconciliation*
Social Organization—*Territoriality*

Further Resources

Connor, R. C., Wells, R., Mann, J. & Read, A. 2000. *The bottlenose dolphin: Social relationships in a fission–fusion society*. In: *Cetacean Societies: Field Studies of Whales and Dolphins* (Ed. by J. Mann, R. C. Connor, P. Tyack & H. Whitehead), pp. 91–126. Chicago: University of Chicago Press.
Harcourt, A. H. & de Waal, F. B. M. 1992. *Coalitions and Alliances in Humans and other Animals*. Oxford: Oxford University Press.

Richard Connor

■ Social Organization
Cooperation and Aggression among Primates

It is commonly argued that a consequence of social group living is an expected increase in feeding competition and in the rate of aggressive interactions among conspecifics (members of the same species). Males are expected to compete for access to reproductive partners, and females are expected to compete for access to important feeding sites. It is thought that there are strong evolutionary advantages in forming affiliative bonds with group members, especially kin, if this helps your group outcompete individuals in other groups for access to resources. However, individual group members also must compete for resources and sexual partners within their social group. Opportunities for such competitive interactions may

Two group-living ringtailed lemurs greeting one another; an example of cooperative behavior among primates. Collars are used to identify individuals in long-term ecological studies in Madagascar.
Courtesy of Robert W. Sussman.

occur on an hourly or daily basis, and threaten to weaken social bonds required for coordinated and cooperative group-level behavior. It is reasoned, therefore, that reconciliatory behaviors evolved to help reestablish social bonds fractured by within group aggression and competition.

The framework described above has been used to interpret the social and mating systems of many primate species. However, sufficient data required to substantiate the basic assumptions of this model have not been collected, and alternative theories on the causes of aggression and cooperation have not been adequately investigated. Affiliative and aggressive behaviors of male and female primates vary depending on the species, the context, and the individual, and understanding of these variations must await a clearer appreciation and investigation of the various social contexts in which social interactions occur.

It seems, however, that some authors may have accepted this competition-based paradigm as a default explanation without critically evaluating its assumptions or appropriately testing alternative hypotheses. There can be considerable advantages to both kin and non-kin group members in developing dyadic, polyadic, and group-level affiliative and cooperative behaviors in which partners receive collective benefits. Theories on the importance of mutualism and other forms of social cooperation are generally lacking from the discussion of primate sociality.

Just how frequent is aggressive behavior in the normal activity cycle of primates. In a recent review of activity budgets of free-ranging primates (86 studies, 23 genera, and 42 species) little time was spent in social behavior in contrast to overall maintenance behavior (such as feeding, resting, or moving). Generally, less than 11% of the time was spent in any type of active social interaction. Aggressive behavior accounted for less than 1% of the daily activity, usually much less. This was true in prosimians, monkeys, and apes.

Diurnal primates live in social groups. Most primate social interactions are affiliative. Aggression and affiliation are necessary consequences of social life. However, if an individual's survival is enhanced by the collective advantages of living in a cohesive and socially integrated behavioral unit, than an understanding of the ability to maintain affiliative and coordinated behaviors and to minimize agonistic and eccentric behaviors is likely to provide critical insight into the evolution of sociality and group living in primates. It appears that affiliation may be a major governing principle of primate sociality, and that aggression and competition represent important but secondary features of daily primate social interaction.

There is evidence that the benefits of cooperation in vertebrate societies, generally, may show parallels to those in human societies, where cooperation between unrelated individuals is frequent, and social institutions are often maintained by generalized cooperation and reciprocity. Cooperation and affiliation represent behavioral tactics that can be used by group members to obtain resources, maintain or enhance their social position, increase reproductive opportunities, and reduce the stress of social isolation. Many affiliative and cooperative behaviors can be explained by individual actions that may benefit several individuals. In acts of cooperation, both participants may receive immediate benefits from the interaction.

Coordinated behaviors, such as joint resource defense, range defense, cooperative hunting, alliance formation, cooperative food harvesting, and predator vigilance, can be explained in terms of immediate benefits to participating individuals. Acts that appear to benefit recipients may also benefit actors. These benefits need not be equal for each individual. If the cost to the actors of affiliative behavior is low, even if the rewards are low and/or variable, we should expect affilation and cooperation to be common. This intraspecific mutualism may help to explain observations that nonhuman primates live in relatively stable social groups and solve the problems of everyday life in a generally cooperative fashion.

Recently, mathematical and genetic models have been developed illustrating how the evolution of cooperation among nonrelated and nonreciprocating individuals can occur. Furthermore, there is growing evidence that endocrinological and neurological mechanisms exist that help sustain, maintain, and positively reinforce social cooperation among mammals. Finally, there is evidence that when resources are distributed heterogeneously in time and space, feeding competition and group living might be less costly than previously thought. This has been referred to as the *resource dispersion hypothesis* (RDH). It asserts that resources might be distributed such that the territory that will support a primary pair (or minimum social unit) might also support additional individuals without any requirement for competition among them.

Given these new findings, we might speculate on an evolutionary scenerio that does not necessitate constant competition or complex calculations of kin recognition. In experiments using MRI scans, mutual cooperation has been associated with consistent activation in two broad brain areas linked with reward processing—the anteroventral striatum and the orbitofrontal cortex. It has been proposed that activation of this neural network positively reinforces cooperative behavior. Furthermore, both of these brain areas are rich in neurons able to respond to dopamine, the neurotransmitter known for its role in addictive behaviors. The dopamine system evaluates rewards—both those that flow from the environment and those conjured up by the mind. When something good happens, dopamine is released which makes the individual take some action. The dopamine system works unconsciously, providing guidance for making decisions when there is not time to think things through. With these systems, investigators believe that they have identified a pattern of neural activation that may be involved in sustaining cooperative social relationships, perhaps by labeling cooperation as rewarding.

Another physiological mechanism associated with affiliation and nurturing is the neurological circuitry related to maternal responses in mammals. The hormone oxytocin, for example, controls a broad suite of maternal responses. Oxytocin has been related to every type of animal bonding: parental, fraternal, sexual, and even in the capacity to sooth one's self. It has been suggested that, although its primary role may have been in forging the mother–infant bond, oxytocin's ability to influence brain circuitry may have been coopted to serve other affiliative purposes that allowed the formation of alliances and partnerships, thus facilitating the evolution of cooperative behaviors. These neuroendocrinological systems could explain the evolution not only of cooperation among nonrelatives but also of "nonselfish" altruistic behavior.

Of course, aggression and competition are important in understanding primate social interactions. However, it seems that affiliation, cooperation, and social tolerance associated with the long-term benefits of mutualism form the core of social group living and that, in most instances, aggression and competition are better understood as social tactics and individual adjustments to the immediate and ephemeral conditions of particular social situations. Primate sociality, and affiliative, cooperative, and agonistic behaviors are best understood in terms of the individual benefits and mutually collective advantages that individuals obtain as members of a functioning social unit.

See also Cognition—*Social Cognition in Primates
       and Other Animals*
Cooperation
Play—*Dog Minds and Dog Play*
Play—*Social Play Behavior and Social Morality*
Social Evolution—*Optimization and Evolutionary
       Game Theory*

Further Resources

Clutton-Brock, T. 2002. *Breeding together: Kin selection and mutualism in cooperative vertebrates.* Science, 296, 69–72.

Dugatkin, L. A. 1997. *Cooperation Among Animals: An Evolutionary Perspective.* New York: Oxford University Press.

Rilling, J. K., Gutman, D. A., Zeh, T. R., Pagnoni, G., Berns, G. S., & Kilts, D. 2002. *A neural basis for social cooperation.* Neuron, 35, 395–405.

Strum, S. C. 2001. *Almost Human: A Journey into the World of Baboons.* Chicago: University of Chicago Press.

Sussman, R. W., & Chapman, A. (Eds.). 2004. *The Origin and Nature of Sociality,* New York: Aldine De Gruyter.

Robert W. Sussman

■ Social Organization
Dispersal

Animals are born in the home range of one or both of their parents. As they approach reproductive maturity, juveniles face a choice: remain in their natal home range or leave and seek a new home range elsewhere. A permanent movement away from the current home range to a new home range is called *dispersal*. This behavior is most prevalent among juveniles, and dispersal before reproductive maturity is termed *natal dispersal*. But adults also sometimes disperse, and dispersal after an animal has successfully reproduced is termed *breeding dispersal*. Dispersal occurs in almost all species of birds and mammals, including humans, and many other animals as well.

Dispersal has long fascinated biologists because of the distances dispersers sometimes move and the risks they seem to take. Animals that remain in the same home range benefit from being familiar with essential resources there; they know the location of food, nest or den sites, and dense cover to escape from predators. Once a disperser leaves its home range and enters unfamiliar terrain, it loses the benefits of familiarity. Hence, dispersers are thought to face numerous risks; they may starve from lack of food, suffer predation when surprised by a predator away from cover, or be unable to find suitable shelter from inclement weather. Further, they face the energetic cost of movement during dispersal, and they may fail to find suitable habitat or mates. Animals that live in social groups may entail additional costs. Group-living animals benefit from social bonds with other members of the group, who are often close relatives, and leaving that group means the loss of those benefits, as well as the risk of aggression when encountering members of the same species who are unfamiliar or unrelated.

If dispersal is so risky, then why is this behavior so common? This question has perplexed biologists for decades. However, recent research has shown that the cost of dispersal may be lower than previously thought. For example, dispersers usually move rapidly, thereby reducing the time they are exposed to predators. Further, remaining in the natal home range, called *philopatry*, also has costs, and animals probably disperse because the costs of philopatry are greater than the costs of dispersal. Several costs of philopatry have been identified. One is competition for resources. A juvenile approaching maturity has benefited from living in suitable habitat, but that habitat is also occupied by members of the same species—one or both parents, and often other individuals as well. Hence, the juvenile must compete with these adults for resources such as food, nest sites, or den sites. Since juveniles are usually inferior to adults as competitors, dispersal in search of unoccupied habitat may be the best choice. Another cost of philopatry is competition for mates, which can be fierce, depending on the mating system. For example, in a polygynous mating system, in which one male mates with several females, a dominant male will attempt to exclude subordinate males, thereby depriving subordinates of the opportunity to mate. Dispersal may be the best option for acquiring mates. A third cost of philopatry is the potential for inbreeding depression. Mating opportunities in the natal area may be limited to parents and other close relatives, and mating with close relatives can lead to lower offspring viability, termed *inbreeding depression.* Consequently, dispersal to find unrelated mates may improve reproductive success.

The costs and benefits of dispersal often affect males and females differently, resulting in a sex bias in dispersal tendency in many species. In most bird species, females are the predominant dispersers, and in most mammal species, males are the predominant dispersers. Although the cause of this difference in sex-biased dispersal is not fully known, two explanations seem likely. One is inbreeding avoidance; one sex disperses to avoid inbreeding with the other, although this explanation does not indicate which sex should be the predominant disperser. Another explanation is the mating system. Mammals typically have a mate-defense mating system, in which males compete with each other for access to females, and a dominant male may exclude several subordinates; thus, males disperse to find undefended females. Birds usually have a resource-defense mating system, with males competing with each other for high quality breeding territories to attract females; thus, males remain philopatric because familiarity confers an advantage in territorial defense. Females require adequate resources for reproduction and make decisions accordingly. Female mammals benefit from remaining philopatric in a home range with sufficient resources, whereas female birds benefit from dispersing to seek out a male defending a high quality territory.

Dispersal is inherently difficult to study because this behavior is rarely observed in its entirety. Dispersers often leave abruptly and move rapidly in an unpredictable direction, and they may travel long distances before settling. In the early years of research on dispersal, most information was obtained by capturing or observing marked animals away from their home range. However, unless the disperser settled within the study area of the researcher, locating an animal after dispersal required a combination of intensive searching and good fortune, and many dispersers were never recaptured or resighted. With the development of radiotelemetry, in which an animal is equipped with a small radio transmitter that allows determination of its location, researchers have been able to track dispersers from before emigration until after settlement in their new home range. Recent advances in genetic techniques, especially those for studying DNA, have added additional information about dispersal.

Dispersal in Marmots

Dirk H. Van Vuren

Dispersal plays a central role in the behavior of the yellow-bellied marmot, a large ground-dwelling squirrel that lives primarily in high elevation areas throughout western North America and feeds on vegetation. Marmots live in meadows that have rock outcrops or talus, because burrows within the rocks are their primary escape cover from predators. Rocky meadows are patchily distributed, and marmots living on these habitat patches form social groups called colonies, which consist of several related adult females (typically mother–daughter and sister–sister relationships), one adult male, and juveniles. These kin groups form by adult females recruiting their own daughters. The mating system is polygynous, with the one adult male excluding all other adult males. Females benefit by living with relatives because they alert each other to the presence of a predator by alarm calling, and because they exclude unrelated females from resources. However, a cost of group living is competition for resources within the group, and one outcome is reproductive suppression, in which a dominant female prevents subordinates from reproducing.

Natal dispersal is common and usually occurs before reproductive maturity is reached at the age of 2 years, but breeding dispersal is rare. Marmots show the typical mammalian pattern of male-biased dispersal: All males disperse, but only about half of females disperse, with the other half remaining philopatric and being recruited as members of the kin group. Males disperse to avoid competition for mates with the dominant male, or possibly to avoid inbreeding with females in their natal colony, all of which are usually relatives. Males that delay dispersal much beyond 1 year of age may be attacked and bitten. Female dispersal is influenced by adult female behavior. Amicable behavior from the mother favors juvenile female recruitment, while agonistic behavior from the mother, or the mother's absence, favors dispersal. Females that emigrate face the costs of dispersal, but they escape aggression from adult females and the possibility of reproductive suppression. Inbreeding probably is not a factor in dispersal of females because the tenure of the dominant male is so short that their father has usually been replaced by an unrelated male before they reach reproductive maturity.

Dispersing marmots emigrate in all directions, but there is a tendency to follow linear features in the landscape, such as rock scarps and rocky canyons, that provide suitable cover from predators. Some marmots, however, traverse large expanses of habitat that provides no cover. Most marmots move a relatively short distance before finding a place to settle. However, some move distances exceeding 15 km (9 mi), crossing mountain passes and swimming rivers in the process, a remarkable feat for an animal that weights only 2 kg (4.4 lb), has legs only about 10 cm (4 in) long, and lives in a home range only 100 m (328 ft) across. As is common for mammals, males disperse farther than females. A few marmots disperse to an adjacent home range when a vacancy occurs, but most locate a more distant site to disperse to based on information gained from exploratory excursions up to 1,500 m (5,000 ft) long, or they simply emigrate abruptly, moving rapidly until a settlement site is encountered. Surprisingly, the survival cost of dispersal for marmots is relatively low. About three fourths of dispersers survive until they find a new home range, and predation is the only cause of mortality during dispersal.

Tracking dispersers through radiotelemetry has revealed much about the process of leaving home. Dispersers may leave in any direction, and some move in a relatively straight line, whereas others follow a more circuitous path. Dispersers often select good quality habitat for traveling, but sometimes they cross expanses of habitat known to be unsuitable for the species. Distances traveled by dispersers before settling are usually right-skewed, meaning that most settle within a relatively short distance although a few travel long distances. In general, dispersal distance is related to body size; larger species tend to disperse farther than smaller species.

Radiotelemetry also allows identification of the ways that dispersers locate a new home range. Some animals detect a vacancy adjacent to their current home range and disperse through a local shift in location. Sometimes this shift occurs through a gradual extension of activity into the new home range, coupled with diminishing activity in the former home range. Animals dispersing through a local shift in home range benefit from having information about prospects in the new home range before dispersing. Other animals disperse longer distances after discovering opportunities beyond their home range through exploratory excursions, round-trip forays that allow assessment of prospects elsewhere before eventual dispersal. The distances traveled in these exploratory excursions may be several times greater than the distance across the animal's home range. Yet other dispersers discover opportunities for settlement by chance encounter during one-way movements through previously unknown areas. These animals often emigrate abruptly, usually move rapidly, and may cover long distances before encountering suitable habitat for settlement.

Dispersal is a behavioral attribute of certain individuals, with important implications for survival and reproductive success, thus it plays a major role in animal behavior. But dispersal also has consequences at the population level, hence it is important in other scientific disciplines as well. Ecologists have studied dispersal because of its influence on population structure and population size. For example, movement of dispersers into or out of a population affects age and sex structure, and a net movement of dispersers out of a population at high density may be an important factor regulating population size. Geneticists and evolutionary biologists are interested in dispersal because animals that survive dispersal, and subsequently breed, generate gene flow, the movement of genes among populations. Gene flow influences the genetic structure of populations, and it is an important factor in evolution. Biogeographers emphasize dispersal as the process by which species extend their ranges. Finally, conservation biologists have focused on dispersal as an important component in plans to save rare species from extinction. As species are increasingly limited to smaller and smaller fragments of natural habitat because of logging, urbanization, and other human activities, local extinctions are likely to result. Dispersal provides a means of restoring locally extinct populations from intact populations nearby.

See also Social Organization—*Home Range*
Social Organization—*Territoriality*

Further Resources

Clobert, J., Danchin, E., Dhondt, A. A. & Nichols, J. D, (Eds.). 2001. *Dispersal*. Oxford: Oxford University Press.

Gaines, M. S. & McClenaghan, L. R., Jr. 1980. *Dispersal in Small Mammals*. Annual Review of Ecology and Systematics, 11, 163–196.

Greenwood, P. J. 1980. *Mating systems, philopatry and dispersal in birds and mammals*. Animal Behaviour, 28, 1140–1162.

Holekamp, K. E. & Sherman, P. W. 1989. *Why male ground squirrels disperse.* American Scientist, 77, 232–239.

Johnson, M. L. & Gaines, M. S. 1990. *Evolution of dispersal: theoretical models and empirical tests using birds and mammals.* Annual Review of Ecology and Systematics, 21, 449–480.

Van Vuren, D. 1998. *Mammalian dispersal and reserve design.* In: *Behavioral Ecology and Conservation Biology* (Ed. by T. Caro), pp. 369–393. New York: Oxford University Press.

Dirk H. Van Vuren

■ Social Organization
Eusociality

There are many and diverse social systems among animals. Perhaps the most extreme social system is eusociality. Eusocial species are distinguished from other social groups by three characteristics:

1. cooperative brood care, in which the group as a unit rears young together;
2. overlap of generations, in which adult offspring stay in the colony and help parents rear additional offspring; and
3. the possession of reproductive division of labor, so that only one or a few individuals reproduce, while the others remain functionally or physically sterile throughout their lifetimes.

For a species to be considered eusocial, all three conditions must be met (Wilson 1971). Systems that meet one or two of the criteria (generally the first and second) are considered *presocial*. As an example of a presocial system, a number of bird species cooperatively care for brood, but individuals assisting in brood care are not necessarily sterile and are not morphologically different from reproducing individuals.

Classic examples of eusociality are found in the social insects, especially the ants, honeybees, and termites, where colonies may contain thousands to millions of morphologically sterile workers and a single reproductive queen (ants and honeybees) or queen and king (termites). Workers in some species of ants and termites show additional morphological differentiation into various *castes*, which are groups that perform behaviorally distinct tasks in the colony. However, eusociality does not necessarily involve absolute sterility of workers. In primitively eusocial species, such as many wasp species, nonreproductives are often physiologically capable of reproduction, but remain functionally sterile throughout their lifetimes; they neither mate nor produce offspring, and in many cases their ovaries do not fully develop. Eusociality via functional sterility also occurs in vertebrates, most especially the social mammals, the most pronounced example of which are the naked mole-rats (Sherman et al. 1991).

The degree to which species show eusociality can be determined via reproductive *skew*, or the relative difference in reproductive output among individuals in the group, where groups with higher levels of reproductive skew are considered to be more highly eusocial. However, the question of precisely how much reproductive skew is necessary for a group to be considered eusocial has been the subject of debate (Sherman et al. 1995; Crespi & Yanega 1995). Regardless, categorization as a fully eusocial species is generally reserved for species in which the overwhelming majority of individuals remain sterile throughout their lifetimes, but cooperate to rear the brood of one or a few individuals.

Eusociality provides an evolutionary paradox; despite the extreme success of eusocial species, as measured by their relative biomass, very few groups have evolved to be truly eusocial. For example, sociality has evolved multiple times within the social insects (order Hymenoptera), and yet eusocial species make up only a small fraction of the order. The relative rarity of this social system is likely related to the fitness costs for nonreproductives, which pass up the opportunity to reproduce and instead help rear another individual's offspring.

Explanations of the origin of eusociality are typically based on the roles of two main factors: kin selection and ecological characteristics. The concept of kin selection argues that the indirect fitness that an individual gains by assisting in the care of siblings can outweigh the direct fitness that individual would gain if it were to reproduce on its own (Hamilton 1964). In support of this explanation, individuals within eusocial groups are often closely related to each other, and in many cases are more closely related genetically to the sibling offspring they help rear than they would be to their own offspring. Extremely high relatedness within eusocial Hymenopteran colonies is generated by an unusual genetic system, haplodiploidy, in which females are diploid, while males are haploid and are produced as unfertilized eggs. In this system, females, but not males, are highly related to each other (supersisters), and only females become sterile workers.

There are also many diploid eusocial organisms; the two most studied are the termites and naked-mole rats (Shellman-Reeve 1997, Sherman et al. 1991). In most known diploid eusocial species, ecological conditions generate extremely low dispersal rates, resulting in highly inbred, and thus highly related individuals within colonies. In these systems, the risk of dispersing to a new location is often so great that it is more beneficial for individuals to remain within the colony and help closely related kin rear offspring. Additionally, increased numbers of individuals are better able to defend a lucrative location against parasites, predators, and usurpers. These ecological characteristics, along with kin selection, may be the primary determinants of the evolution of eusociality.

See also Cooperation
Naked Mole-rats
Sociobiology

Further Resources

Crespi, B. J., Yanega, D. 1995. *The definition of eusociality*. Behavioral Ecology, 6, 109–115.
Hamilton, W. D. 1964. *The genetic evolution of social behaviour*. Journal of Theoretical Biology, 7, 1–52.
Shellman-Reeve, J. S. 1997. *The spectrum of eusociality in termites*. In: *Social Behavior in Insects and Arachnids*. (Ed. by J. C. Choe, & B. J. Crespi, pp. 52–93. Cambridge: Cambridge University Press.
Sherman, P. W., Lacey, E. A., Reeve, H. K., Keller, L. 1995. *The eusociality continuum*. Behavioral Ecology, 6, 102–108.
Sherman, P. W., Jarvis, J. U. M., Alexander, R. D. (Eds.). 1991. *The Biology of the Naked Mole-Rat*. Princeton, NJ: Princeton University Press.
Wilson, E. O. 1971. *The Insect Societies*. Cambridge, MA: Cambridge, MA: The Belknap Press of Harvard University Press.

Rebecca Clark & Jennifer Fewell

■ Social Organization
Herd Dynamics: Why Aggregations Vary in Size and Complexity

From Individuals to Groups

Apart from coming together to mate, most animals live solitary lives. Such a preponderance of asocial behavior should not be too surprising since as Alexander (1974) first noted, there are no automatic benefits of living in groups, only automatic costs. Typically disease and parasite transmission increase when individuals spend time together, as has been demonstrated in species from swallows to horses. Yet groups do form, and many are long lasting. Clearly, there are compensating benefits that can offset the automatic costs associated with disease and other likely detriments, such as increased competition for resources, including mates, and increased detectability by predators. Benefits mostly come in three forms. First, animals can develop types of social behavior specific to stable groups that directly compensate them for specific costs of group living. One such example is forming mutual grooming partnerships as do olive baboons (*Papio anubis*) to lower disease and parasite transmission. Second, by forming groups, animals can enhance foraging by being better able to find, acquire or defend food. Examples include colonial cliff swallows (*Hirundo pyrrhonota*) that transfer information about the locations of rich but ephemeral feeding sites or troops of monkeys who drive smaller troops away from feeding trees, or even lions (*Felis leo*) that defend kills to keep competitors away. And third, animals in groups can reduce their risk of being preyed upon by either increasing the likelihood of detecting predators, diluting their personal risks or by decreasing the likelihood that predators can make a kill; confusion and cooperative defense are mechanisms that provide such antipredator benefits. Examples include the scattering of fish in schools, tail flashing by white tail deer (*Odocoileus virginianus*), stotting by gazelles, or the gathering of young inside a ring of adult musk oxen (*Ovibos moschatus*) with upturned horns facing outwards toward approaching predators.

Whenever the sum total of all benefits exceed costs, group living will be favored by natural selection because such social individuals will sire more offspring and rear them to independence then asocial individuals. As genes for being social increase in frequency in the population, sociality will necessarily become the norm. In fact, groups will tend to be of similar size in similar environments in part because the most favored groups will be those where benefits maximally exceed costs. But such groups are not inherently stable since any new member joining the group will lower the net benefits for all. Yet, given that the resulting net gain for each group remaining a group is still greater than that of an individual living alone, joining will continue until the difference between the net gain of living alone and living in a group of this particular size vanishes. At this size, the group becomes stable in

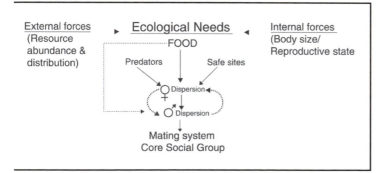

A framework showing how ecological and internal constraints determine mating systems and thus the structure of core social groups by first affecting distributions and associations of females which in turn affect distribution and associations of males.

Courtesy of Daniel I. Rubenstein

terms of numbers. Although "stable" groups are always larger then "optimal" sized groups, both hypothetical group types scale in similar ways as environmental conditions change, so either group type can be used to examine how environmental circumstances shape the dynamics of group living.

Groups take on particular structures depending upon the nature of the relationships that develop among their members (Rubenstein 1986). Particular distributions of food, water, predators, and safe sites produce different core social structures. Since natural selection favors females that leave more progeny then their peers, female distributions and abundances on particular landscapes will change as the distribution and abundance of key resources change. Such flexibility comes about because different ecological settings often select for specific types of behaviors that best enable females to satisfy their needs at the lowest possible costs. Once females have responded, their associations, and the distribution that develops on the landscape in turn, constrain what males can do. As males respond to the actions of females, with perhaps females adjusting their initial response to the subsequent actions of males, mating relationships develop and the core of a social system is born. (See the graph below.)

In environments in which resources are abundant, especially when they are evenly distributed, competition is often low enough to permit females to aggregate. Conceptually, if sufficiently large foraging or antipredator benefits can be derived by females that aggregate, and the groups that form are not too large, then these groups can be defended by single males and so-called "Harem Defense Polygyny" results. However, if resources are more patchily distributed so that competition among females intensifies, then female group sizes will vary, and female associations may become more transitory. Rather then defending unstable groups of females, males instead attempt to defend in advance of females arriving, resource patches sought by females (dotted line in the graph). Typically, in these systems of "Resource Defense Polygyny," the most able males defend the best patches and thus gain access to the largest number of females for the longest periods of time. If resources are not only patchily distributed, but the patches are large, widely separated, and fluctuate seasonally in abundance, then competition among females often becomes so low that the formation of large aggregations is even more likely, provided that females can range widely and follow the shifting locations of peaks in food abundance. Males will thus be forced either to follow these large groups competing for, and then tending, individual reproductive females one at a time (*wandering*), or to position themselves at the intersection of female migratory routes waiting for females to visit them (*lekking*). In either case, intense male–male competition generates a mating system based on "Male Dominance Polygyny," and in the latter case, females are afforded an exquisite opportunity to compare many males simultaneously before choosing with which one to mate! Whenever resources are sparsely, but somewhat evenly

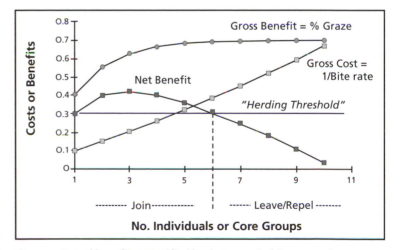

Gross costs and benefits as typified by decreases in bite rate or increases in percentage of time spent grazing, and net benefits as computed by the difference between gross costs and benefits as a function either of number of individuals in a group or the number of core social groups in a multilevel herd.

Courtesy of Daniel I. Rubenstein.

distributed, high levels of competition prevent females from forming groups. As a result, individual females defend territories, thus insuring a regular supply of a renewing resource. Since solitary individuals searching for members of the opposite sex, especially if small-bodied, will face increased predation risk, pairs often share territories, and monogamy results.

This conceptual framework can account for the diversity of core social groups that form for many different taxa, including insects, fish, reptiles, rodents, and many varieties of birds, ungulates, carnivores, marine mammals, and primates. One of the best illustrations showing how environmental forces interact with physiological constraints to shape a species' social system emerges from Peter Jarman's (1974) classic study of African antelopes. By showing how body size affected the ways in which different species perceived, and then responded to, the distribution and abundance of forage and predators of grasslands, Jarman showed why particular social systems increased survival and reproductive prospects for particular species. He argued that the smallest-bodied species, such as did-dik (*Madoqua kirkii*), duikers (*Cephalophus spp., Sylvicapra spp*), suni (*Neotragus moschatus*) and klip-springers (*Oreotragus oreotragus*) require limited amounts of high-quality vegetation. But given their small size, such food items often appear as if they are widely scattered. Faced with high levels of competition and intensified risks of predation, territoriality and monogamy appear to be the best strategies. Pairs generally live in wooded or shrub-rich areas where moisture enables vegetation to grow and renew itself well into the dry season. By signaling territorial ownership via scent rather than by means of sound or visual display, these small-bodied species reduce the chances that any of a large number of carnivores will prey upon them.

As species increase in body size, both physiologically determined dietary needs and the way acceptable forage becomes distributed on the landscape changes. Since "crypsis" (camouflage) becomes an untenable antipredator strategy for larger and more widely ranging species, forming groups becomes the best strategy for larger species to lower predation risk. Fortunately, with larger size also comes an ability to subsist on more abundant, lower-quality vegetation. When it is patchily distributed, as it is for impala (*Aepyceros melampus*), reedbuck (*Redunca spp.*), and some gazelles (*Gazella spp*), males defend the best patches that females prefer. When the vegetation is more evenly distributed, which often results simply from the fact that larger species such as eland (*Taurotragus oryx*) and cape buffalo (*Syncerus caffer*) can utilize even the lowest-quality items, larger groups form. Because the largest species view all types of vegetation populating continuous swards of a landscape as acceptable, competition is virtually eliminated, and many males associate with many females. With such high levels of male–male competition, defense of a small subgroup of females becomes impossible, and dominance defense systems develop.

The same sorts of connections between changing ecological circumstances shape the types of sociality exhibited in other taxa. But, because ecological circumstances that are not too different from those described above generate novel social variants, it appears that strong biological interactions among internal and external constraints themselves become important determinants of sociality. Among the equids, for example, the close association between food and water enable horse (*Equus caballus*) and plains zebra (*Equus burchelli*) females of different reproductive states to associate permanently. Thus males are able to defend such groups and so-called "harems" form. When these two resources are widely dispersed, as for Grevy's zebra (*Equus grevyi*) and the Asiatic wild ass (*Equus hemionous*), females of different states are precluded by metabolic constraints from foraging together. As a result, males compete for territories along traveling routes that take females from feeding areas to watering points (Rubenstein 1994). Differences in body size and the way they alter patterns of water dependency appear to be ultimately responsible for why two similar groups of species exhibit such fundamental differences in the strength of social relationships and hence core patterns of social organization.

From Groups to Complex Societies

Core social groupings that are built mostly upon mating relationships do not fully explain the diversity of social structures that develop among species. For many species, social relationships are limited to those with core social units. Consequently, such groups remain apart from other such units. Yet for other species, interactions and relationships develop among different classes of individuals beyond these basic units. As a result, some species live in large fission–fusion networks where either the nature or strength of relationships among individuals, or core social groups within which they live, change over time. In the latter case, multilevel societies develop in which core groups aggregate to form larger and more complex social networks. Some of the best examples of hierarchical social structuring are displayed by elephants (*Loxodonta africanas*) where female-led family groups aggregate during the rains but split during the dry season, or by gelada baboons (*Theropithicus gelada*) where herds of harems aggregate on the tops of plateaus where predation risk is high, but fragment on steeper slopes where the mobility of predators is more limited. Yet it is in the equids where the relative simplicity of the multifaceted social networks of both the plains and Grevy's zebras reveals how both natural and sexually selective forces interact to produce a multiplicity of social interactions that, in some circumstances, lead to multileveled societies.

A tightly-knit herd of zebras grazing while walking.
Courtesy of Corbis.

As with elephants and gelada baboons, natural selection to maximize food acquisition and minimize risk of predation play important but small roles in determining whether large cohesive herds of harems and bachelor groups form in plains zebras (Rubenstein and Hack 2004). Populations are often large in areas where predators are absent, and where zebra density is high, the size of herds that form tend to be large. Similarly, as food abundance increases, so do herd sizes. But these effects are weak and pale when compared to the forces of sexual selection: Cuckolding risk brought about by the presence of bachelor males is hard to reduce when alone on a landscape. In areas where bachelor groups are large and the groups are cohesive, zebra herd sizes grow to be quite large. It appears that males tend to bring their harem groups together to insure that, if discovered by bachelor males, individual harem-"owning" males can more effectively drive the bachelors off before they can infiltrate his group and mate with those of his females in heat. Herds tend to be significantly larger where the relative number of bachelor males per breeding male is high. Interestingly, males are successful at forming these alliances only when females suffer no significant foraging or harassment costs when in herds. Since this occurs only when a moderate number of breeding stallions band together, limits to growth of herds exist. Thus by avoiding the creation of any intersexual conflicts, males are able to organize themselves to reduce the risk of losing exclusive control of mating. In essence, a second layer of sociality appears to have evolved to solve simultaneously both female and male problems that the fundamental set of relationships characterizing the core social group could not.

Although Grevy's zebras do not structure their extensive social relationships in simple multilevel tiers, the existence of multiple networks of relationships that develop, and that individuals modulate as circumstances vary, underscores the notion that complex societies develop to solve problems that simple one-dimensional social structures can not. Grevy's zebra females in particular have the options of bonding with many different classes of females as well as males. The details of how their oscillations among relationships as a function of "top-down" predatory, or "bottom-up" foraging, pressures change is only now emerging. But variation in the diversity of relationships is responsive to changes in population density and predation pressure. Thus for Grevy's zebras, individual decision making remains at all times more flexible then in plains zebras where, although females solve foraging and antipredator problems at the level of the core group, males can only solve problems of postmating competition at the level of the herd. In Grevy's zebras, females appear to solve their multiplicity of problems by changing the nature and strengths of their relationships at the primary level, in part because resource dependency is not similar for all individuals and thus constrains their ability to simultaneously solve different types of environmental problems with a society structured hierarchically.

From Phalanxes to Waves

Although group living, herd structure, and overall patterns of social organization are best accounted for by functional or adaptive explanations associated with ultimate consideration of reproductive success or even more proximate measures of more short-term behavioral costs and benefits, the actual shape of herds is best described by more mechanistic processes. Typically, models are constructed that imbue individuals on a landscape with particular attributes, and then these individuals act as particles or cellular automata responding to the actions of other such particles on landscapes with hypothetical environmental features. In most such models, forces of repulsion associated with competition and violation of personal space are combined with forces of attraction that reduce separation and the risks of predation. Debate exists over whether individuals are "aware" of all neighbors or are attending to only one or just a few neighbors located in detection "zones." Issues of detection ability get to the heart of how animals "know" what they know and provide an opportunity to enrich the study of animal decision making with important insights emerging from cognitive psychology.

Even with only simple assumptions about animal reasoning abilities, however, the results of simulations can predict the shape of herds that form on real landscapes and show how different modes of detection can alter energetic efficiency of herd dynamics (Gueron et al. 1996). For example, in cases where individuals exhibit hierarchical decision-making patterns in which individuals in particular zones evoke particular movements—first, moving away from neighbors that get too close ("stress" zone), second, moving freely according to idiosyncratic tendencies when repulsion has been overcome and a nearest neighbor is neither too near nor too far ("neutral" zone), and third, moving to a nearest neighbor when a nearest neighbor is not close enough for diluting the risk of predation ("attraction" zone)—stable herds form as long as all individuals are of similar phenotype. Moreover, energetic savings accrue to individuals with "neutral" zones since acceleration and deceleration rates are low, and overall, the shape of these herds become wavy-front-like when moving slowly and columnar or phalanx-like when moving quickly, as is the case when wildebeest or zebras move across African savannas. However, for stability to be achieved when individuals vary markedly in phenotype, which can occur when females have different dietary or physiological needs, for example, then a fourth rule is required. Since different needs often lead to temporary separation, only by invoking a "stop" until discovered or a "speed-up" until a peer is found "rule," can long-term cohesion be maintained. While the use of such "toy models" to explore the ways in which global patterns of behavior can emerge from local interactions is only beginning, the ability to account for the complexity of herd shape and the energetics of group living illustrates the power of incorporating functional with mechanistic explanations to unravel the structural and shape dynamics of complex animal groups.

Conclusions

The dynamics of core groups and higher levels of sociality are driven by individuals joining together to minimize costs and maximize benefits. Such costs and benefits may be in similar or different currencies but the net result of forming groups is that individuals are better able collectively to solve problems posed by the environment. Female relationships in particular develop as they try to maximize the number of surviving offspring. Since male reproduction is much less limited by meeting ecological needs and much more governed by their need to sire more offspring then other males, if males can be enticed by females to help solve their most pressing problems, then it is likely that social relationships will be structured. In such societies the needs of all will be *simultaneously* satisfied as long as conflicts of interest among the sexes do not develop. When the limited, but strong, bonds of core networks cannot solve the sexually selected problems facing males, or even all the environmental problems confronting females, then the core social network itself may become more diversified, and the relationships more complex. In such a circumstance, individuals may modulate their relationships on condition-dependent bases in order to solve *sequentially* conflicting problems posed by complex physical and social environments.

Complexity of herds also emerges from the ways their shapes change in response to individual perceptions about landscapes. Not surprisingly, the decision making of individuals that leads to shape change is governed by "rules" that emerge from the forces of natural and sexual selection. Perhaps somewhat surprising is the fact that only a limited number of such rules can account for the emergence of complex patterns that characterize the shapes and energetics of real herds. Exploring the extent to which cellular automata type models can account for global patterns of behavior emerging from local interactions will provide important insights into herd structure and the dynamics of social organization.

See also Feeding Behavior—*Scrounging*
Feeding Behavior—*Social Foraging*
Social Evolution—*Optimization and Evolutionary
   Game Theory*
Social Organization—*Alliances*
Social Organization—*Home Range*
Social Organization—*Territoriality*

Further Resources

Alexander, R. D. 1974. *The evolution of social behavior.* Annual Review of Ecology and Systematics, 5, 325–83.

Gueron, S., Levin, S. A. & Rubenstein, D. I. 1996. *The dynamics of herds: From individuals to aggregations.* Journal of Theoretical Biology, 182, 85–98.

Jarman, P. J. 1974. *Social organization of antelope in relation to their ecology.* Behaviour, 28, 215–67.

Rubenstein, D. I. 1986. *Ecology and sociality in horses and zebras.* In: *Ecological Aspects of Social Evolution.* (Ed. by D. I. Rubenstein & R. W. Wrangham), pp. 282–302. Princeton, NJ: Princeton University Press.

Rubenstein, D. I. 1994. *The ecology of female social behaviour in horses zebras and asses.* In: *Animal Societies: Individuals, Interactions and Organisation,* (Ed. by P. J. Jarman & A. Rossiter), pp. 13–28. Kyoto, Japan: Kyoto University Press.

Rubenstein, D. I. & Hack, M. 2004. *Natural and sexual selection and the evolution of multi-level societies: Insights from zebras with comparisons to primates.* In: *Sexual Selection in Primates: New and Comparative Perspectives.* P. M. (Ed. by Kappeler & C. P. van Schaik), pp. 266–279. Cambridge: Cambridge University Press.

Daniel I. Rubenstein

■ Social Organization
Home Range

The term *home range* is strictly connected to the study of local movements and space use of animals during their usual daily activities. The home range may be defined as "that area traversed by the individual in its normal activities of food gathering, mating and caring for young." It should be underlined that home range is not *all* the area an animal traverses, but rather the area in which it normally moves; accordingly, occasional excursions outside its normal area should not be considered as part of the home range.

Unfortunately, home range is not the sole concept considering the local space use of animals. We have also to consider the concept of *territory*, that may be defined as "a more or less exclusive area *defended* by an individual or group." However, it is often very difficult to ascertain the exclusive use of an area and to detect the keeping-out signals displayed. Moreover, the same species may be territorial in certain ecological contexts and nonterritorial in others. The degree of overlap between adjacent territories is also variable, and the very concept of territorial exclusiveness is far from adequate. Accordingly, the borderline between these two spatial systems is not as clear-cut as may appear from the above-mentioned definitions. However, the home range concept is used in a more general and comprehensive way than that of territory. In fact, spatial systems, whose exclusiveness has not been ascertained, are usually referred to as home ranges even though they may really be territories.

Radiotelemetry is a useful technique for the study of home ranges of vertebrates: Animals are caught, tagged with small transmitters, and tracked by researchers who detect signals by means of receivers. Transmitters can be radio-located using antennas to locate animals. The exact position occupied by a radiotagged individual at a given time is called *fix*. A major advantage of radio telemetry is that this technique records events in relation to time; radiotracking can show not only how often an individual visits each part of its range, but also what time of day it went there. Movement records from radiotagged animals fall into two main types. There are the short-term movements that animals make daily during foraging, courting, and general maintenance activity, and there are the long-term movements that are on a seasonal or even a lifetime basis. Note that while daily movements are generally useful to define individual home ranges, long-term movements, when applied to migration or dispersion, are not.

The analysis of radiotracking data may be carried out through different theoretical approaches using different computer programs. The home range estimation, in particular, can give rise to home ranges whose graphic output is an ellipse, a convex polygon, or a figure whose boundaries are curvilinear countours. Another method is that of plotting home ranges as an accumulation of grid cells, usually grid squares. The first aim of these analyses is to calculate the *size* of the home range (i.e., its area, usually in hectares.). When a home range size is estimated, the measured size increases very rapidly when fix recording starts, but as more and more positions are recorded, a point of "sampling saturation" is reached. It is generally assumed that beyond this point the animal's range does not increase significantly. Home ranges of erratic individuals or *floaters* are obviously expected to continue increasing, but the presence of a true home range for these individuals can be questioned. However, since many studies have shown continuous range size increases even in apparently sedentary animals, the idea of a stable home range has recently been challenged. *Shape* may be important for analyzing how a home range is placed in the landscape as indicator of resource and security requirements; *structure* within a home range is also relevant to study

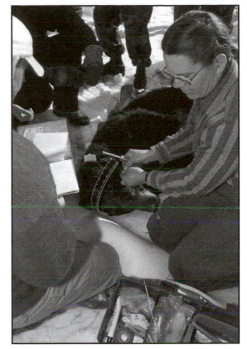

Scientists fitting a tranquilized black bear with a radio collar to track its territory.
© *Ned Therrien / Visuals Unlimited.*

how the intensity of using different areas relates to habitat and sociality. Estimations have shown that home ranges are highly diversified; that is, individuals may inhabit areas of different sizes, use certain portions of their area more often than others (i.e., *core areas*), frequent two or more distinct and separated areas (i.e., *disjointed ranges*), and overlap to different degrees in the use of the same area (*home range overlap*). Different species may present different kinds of home ranges (larger species, for instance, usually have larger home ranges), but individuals of the same species may differentiate their home ranges as well. The red fox (*Vulpes vulpes*), for instance, is highly flexible and can change its ranging behavior according to local environmental and social conditions.

Local movements are under the control of several determinants. From this point of view, it is important not to mix *biological processes* with *spatial patterns* (i.e., home range size and shape, core areas, home range overlap, movements, etc.) they generate. The topic of habitat selection is central in vertebrate space use. At a local level, all decisions taken by individuals to better use their habitat fall into the habitat selection process. Habitat selection decisions may significantly influence individual movements and spatial distribution, and thus it can be assumed that habitat selection is an important process in determining spatial patterns. However, home range spatial patterns may also be controlled by processes that are independent of habitat selection. For instance, the presence of certain mating systems may significantly influence animals movements, in which case spatial patterns are controlled by the process of mating. There are a lot of *factors* affecting such processes. Many of them are *direct* in the sense that they directly influence the processes (e.g., food availability directly affects habitat selection which affects movements). Others are *indirect* because they affect other factors which then affect processes (e.g., human disturbance may change food availability which then affects movements) or *false* in the sense that they seem to affect processes while, in reality, other factors do so (habitat type and structure seem to affect animals' decisions, where they might not choose the habitat for its type or structure but rather for its food resources).

Hereafter processes and factors influencing home ranges in birds, extensively studied by radiotelemetric techniques for more than two decades, will be discussed, because most of these factors are expected to play the same role in other vertebrates (in mammals, in particular) as well. The most important factor affecting home ranges in birds is likely *food availability*. Typically, when food is abundant, there is no need to increase movements, whereas when it is scarce, individuals are forced to amplify movements to find new resources and, consequently, home range size increases in response to declining resource abundance. Ranging behavior may suddenly change when a new food source becomes available. Movements of the nutcrackers (*Nucifrga caryocatactes*), for instance, become frenzied when hazel nuts or pine seeds ripen at the end of the summer. *Nesting site availability*, *climatic conditions*, and *topography* (at least when the occurrence of steep slopes significantly affect the birds' energy requirements) are other direct factors.

Remaining factors sometimes act directly, sometimes indirectly, or are false. *Habitat type and structure*, for instance, may be a false factor, since the birds may not use habitat features but rather the food resources, nesting sites, or climatic conditions found in it. Similarly, about half of the studies dealing with *predator–prey interactions* demonstrate that predators select habitats where prey are available (thus food availability is the true direct factor) or where they are able to hunt (in this case, habitat type and structure is the true direct factor). The other half of predator–prey interaction studies demonstrate that prey select habitats that offer effective or potential protection from predators; that is, the predators directly affect the habitat selection of the prey. However, certain habitats that are positively selected because of their potential protection may be extremely dangerous for prey because they are often visited by predators. This is the case of the pheasant (*Phasianus colchicus*) hens which take cover in hedges where foxes hunt efficaciously. *Habitat fragmentation* may be another indirect factor because it induces a change in food availability that modifies habitat selection processes and, consequently, spatial patterns. *Human disturbance* usually directly affects home range, but in a few cases it works as an indirect factor, modifying food availability or habitat structure. In many instances *age* directly affects home ranges. It is well known, for instance, that home ranges of adults are significantly smaller than those of juveniles, which are usually floaters and move a lot looking for a site in which to settle down. *Sex* and *social status* may also sometimes determine differences among home ranges.

Among all these factors, food availability is the only one that always directly affects home ranges (the effect of topography, climatic conditions, and nesting site availability are more likely marginal factors). The other factors may be direct, but in many instances are false or indirect, often because food availability is the only direct factor. Moreover, although many other factors affect only certain species (predators affect the home range only of their prey) or habitats (e.g., only fragmented forests), and sometimes do not work at all (e.g., age, sex, and social status), food availability affects all individuals of all species, everywhere, and at all times. All the above clearly suggest that food availability is the most important factor controlling home ranges in birds and that the other factors are secondary. Many of the identified factors affect the process of habitat selection. In particular, food availability, habitat type and structure, habitat fragmentation, predator–prey interactions, human disturbance, nesting site availability, climatic conditions, age, and sex significantly and directly affect the way birds use their habitat. In turn, habitat selection choices affect the way birds move locally.

Breeding is another process affecting home range spatial systems of breeding individuals during the reproductive period. It may be affected by the sex, age, and social status of the birds. Other processes that may also occasionally affect home range spatial patterns of birds are *mating* (influenced by sex and age) and *flocking*. All these processes are obviously not mutually exclusive since spatial patterns may be contemporaneously determined by more than one process. For instance, movements of breeding birds may be constrained by the need to defend eggs or nestlings, while at the same time they also depend upon the habitat selection decisions regarding food finding, predator avoidance, and so on.

Functional relationships among factors, processes, and patterns are hierarchical in the sense that home range spatial patterns are determined by ecological and biological processes, which in turn are affected by factors in a direct or indirect way. There are numerous relationships among factors, between factors and processes, and between processes and spatial patterns. Although it is obviously not possible to weigh precisely all the factors and processes to obtain precise hierarchies, it can be suggested that the most important process affecting home range spatial patterns in birds is habitat selection, which in turn is mostly affected and controlled by food availability and location. Hence, the main conclusion is that food availability is the primary determinant of the home ranges in birds, and all the other factors are secondary. After all, the most impelling need of animals is feeding.

See also Social Organization—*Dispersal*
Social Organization—*Herd Dynamics;*
Why Aggregations Vary in Size and Complexity
Social Organization—*Territoriality*

Further Resources

Cavallini, P. 1996. *Variation in the social system of the red fox.* Ethology Ecology & Evolution, 8, 323–342.
Gautestad, A. O. & Mysterud, I. 1995. *The home range ghost.* Oikos, 74, 195–204.
Kenward, R. E. 2001. *A manual for wildlife radio tagging.* San Diego: Academic Press.
Mace, G. M. & Harvey, P. H. 1983. *Energetic constraints on home-range size.* American Naturalist, 121, 120–132.
Rolando, A. 1998. *Factors affecting movements and home ranges in the jay* Garrulus glandarius. Journal of Zoology, London, 246, 249–257.
Rolando, A. 2002. *On the ecology of home range in birds.* Revue d'Écologie (la Terre et la Vie), 57, 53–73.
Rolando, A. & Cariso, L. 1999. *Effects of resource availability and distribution on autumn movements of the Nutcracker* Nucifraga caryocatactes *in the Alps.* Ibis, 141, 125–134.
White, G. C. & Garrott, R. A. 1990. *Analysis of wildlife radio-tracking data.* San Diego: Academic Press.

Antonio Rolando

■ Social Organization
Monkey Families and Human Families

Although family life is common among most human societies, it is rare to find other mammals living in families. In some species, mothers rear infants alone with no help from fathers and other relatives. Other species live in large multimale, multifemale groups where males defend the group as a whole, but are not directly involved in infant care. However, one group of monkeys, the marmosets and tamarins, found in Central and South America are called cooperative breeders because they live in small groups with only one pair (typically) reproducing while others in the group (often older siblings) do not reproduce but help with rearing infants.

Studies both in the field and in captivity have shown that fathers and other helpers play an important role in infant care and are essential for infant survival. For example, in the cotton-top tamarin, an endangered species found only in northern Colombia, a family size of five is essential for the survival of all infants born. That is, infant survival increases with increased group size up to five. Amazingly, we found the same results in captivity where the monkeys have all the food they can eat, do not have to travel far to find food, and have no predators. Why are fathers and other helpers so important for infant survival?

All marmosets and tamarins usually give birth to twins. The twins weigh about 20% of their mother's weight at birth (the equivalent of a 130 lb—59 kg—woman giving birth to twins each weighing 13 lbs—6 kg). Furthermore, when food resources are good, mothers become pregnant within a month of giving birth. Thus mothers are nursing large infants and are pregnant at the same time. These monkeys do not build nests or have permanent shelters, so the heavy infants must be carried around all day as the family forages for food while climbing up and down trees. The mother cannot do it all!

A cotton-top tamarin father carrying an infant.
Courtesy of Carla Y. Boe.

Fathers and other helpers carry infants on their backs until the infants are independent (about 8 weeks), and they share food with infants at the time of weaning. In captivity, fathers lose up to 10% of their weight during the first 8 weeks of infant care, but with other helpers around to share infant care, fathers (and helpers) lose only 1–2% of body weight during infant care. In addition to carrying infants, group members must be vigilant toward predators and find food for the whole group. Cooperation is an essential strategy for raising healthy infants.

But cooperative breeding creates problems for the parents. First, since infant survival depends upon having a male willing to take care of infants, mothers must make good choices of mates before breeding. But fathers are at risk because they lose weight and are more vulnerable to predators while carrying infants, so they too must make a careful choice of mates. A father must be sure his mate is faithful so he does not spend his energy caring for another male's infants. Finally, both parents must recruit and maintain helpers while preventing them from reproducing.

Marmosets and tamarins have several ways to form and maintain close pair bonds. When pairs first meet, they copulate at high rates (up to 14 times a day) and spend a lot of time sitting close and grooming each other. In wild common marmosets, nearly 14% of the day is spent in grooming, with the breeding pair doing most of the grooming. Interestingly, reproductive males groom reproductive females much more than vice versa, and reproductive females groom subordinates much more than the reverse. Grooming produces the release of natural opiates, called endorphins, in the brain, providing a feeling of pleasure for the one being groomed.

Mates show emotional responses to each other when separated briefly, seen both in increased rates of calling to each other while separated, and by increased rates of sex and grooming when they are reunited. Other challenges to the relationship—the odor of a novel female or the loss of infants—lead to increased grooming and closeness between the pair.

Using noninvasive methods to measure hormones in urine, we know that subordinate females never ovulate while they are living with a dominant female, but within 8 days of being paired with a novel male, the same female can be pregnant. Scents from the reproductive female can inhibit ovulation (*social contraception*), but even the sight and smell of a novel male is enough to induce a female to ovulate. Measurements of the stress hormone, cortisol, show very low levels in helpers, suggesting that stress is not involved in inhibiting reproduction. Indeed we rarely see much conflict or fighting between parents and offspring. The helpers benefit in two ways: They are often related to infants and thus help the survival of their kin. However, monkeys also need to learn parental care skills, so helping care for someone else's infants prepares monkeys to be better parents themselves.

Fathers display some of the same hormones as nursing mothers. Prolactin, which is necessary for milk production and, in rodents, is critical for good maternal care, is seen in marmoset and tamarin fathers both during the time of infant care, but also even 6–8 weeks before the infants are born. The father begins to take care of infants from the moment of birth, so being prepared ahead of time makes sense, but fathers do not experience the hormonal changes that pregnant mothers do. We suspect that a combination of changes in female odors and increased closeness during pregnancy lead fathers to the hormonal anticipation of infant births.

Marmosets and tamarins coordinate behavior closely and have a wide range of vocal, visual, and chemical signals to coordinate behavior. They are the only primates, other than great apes, to show teaching behavior toward young (using food sharing to help infants learn about novel or rare foods and avoid noxious foods) and to exhibit true imitation. Cooperative infant care leads to cooperation in other behavior as well.

Humans are also cooperative breeders, according to anthropologist Sarah B. Hrdy, but for different reasons. Human infants are relatively small, with much of the growth in brain size and structure occurring after birth. Furthermore, human children generally are not independent of their families until they are at least 15. Human mothers need helpers, whether from fathers, aunts, friends, daycare providers, or older siblings, to provide economic security and adequate care to rear a child successfully.

Several lessons emerge from studies of cooperatively breeding primates that are relevant to human families:

1. It is important to choose a mate carefully. Trust and commitment are important.
2. Maintaining a relationship takes effort, and both partners must work to provide rewards to each other and provide reassurance at times when the relationship is strained.
3. Humans probably need to learn parenting skills at least as much as monkeys.
4. It is important to delay having children until parents have adequate economic and social resources to support their children.
5. Developing good communication skills is essential to coordinate infant care, and to minimize and resolve conflict and stress within families.

See also Lemurs—*Behavioral Ecology of Lemurs*
            Social Organization—*Socioecology of Marmots*

Further Resources

Hrdy, S. B. 1999. *Mother Nature*, New York: Ballantine.

Snowdon, C. T. 1996. *Infant care in cooperatively breeding species*. In *Parental Care: Evolution, Mechanisms, and Adaptive Significance* (Ed. by J. S. Rosenblatt & C. T. Snowdon), pp. 643–689, San Diego: Academic Press.

Snowdon, C. T. 2001. *Social processes in communication and cognition in Callitrichid monkeys: A review*. Animal Cognition, 4, 247–257.

Solomon, N. G. & French, J. A.1998. *Cooperative Breeding in Mammals*, New York: Cambridge University Press.

Charles T. Snowdon

■ Social Organization
Pika Behavior Studies

Pikas are small relatives of rabbits and hares. Two of the 12 living species live in the United States, Canada, and Alaska where they inhabit talus slopes in the mountains. They are restricted to higher elevations and mountain tops because they cannot tolerate warm temperatures. In fact, they are very vulnerable to global warming and may serve as an important indicator of increases in temperature. They are great subjects for studies of animal behavior because they are territorial and active during the day.

Pikas collect vegetation in the summer and bring it back to haypiles that they establish in their territories. With individual territories, haypiles worth guarding, and a finite amount of open talus, pikas have to defend their territories from other pikas. Young must establish their own territories, whether squeezing in between adults or taking over a territory from an animal that has died. They can be quite aggressive toward each other, and use calls and scent marks—they have a cheek gland located just below the jawbone—to defend their territories; sometimes they resort to chases and fights. Pikas will steal hay from each other's hay piles!

I was interested in understanding where and how often they scent marked, and how this behavior might play a role in territorial defense. I also was curious how animals that defended their territories so staunchly could ever let their guard down enough to find mates and breed. The first thing I had to do was mark each animal in the field and then watch them individually. This is called *focal animal sampling*. Furthermore, I needed to create a grid on the landscape so that I could mark where each animal was during observations. I live trapped all the pikas on my study site and put ear tags on them in the form of colored plastic disks. Thus "RY" was the animal that had a red ear tag and a yellow ear tag, and "BO" had a blue and an orange ear tag. During an observation period, I recorded a dot for my subject animal's location every minute on the minute. And I recorded an "x" on my data sheet for the location of every cheek mark.

A pika stops for a brief moment with a mouthful of hay on her way back to her haypile. The haypile serves as a source of food.
Photo by William Ervin.

With this approach, I learned that pikas scent mark more frequently during the breeding season, especially the males. Both males and females place most of their scent marks in the center or core of their territory and in the area of overlap with their neighbor. I also found that they eliminate urine and feces on rocks right by their haypiles. The urine forms a white "pinnacle" on the rock, and an orange nitrogen-loving lichen grows on these urine pinnacle rocks. This serves to mark and defend their haypiles, which are typically located at the center of their territory.

I also brought pikas into a laboratory setting to find out whether females were more likely to mate with males whose odors were familiar. I rubbed a clean rock across a male's cheek gland and put the rock in the female's enclosure, repeating this three times during one week. At the end of the week I created a connection between the female's enclosure and the male's whose odors were familiar to her. I also connected her enclosure with a male's whose odors were not familiar, and alternated which male she was connected with first. I did find that the two pikas had more amicable behaviors, fewer fights, and more copulations when the female was connected with the enclosure of a male whose odors were familiar compared to a male whose odors were not familiar. The results of this study suggested to me that pikas use scent marks to defend their territories and facilitate breeding behavior, and that they use urine and feces to mark their haypiles.

See also Social Organization—*Social Dynamics in the Plateau Pika*

Further Resources

Beever, E. A., Brussard, P. F. & Berger, J. 2003. *Patterns of apparent extirpation among isolated populations of pikas* (Ochotona princeps) *in the Great Basin*. Journal of Mammalogy, 84 (1), 37–54.

Fitzgerald, J. P., Meaney, C. A. & Armstrong, D. M. 1994. *Mammals of Colorado*. Boulder: University Press of Colorado and Denver Museum of Natural History.

Meaney, C. A. 1987. *Cheek-gland odors in pikas* (Ochotona princeps): *Discrimination of individual and sex differences*. Journal of Mammalogy, 68(2), 391–395.

http://ens-news.com/ens/feb2003/2003-02-25-06.asp
    The negative effects of global warming on pika populations.

Carron A. Meaney

■ |Social Organization
|Reconciliation

Social conflict is an unavoidable part of life for animals living in stable social groups. Animals have behavioral mechanisms to prevent social conflict from escalating into aggression, but sometimes such escalation occurs. Aggression has negative consequences, including bodily harm, stress, and the potential disruption of beneficial social relationships. Aggression has the potential to force animals to leave their social group, and thereby reduce the benefits of group living for all concerned. Consequently, social animals should be under strong selective pressure to reduce social conflict and mitigate the negative consequences of aggression. One technique animals use to repair the damage caused by aggression is reconciliation.

Reconciliation is defined as affiliative (i.e., friendly) contact between former opponents shortly after a fight. Reconciliation is not something that can be observed; rather it is inferred when affiliative behavior occurs more quickly after a fight than would otherwise be expected. Reconciliation is a functional rather than a descriptive term, and some scientists prefer the term *friendly reunion* until the consequences of such reunions are established and the term reconciliation justified. In this essay I will use the term reconciliation, and will only use the term friendly reunion to emphasize when the function is still unclear.

A female bonnet macaque gives a subadult male an open-mouth threat. The subadult male holds his ground and returns the threat.
Courtesy of Matthew Cooper.

Reconciliation was first described in a captive colony of chimpanzees in 1979 by Frans de Waal and Angeline van Roosmalen. They noticed that a few minutes after a fight chimpanzees sometimes approached, embraced, and kissed their former opponents. While reconciliation has been studied in a variety of species since 1979, the majority of research has been done with nonhuman primates. The type of behavior used to reconcile depends on what constitutes friendly behavior for that species. Some species use specific types of behavior to reconcile, whereas other species lack specific behavior for reconciliation. Macaques and baboons, which are medium-sized terrestrial monkeys from Africa and Asia, reconcile using a variety of friendly behavior including social grooming, mounting, or briefly sitting in contact. In baboons, reconciliation is often accompanied by friendly vocalizations such as grunts. In spotted hyenas postconflict friendly reunions include licking and sniffing while making affiliative vocalizations. Dolphins engage in postconflict friendly reunions by rubbing against the former opponent's flipper and swimming in contact. In patas monkeys, friendly reunions are more subtle, and simply reestablishing proximity to a former opponent after a fight may be enough to reconcile.

The rate of reconciliation depends on the type of species. Chimpanzees reconcile between 18–47% of their fights. In some species of macaques, as few as 7% of fights are reconciled, whereas in other macaque species, 53% of fights are reconciled. Researchers have reported that olive baboons reconcile 16% of their fights, and gelada baboons reconcile between 30–45% of their fights. Variation in the rate of reconciliation is also observed in nonprimates. Domestic goats reconcile 17% of their fights, and spotted hyenas reconcile 15% of their fights. In some species, researchers have failed to find reconciliation. For example, there is no evidence of reconciliation in ringtailed lemurs, which are prosimians from Madagascar, or in red-bellied tamarins, which are small arboreal monkeys from South America. While reconciliation is not expected to occur in every animal species, it is likely to occur in species that experience aggression and live in stable social groups where animals recognize each other individually.

Different groups of the same species also differ in the rate of reconciliation. Researchers have found that factors such as the intensity of aggression, the context of aggression, the decidedness of its outcome, the mating season, and the distance between opponents after aggression can influence the likelihood of reconciliation. Other studies have found that fights are rarely reconciled when they occur during feeding. After a fight over food, the winner is often busy eating the contested food item, and the loser may not wish to approach the winner in this high-risk situation. Alternatively, fights over food might not disrupt the social relationship between opponents.

Factors that most consistently influence the rate of reconciliation are related to the social relationship between opponents. In some macaque species, closely related females maintain strong social bonds throughout their entire lives, resulting in kin-biased patterns of affiliation and support during aggression. In these species, kin reconcile more often than do non-kin. In societies that are less kin oriented, reconciliation tends to occur equally among kin and non-kin. The amount of reconciliation is also related to the amount of friendly behavior that animals normally exchange. For example, unrelated macaque females that groom one another frequently may reconcile as frequently as do kin. In chimpanzees, males remain in the same social group and form alliances whereas females transfer to new groups at puberty. Not surprisingly, male chimpanzees reconcile their fights more often than do females. Likewise, in gorillas the strongest social relationships exist between females and the silverback male, and reconciliation is most common between these dyads.

A female bonnet macaque grooms another individual while other macaques huddle next to them. Social grooming is a very common form of reconciliation among macaques.
Courtesy of Matthew Cooper.

As discussed earlier, reconciliation does not occur after every fight, suggesting that approaching a former opponent shortly after a fight can be costly. In many primate species, recipients of aggression are at increased risk of renewed aggression from their former opponents shortly after the initial conflict. Likewise, third parties are more likely to attack recipients of aggression shortly after the original conflict than they are during control periods. Moreover, chimpanzees have been observed to invite a former opponent to approach shortly after a fight only to attack the individual at the last moment. Thus, the possibility of renewed aggression can keep former opponents separated, and scientists expect reconciliation to occur only when the benefits outweigh the costs.

So why should opponents reconcile? We can discuss the benefits of reconciliation at two levels, the ultimate and the proximate level. *Ultimate level explanations* concern the long-term evolutionary benefits and functions of behavior. Ever since reconciliation was first described in chimpanzees, researchers thought that reconciliation might function to repair relationships damaged by aggression. Since that time researchers have found evidence supporting this idea. For instance, in a variety of primate species reconciliation reduces the probability of renewed aggression. Recently, Sofia Wahaj and colleagues found in their 2001 study on spotted hyenas that friendly reunions are associated with reduced rates of aggression between former opponents. Also, reconciliation has been shown to restore tolerance between the aggressor and the victim. In a 1992 study by Marina Cords on pairs of long-tailed macaques, the aggressor was less likely to exclude the former victim from a preferred resource, and the victim was less hesitant to approach and feed in proximity to the former aggressor after reconciliation.

In her 1996 study on baboons, Joan Silk and colleagues suggested that postconflict friendly reunions signal the end of hostilities rather than an attempt to repair relationships damaged by aggression. In practice, these functions may be indistinguishable, insofar as signaling the termination of hostilities allows for future affiliative behavior and thus restores interactions to baseline levels. At issue is how we might determine that aggression damages social relationships. In a 2001 study on Japanese monkeys, Nicola Koyama demonstrated that failure to reconcile fights has long-term negative consequences. She found that aggression levels were higher, and grooming, proximity, and approach levels were lower compared to baseline levels during the 10 days following nonreconciled fights, whereas aggression, grooming, proximity, and approach levels were restored to baseline following reconciled fights. This study supports the conclusion that aggression disrupts social relationships, and reconciliation restores them.

Social relationships can be viewed as investments, and by extension, social partners have a certain value to others with whom they interact. Animals may be valuable to one another for a variety of reasons. For example, animals may frequently groom one another, they may aid one another during fights with other group members, they may help protect each other from external threats, they may be sexual partners, or they may be kin. If aggression damages social relationships, then animals should seek to repair this damage. Animals should be particularly motivated to repair relationships with valuable partners because they risk losing the benefits associated with that relationship. Numerous studies on macaques have demonstrated that dyads with good social relationships reconcile more than do dyads with bad social relationships. These studies tend to use a composite measure of high affiliation and low aggression to define "good" or "friendly" relationships. Experimental evidence also supports this assertion. In a 1993, study by Marina Cords and Sylvie Thurnheer, pairs of longtail macaques were trained on a task in which they needed to cooperate to obtain food. The researchers found that reconciliation increased three-fold after the pair learned to cooperate. Thus, these monkeys modified their level of reconciliation in response to changes in their partner's value.

A male liontail macaque mounts another male. Among male macaques mounting is used as a form of greeting and as a means to reconcile after a fight.
Courtesy of Matthew Cooper.

Variation in the rate of reconciliation within different groups of the same species is also related to relationship value. In 1996, Castles and coworkers studied two groups of pigtailed macaques and found that reconciliation was more common in the group in which individuals had a few strong affiliative bonds than in the group with numerous weak affiliative bonds. The researchers concluded that in a society in which some individuals have strong affiliative relationships, reconciliation occurs frequently because each of these relationships is valuable. Species differences in the rate of reconciliation have also been attributed to the value of social relationships. In some macaque species, such as rhesus, Japanese, and longtail monkeys, animals reconcile at low rates and reconciliation occurs more often among kin than non-kin. In other macaque species such as stumptail,

Barbary, and Tonkean monkeys, reconciliation is common and occurs equally among kin and non-kin. It has been suggested that in these latter species, cooperation is important in group defense against predators, neighboring groups, or infanticidal males, whereas it is less important for the former species. The importance of cooperation would have the effect of increasing the value of the average group member. While intriguing, this explanation remains speculative and requires additional research.

Proximate level explanations concern the immediate cause of behavior. One possible reason animals reconcile is to reduce their postconflict anxiety. Displacement activities such as scratching, yawning, and body shaking have been used as a noninvasive measure of anxiety-like emotion in animals. In macaques, baboons, chimpanzees, and domestic goats, displacement activities are often elevated following aggression, and are reduced by reconciliation. Similarly, heart rate increases following aggression, is reduced by postconflict affiliation, and this reduction is most pronounced when affiliation occurs between former opponents. Filippo Aureli and colleagues theorize that the risk of renewed aggression causes postconflict anxiety, and that reconciliation reduces anxiety by alleviating an animal's uncertainty about the future behavior of its opponent. Further support for the uncertainty-reducing function of reconciliation comes from playback experiments in baboons. In a series of studies, Dorothy Cheney and colleagues have shown that baboons use grunts when they reconcile, that grunts reduce the uncertainty experienced by former recipients of aggression, and that reconciliation is more likely to occur when grunts are played back to former opponents than when they are not. Initial studies also suggest that individual variation in postconflict anxiety could account for variation in the level of reconciliation. For instance, in long-tailed and Japanese macaques, dyads with strong affiliative bonds show high levels of scratching after fights. These dyads frequently reconcile their fights, and reconciliation reduces their scratching.

Anthropological studies show great cultural variation in the types of techniques humans use to manage social conflict, and some similarities with other animals' reconciliation exist. During childhood and adolescence, conflict management is closely related to friendship relations. Children manage social conflict differently with friends and nonfriends. For instance, conflicts are usually less intense among friends than among nonfriends. Also, children use more constructive types of conflict resolution with friends than with nonfriends, such as disengagement and negotiation as opposed to standing firm and subordination. Among adults, conflict resolution is often mediated by third parties, and friendships continue to play an important role in the process. Specifically, conflicts are most likely resolved when social relationships are important, and new relationships are difficult to establish.

In conclusion, among nonhuman primates, reconciliation appears to repair social relationships damaged by aggression. Whereas the function of postconflict friendly reunions is less well researched in other animals, initial studies on domestic goats and spotted hyenas suggest a similar function for reconciliation. Animals tend to reconcile more often with those with whom they have strong social bonds than they do with others. This is consistent with the notion that humans and other animals are particularly motivated to reconcile with valuable social partners. Furthermore, initial studies on primates suggest that the anxiety and uncertainty following conflict causes animals to reconcile. In animals we see evidence for an evolved predisposition to resolve conflict and the biological precursors for our own style of conflict resolution.

See also *Cooperation*
       Empathy
       Social Organization—*Monkey Families and Human*
         Families

Further Resources

Aureli, F. & de Waal, F. B. M. 2000. *Natural Conflict Resolution*. Berkeley, CA: University of California Press.

Aureli, F., Cords, M. & van Schaik, C. P. 2002. *Conflict resolution following aggression in gregarious animals: A predictive framework*. Animal Behaviour, 64, 325–343.

Castles, D. L., Aureli, F. & de Waal, F. B. M. 1996. *Variation in conciliatory tendency and relationship quality across groups of pigtail macaques*. Animal Behaviour, 52, 389–403.

Cheney, D. L., Seyfarth, R. M. & Silk, J. B. 1995. *The role of grunts in reconciling opponents and facilitating interactions among adult female baboons*. Animal Behaviour, 50, 249–257.

Cords, M. 1992. *Post-conflict reunions and reconciliation in long-tailed macaques*. Animal Behaviour, 44, 57–61.

Cords, M. & Killen, M. 1998. *Conflict resolution in human and nonhuman primates*. In: *Piaget, Evolution, and Development* (Ed. by J. Langer & M. Killen), pp. 193–218. Mahwah, NJ: Lawrence Erlbaum Associates.

Cords, M. & Thurnheer, S. 1993. *Reconciliation with valuable partners by long-tailed macaques*. Ethology, 93, 315–325.

Koyama, N. F. 2001. *The long-term effects of reconciliation in Japanese macaques* (Macaca fuscata). Ethology, 107, 975–987.

Schino, G. 1998. *Reconciliation in domestic goats*. Behaviour, 135, 343–356.

Silk, J. B., Cheney, D. L. & Seyfarth, R. M. 1996. *The form and function of post-conflict interactions between female baboons*. Animal Behaviour, 52, 259–268.

de Waal, F. B. M. 1989. *Peacemaking among Primates*. Cambridge, MA: Cambridge University Press.

de Waal, F. B. M. & van Roosmalen, A. 1979. *Reconciliation and consolation among chimpanzees*. Behavioral Ecology and Sociobiology, 5, 55–66.

Wahaj, S., Guse, K. R. & Holekamp, K. E. 2001. *Reconciliation in the spotted hyena* (Crocuta crocuta). Ethology, 107, 1057–1074.

Matthew A. Cooper

■ Social Organization
Shoaling Behavior in Fish

Fish gain numerous reproductive fitness benefits from joining social groups known as shoals. This essay will focus on these benefits and analyze the complex decision-making processes involved in a fish's choice of shoal mates.

Terminology

Unique names are given to groups of specific animals. For instance, a group of whales is known as a pod, a group of lions is a pride, and a group of geese is a gaggle (see the figure on p. 1013 for more).

People often refer to a group of fish as a *school* although this term is not precisely correct. A *school* is a group of fish that swim together in a synchronized and polarized fashion. Schooling fish move in the same direction, at the same speed, and turn simultaneously. In addition, schools typically display a consistent *nearest neighbor distance (NND)*, in which each individual fish maintains a constant distance between itself and adjacent fish. To be called a school, a group of fish must demonstrate all of these rather complex behavioral patterns.

The word *shoal*, on the other hand, is the proper term for any simple social grouping of fish. Fish in shoals can be randomly oriented, exhibit variable NND, and move freely without regard to the movements of the other fish. In short, any group of fish is a shoal, but not all shoals demonstrate the properties of a school.

Some Terms for Groups of Animals

An **Ambush** of Tigers	A **Crash** of Rhinoceros	A **Prickle** of Hedgehogs
A **Bale** of Turtles	A **Family** of Otters	A **Murder** of Crows
A **Bevy** of Quail	A **Grist** of Bees	A **Romp** of Otters
A **Bloat** of Hippos	A **Herd** of Elephants	A **Shoal** of Fish
A **Clouder** of Cats	A **Knot** of Toads	A **Shrewdness** of Apes
A **Congress** of Baboons	A **Muster** of Peacocks	A **Team** of Horses
A **Conspiracy** of Ravens	A **Pod** of Whales	A **Warren** of Rabbits

Benefits Associated with Shoaling

Shoaling behavior offers fish two critical benefits that improve fitness: increased success in finding food and increased protection from predators.

Food Acquisition

A constant concern for fish, as with all species, is the need to locate sources of food. Since food sources may be patchy, searching for food can be energetically costly and potentially risky. It has been shown that fish in shoals find food faster and spend more time actually feeding. This benefit probably stems from the simple fact that a large group of fish has a greater chance of finding food than does a single individual. For this reason the size of the shoal matters, with larger shoals gaining more benefit than smaller shoals. Once food is found, information about the food source is transferred quickly among the members of the shoal. This may be one of the reasons why fish are more likely to join large shoals than small shoals.

Protection from Predators

Fish in shoals are at less risk of predation than fish that are swimming alone. Therefore, it is not surprising that individual fish often join shoals when threatened by a predator. Once in the shoal each fish benefits from the *numerical dilution effect*, which suggests that simply being part of a group reduces the chances of being killed. For instance, in a shoal of 100 fish, each individual has only a 1 in 100 chance of being killed during an attack. This benefit, as with foraging success, increases in value as the size of the shoal increases. This may be yet another reason why fish will choose larger shoals when given the choice.

Once an individual fish has chosen a shoal, its position within the shoal may have strong effects on the benefits it incurs from shoaling. Although fish at the periphery of a shoal may experience enhanced feeding opportunities since they will be the first to encounter new food sources, they are also at the most risk for being killed by a predator. This situation leads to a phenomenon known

Blackbar soldierfish shoaling in the Caribbean.
© *Brandon Cole / Visuals Unlimited.*

as the *selfish herd theory*, which predicts that individuals within any social group will move toward the center of the group when threatened by a predator.

Perhaps the most important antipredator benefit associated with shoaling is the *confusion effect*. This phenomenon is associated with the high degree of similarity between members of a shoal. In theory, the confusion effect suggests that a predator, confronted with a large group of similar-looking individuals, experiences perceptual confusion, which slows the attack. This may occur because the predator has difficulty settling on a single individual to attack. Or, similarly, the visual image of many fish of similar shape, size, and coloration may make it difficult for the predator to identify and target a single individual fish. In either case, the confusion effect relies heavily on conformity within a shoal. In a related phenomenon, known as the *oddity effect*, an individual fish that does not resemble the other members of the shoal will easily be targeted by a predator. For example, imagine a bright red fish shoaling with a group of black fish. The attention of an approaching predator will be drawn to this red fish because it stands out in stark contrast to the rest of the group.

Other antipredator benefits related to shoaling involve the acquisition of information by fish within a shoal and the transmission of information between shoal mates. Information acquisition may benefit the surviving members of a shoal that have experienced an attack. These fish may learn about the attack strategies of particular predators, they may learn to recognize unfamiliar predators, and they may learn evasive maneuvers from other members of the shoal. Therefore, joining a shoal, especially a large shoal, and surviving attacks may give an individual fish a selective advantage during subsequent encounters with predators.

Transmission of information between shoal mates may also occur during an encounter with a predator. Since it is typical for a shoal under attack to become more compact, the resulting proximity of the shoal mates allows for rapid communication. Pressure waves between individuals (which are detected through the inner ear and lateral lines along the sides of the fish) may help each member of the shoal adjust its swimming speed and direction to avoid the predator. In some cases, the shoal may actually demonstrate coordinated evasive maneuvers such as the *fountain effect*, where the shoal splits in two, moves around the predator, then reforms, or *flash expansion*, where the entire group suddenly breaks up with all individuals fleeing in different directions.

Finally, when assessing the benefits and costs involved in shoaling, it has been suggested that the antipredator benefits outweigh the competition that occurs within a shoal. Although food acquisition may be enhanced, there will certainly be greater competition for the food resources. However, the benefits associated with the confusion effect and oddity effect may outweigh the costs of competition. Simply put, it costs less to miss a meal than to be a meal. For this reason it is thought that the reduced risk of predation may be the primary evolutionary reason involved in the development of shoaling behavior in fish.

Factors Affecting Shoal Mate Choice

Since fish may change shoals on a regular basis, individuals are often faced with the decision of which shoal to join. Joining the "correct" shoal is critical, especially with respect to the confusion effect, which depends on an individual's ability to "blend in" with its shoal mates. Individual fish must, therefore, assess the looks and behavior of the fish within shoals, and only join those shoals in which they will fit in. Joining the wrong shoal, where an individual looks, or acts, differently from its shoal mates can draw the attention of predators, with disastrous consequences.

There is a large body of research dedicated to analyzing the factors that affect the choice of shoals. Factors that have been shown to be important include the species composition of the shoal, the number of individuals within the shoal, the physical appearance of the individuals within the shoal and the familiarity of the fish within the shoal.

Species Composition

Fish typically prefer to shoal with *conspecifics*, that is, fish of the same species. This is probably a choice that leads to success for a number of reasons. First, individuals of the same species are likely to be similar in appearance, thus conspecific shoals may benefit from the confusion effect and avoid problems associated with the oddity effect. Second, individuals of the same species will likely have similar dietary preferences, and joining a shoal of conspecifics may lead to enhanced foraging benefits. And, third, joining a shoal of conspecifics may enhance an individual's chances of meeting potential mates.

Physical Appearance

Since the confusion effect depends on morphological homogeneity within the shoal (that is, all members of the shoal being of similar appearance), individual fish must constantly assess their own appearance and find shoal mates that they resemble. Factors that have been shown to be important in this choice include body coloration, body patterns, body shape, and body size. In each case, fish will choose to shoal with groups of fish that most closely resemble themselves in appearance.

The exact mechanisms by which fish choose shoal mates of similar appearance are unknown. Fish might carry some type of innate, genetically programmed image that causes them to be attracted to fish with similar features. However, such a system would need to be complex enough to take into account changes in appearance that occur throughout the life of the fish. For instance, as a fish grows, its choice of shoal mates needs to change so that it continues to join shoals with fish of the correct length. Numerous studies have shown that this is true, and that body length is an important factor in shoal mate choice. In addition, it appears that even more subtle changes might be taken into account by fish joining shoals. For example, killifish infected with a particular parasite develop black spots on their sides. Uninfected fish appear to take this minor alteration in appearance into account, choosing to shoal with other uninfected fish. Is it possible, therefore, that a fish actually looks at itself, then seeks shoal mates of similar appearance? This question is currently being studied.

Familiarity

Finally, familiarity plays a critical role in a fish's choice of shoal mates. Fish have been shown to be capable of recognizing specific individuals, preferring to shoal with fish they know. The exact mechanisms mediating this phenomenon are not known. However, it has been shown that both olfactory and visual cues can be used to discriminate between familiar and nonfamiliar individuals. Shoaling with familiar fish may reduce the costs of competitive interactions by reducing aggressive encounters between shoal mates. Additionally, familiarity may enable fish to develop cooperative associations associated with predator avoidance and foraging.

The Study of Shoaling Behavior

Studying shoaling behavior in the laboratory is relatively easy. Test tanks can be constructed by dividing an aquarium into thirds by the addition of two panes of glass (see the figure on p. 1016). "Target" shoals of fish can be placed into the side compartments and an individual test fish added to the central compartment. During an experiment, the time the test fish spends near each of the side compartments is measured. Increased time near one

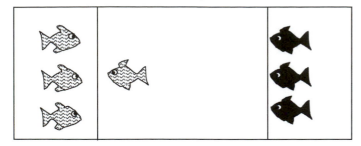

A diagram of a tank used to test shoaling behavior.
Courtesy of Scott P. McRobert.

compartment indicates the test fish's preference for the shoal in that compartment. In this way simple comparisons can be made between shoals containing fish of different species, different sizes, different colors or body patterns, and so on.

Despite the large amount of research that has been completed, many questions associated with shoaling still remain. As noted above, questions remain regarding the exact mechanisms by which fish choose shoal mates. In addition, the vast majority of our information on shoaling has been gathered from laboratory studies. Relatively little is known about the biology of fish shoals under natural conditions. Information regarding the structure of wild fish shoals is needed, as are studies aimed at determining the true advantages of shoaling under natural conditions.

Summary

For individual fish, joining a shoal offers fitness benefits that include increased foraging success and reduced risk of predation. Fish have been shown to select shoal mates based on a number of criteria, including body size, body pattern, and body coloration. By choosing shoal mates with similar features, fish gain advantages associated with the confusion effect and avoid problems associated with the oddity effect. Fish also select shoal mates with whom they are familiar, potentially reducing competitive encounters and benefiting from cooperative behavior. The study of shoaling behavior is an active field of research, with scientists searching for the exact mechanisms underlying the formation of shoals and looking to develop tools to help understand shoaling behavior in the wild.

Further Resources

Godin, J.-G. J. 1997. *Behavioural Ecology of Teleost Fishes*. Oxford: Oxford University Press.

Krause, J., Butlin, R. K., Peuhkuri, N., & Pritchard, V. L. 2000. *The social organization of fish shoals: A test of the predictive power of laboratory experiments for the field*. Biological Review, 75, 477–501.

Pitcher, T. J. 1986. *The Behavior of Teleost Fishes*. Baltimore, MD: The Johns Hopkins University Press.

Scott P. McRobert

■ Social Organization
Social Dynamics in the Plateau Pika

Natural selection favors individuals that survive and reproduce well, and our understanding of the evolution of behavior is enhanced by our ability to discern how particular behavioral patterns allow animals to best achieve these goals. The plateau pika (*Ochotona curzoniae*) is a small (160 gm; 5.6 oz) lagomorph (a rabbit relative) that occupies the vast high-altitude alpine grasslands of the Tibetan plateau. This egg-shaped mammal expresses a variety of intriguing behaviors, especially social behaviors, that enhance its survival and reproductive performance in this rigorous environment.

Plateau pikas live in family groups that defend territories. Each territory (approximately 25 m or 82 ft across) consists of the family's elaborate burrow system and is surrounded by the territories of other families on the open meadow. Throughout summer most family members confine their activities to exclusively within their territory, rarely trespassing onto a neighboring territory. Families are composed of breeding adults and their young produced during the short summer season. Because all adult females breed and produce at least three large litters at 3-week intervals, population density in territories explodes throughout the summer. Cohesive social interactions are common among all family members—up to one social interaction per minute may be observed—making the plateau pika one of the most social of all mammals. They sit in contact, box, wrestle, rub noses, crawl over and groom each other, and follow each other around like a train. These sociable plateau pikas behave more like primates than their lagomorph relatives, most of which are notoriously asocial.

Three plateau pikas keep watch near their burrow.
Courtesy of Andrew T. Smith.

The plateau pika has one of the most varied vocal repertoires of any pika. Males emit a long-call (or song) during the mating season, while all genders and ages produce whines, trills, muffle calls, and alarm calls. The most frequent callers are the youngest pikas, who utter whines and trills that attract older siblings and adults to rush over and interact with them.

Nearly all of these cohesive social behaviors and vocalizations are exchanged among family members (in any combination of age or gender)—rather than with equally accessible pikas living in adjoining territories. It is interesting that adult males are involved in friendly interactions with youngsters far more frequently than are adult females—it is common to observe a line of young pikas following an adult male around the territory. Plateau pikas can be aggressive—normally expressed as long vigorous chases covering several territories or in violent fights with fur flying. Aggressive behavior usually occurs between adult males from different family territories.

A lone plateau pika calls for others.
Courtesy of Andrew T. Smith.

As the summer progresses and successive litters of young pikas are born, density builds within each family territory, and the regional population of pikas may approach up to 300 animals per hectare. The structure of family territories remains intact throughout the fall and winter, although the rigors of the environment and predation erode the population considerably. Few adults live to breed a second time, and most young fail to survive to the next summer.

The population reaches its lowest level in early spring, just before the breeding season commences. At this time, some pikas disperse. *Dispersal* is a common phenomenon in all species—an animal leaving its home to travel to another place to live and breed. The inverse of dispersal is *philopatry*—an individual remaining to breed at the site of its own birth. In most small mammals, juvenile males disperse, and females remain philopatric. In plateau pikas, however, about 60% of juveniles remain philopatric on their family territory while the others disperse; adult dispersal is rare. Both males and females disperse, although dispersal is more common among males. Most dispersers only move to a territory next door, and females rarely move more than two territories from their home. Uncommonly, a male juvenile may disperse as far as three to five territories away from his birthplace. This dispersal shuffle establishes the composition of adults on family territories that will remain unchanged during the upcoming summer breeding period.

Behavioral ecologists have determined that animals may disperse for three reasons: to avoid competition for mates; to avoid competition for environmental resources (most likely food); or to avoid inbreeding depression. Genetic data are not available to determine whether the restricted dispersal (in terms of numbers of dispersers and the short distances traveled by plateau pikas) is sufficient to overcome any potential harm that might be caused by inbreeding. Opportunity for inbreeding exists in some families with philopatric pikas of opposite gender; but normally males or females that disperse have no opposite gender relatives in their new families. Dispersal appears to result in an equalization of density among family territories, thus it is highly suggestive that pikas disperse to avoid competition for food resources.

The final dispersal hypothesis concerns mating opportunities. Most small mammals are polygynous (one male mates with multiple females), and males disperse to avoid competition with their father or other males and to look for new breeding opportunities of their own. Polygyny is the most commonly expressed mating system in mammals because only mammal mothers are physiologically equipped to nurture young (during pregnancy and via provision of milk following birth). Males are thus free to consort with multiple females to spread their genes as widely as possible. However, there are other possible mating system types, namely monogamy (one male paired with a single female) and the rarely observed polyandry (multiple males mating with a single female).

Plateau pikas are unusual among mammals in that any family territory can contain any combination of adults, thus polygynous, monogamous, and polyandrous territories can reside next to one another (some family territories are comprised of multiple males and multiple females—usually two of each—in what can be called a complex family). The percentage of territories of each mating system type varies among years and is determined by the randomness of the number of males and females that happen to survive the preceding winter, coupled with the dispersal shuffle that results in the final composition of adults on a territory prior to the breeding season. Territories rarely contain the same

An adult plateau pika and her two young sitting in contact.
Courtesy of Andrew T. Smith.

mating system type in consecutive years, strong evidence that local environmental features do not play a role in the determination of mating system type.

If plateau pikas behaved like most small mammals, males would disperse to family territories with the most females leading to a primarily polygynous mating system. However, this is not the case. Males do not disperse to gain advantage in competition for mates; they tend to move into families with significantly fewer females. In addition, dispersing females move into families with significantly more males. This strategy appears to pay off for adult females, as more young survive to the next year when there are more males in a family. One explanation for this pattern is that these adult males are real dads; the cohesive behavior engaged in by adult males apparently provides meaningful parental care of young following weaning.

Huddling together in burrows to conserve heat appears to be the most likely survival strategy pikas use to overcome the rigors of the harsh Tibetan winters. Social cohesiveness within families (development of a varied vocal repertoire and frequently expressed social behaviors), coupled with the lack of dispersal during most of the year, ensures that related pikas will be available to huddle together. Restricted dispersal also yields flexibility in formation of low-density mating system groups, thus ensuring food resources are not overtaxed and that all adult males, important providers of parental care, are incorporated into the breeding structure. Behavioral tactics employed by plateau pikas are essential for their survival and successful reproduction in the harsh alpine environment of the Tibetan plateau.

See also Communication—Vocal—*Alarm Calls*
Social Organization—*Pika Behavior Studies*
Social Organization—*Socioecology of Marmots*

Further Resources

Chepko-Sade, B. D. & Halpin, Z. T. (Ed.). 1987. *Mammalian Dispersal Patterns: The Effects of Social Structure on Population Genetics.* Chicago: University of Chicago Press.
Dobson, F. S., Smith, A. T. & Wang, X. G. 1998. *Social and ecological influences on dispersal and philopatry in the plateau pika.* Behavioral Ecology, 9, 622–635.
Kleiman, D. G. 1981. *Monogamy in mammals.* Quarterly Review of Biology, 52, 39–69.
Smith, A. T. & Wang, X. G. 1991. *Social relationships of adult black-lipped pikas (Ochotona curzoniae).* Journal of Mammalogy, 72, 231–247.

Andrew T. Smith & F. Stephen Dobson

■ Social Organization
Social Knowledge in Wild Bonnet Macaques

A very important component of social cognition in primates is the knowledge that individuals might possess with regard to certain attributes of other individuals that they regularly interact with within their social group. In addition to the obvious recognition of each animal as a distinct individual, the possible attributes that such knowledge might encompass could include their dominance ranks and affiliative relationships—factors that seem to influence much of the social behavior displayed in primate societies.

In bonnet macaques, a frequent interaction between females is that of an allogrooming supplant, in which a dominant female displaces one member of a pair of grooming females, both subordinate to her. In the majority of these supplants (about 80%), the most subordinate of the three individuals leaves her grooming partner as soon as she observes the dominant female approaching them—such females are thus clearly aware of their own subordinate status relative to the other two individuals. On about 20% of these occasions, however, it is the other female (the more dominant of the two allogrooming individuals) that leaves—and the factor that most significantly appears to influence this decision is the high social attractiveness of her grooming partner, as defined by the amount and consistency of allogrooming that this individual receives from all the other females in the troop. These dominant retreating females are thus clearly aware of the social attractiveness of their subordinate partners; bonnet macaque females, therefore, appear to be knowledgeable of the social relationships of the other females in the troop.

That individual females might also know the relative dominance ranks of their troop members is revealed by the typical patterns of aggressive behavior and allogrooming choices that occur during such triadic interactions. If neither of the two allogrooming subordinate females retreats when the third dominant female approaches them, for example, the latter usually displays aggression toward the more subordinate of the two. Occasionally, however, she does not display any agonistic behavior but proceeds to directly allogroom one of the two individuals—and, in the majority of these cases, she grooms the more dominant female. Approaching females thus seem to be aware of the relative dominance ranks of the two other females, both subordinate to her.

Further statistical modelling of the decisions made by the females indicate that three factors are taken into consideration when they decide either to remain behind or to retreat during allogrooming supplants: knowledge of the subject's own dominance rank, her rank difference with the approaching dominant female, and rank difference with her grooming companion. Individuals are thus clearly aware, not only of their own positions in the rank hierarchy, but also of that of the other females in the troop. A model which incorporates the absolute dominance ranks of the latter, however, fails to explain the observed behavioral patterns. Knowledge of another individual's dominance rank is, therefore, egotistical in that it seems to be acquired only relative to one's own; a female knows of her rank difference with another female but does not appear to be aware of the absolute position of her adversary in the rank hierarchy.

The observation that rank difference with the approaching dominant female and that with the grooming companion both influence the decision-making process indicates that a bonnet macaque's knowledge system is integrative in nature—females are simultaneously able to process information about all their interacting companions and to use this knowledge effectively during social interactions. The decisions made in this particular situation are, in reality, even more complex: The intermediate female in a grooming supplant chooses to retreat as the approaching individual becomes relatively more dominant to her, whereas her grooming companion becomes comparatively more subordinate (as also more socially attractive).

The knowledge of dominance ranks and social relationships that individual macaques possess appears to constitute a clear example of recognition of individuality and individual attributes by these animals. Furthermore, the decision to retreat or remain behind during allogrooming supplants also depends on the absolute position of the actor in the dominance hierarchy—the more subordinate an individual, the more likely she is to retreat. Clearly then, each bonnet macaque female has knowledge of some of her own individual attributes as well.

Although all of these abilities must obviously call for some form of fairly sophisticated mental representation of particular individuals, including themselves, associated with their specific properties, what remains unclear is how exactly such information is categorized and coded for in the nonverbal cognitive architecture of the macaque mind. It is also important to note that, during triadic interactions, the integrative property of the bonnet macaque's knowledge system allows her to respond appropriately to the relative dominance ranks of the other interacting individuals. It is striking, therefore, that whatever may be the stored imagery of the individual attributes of the two females she is interacting with, it is possible for her to access both these sources and integrate them when finally making a socially complex decision.

See also Cognition—*Tactical Deception in Wild Bonnet
        Macaques*
        Social Evolution—*Social Evolution in Bonnet
        Macaques*
        Tools—*Tool Manufacture by a Wild Bonnet Macaque*

Further Resources

Byrne, R. W. & Whiten, A. (Eds.). 1988. *Machiavellian Intelligence: Social Expertise and the Evolution of Intellect in Monkeys, Apes, and Humans*. Oxford: Oxford University Press.

Cheney, D. L. & Seyfarth, R. M. 1990. *How Monkeys See the World*. Chicago: The University of Chicago Press.

Griffin, D. R. 2001. *Animal Minds: Beyond Cognition to Consciousness*. Chicago: The University of Chicago Press.

Sinha, A. 1998. *Knowledge acquired and decisions made: Triadic interactions during allogrooming in wild bonnet macaques*, Macaca radiata. Philosophical Transactions of the Royal Society, London: Biological Sciences, 353, 619–631.

Tomasello, M. & Call, J. 1997. *Primate Cognition*. New York: Oxford University Press.

Anindya Sinha

■ Social Organization
Social Order and Communication in Dogs

Humans first domesticated dogs when we were still hunter–gatherers, and through our thousands of years of shared history, dogs have adapted to living with us by learning our language. Dogs can be trained to understand simple commands, such as *sit, stay*, and *come*, and to perform relatively complex tasks, but they can also read our body language. Brian Hare, an anthropologist at Harvard University, has discovered that dogs are adept at following our gaze or a pointed finger, and can interpret subtle visual clues that their wild cousins, wolves, and our close relatives, chimpanzees, are unable to understand.

In living with dogs, we have also learned their body language. For example, most people who spend time with dogs are familiar with the play bow and what it means, even if they don't know it by name. When a dog greets another dog by crouching with her forelimbs extended, her hind end in the air, and her tail wagging, she is inviting another dog to play. Most people also recognize happiness in dogs. When my dog Scrap is happy, her

Two Labrador retrievers play with a stick on a beach. The one on the right is in a position resembling a play bow.

© Roy Toft / National Geographic Image Collection.

overall body posture is relaxed, her ears are upright, her mouth is open, and her tail is either down or held out and wagging slightly. If you have watched dogs interact with each other or with humans, you are probably also familiar with the upright, confident stance of an aggressive dog and the cringing of a submissive dog. A dominant aggressive dog will hold his tail and ears upright and will raise the hackles on his back. He will also wrinkle his nose, curl his lips, and show his teeth, and may bark or growl. A submissive dog will lower her body, tuck her tail between her back legs, and fold back her ears. Everything in her body language communicates fear. A submissive dog who is also aggressive has the same body posture, but raises the hackles on her back, wrinkles her nose, curls her lips, and shows her teeth as a way of saying that she may bite if cornered.

Dogs have a wide range of vocalizations, including numerous types of barks, as well as whines, growls, yips, and howls. By paying attention to when and how dogs vocalize, it's possible to figure out not only the general meaning of a bark versus a growl, but also the specific meanings of different types of barks and growls. In my experience, some dogs vocalize more than others. I call Willie my "talking dog" because of his wide range of vocalizations, which he uses to signal joy, contentment, fear, sadness, pleasure, and other emotions. Both Willie and my dog friend Max have a specific sound, an "awwr-roaor-roaor," which starts low and ends high, that they use to greet me when I come home from work. You can teach yourself dog language by paying careful attention to your dog's vocalizations.

By watching dogs interact in your home, if you're the guardian of multiple dogs, or at your local dog park, you can learn about how dogs communicate with each other and form packs. For the past several years, I have been a foster parent to abandoned and homeless

Two dogs running in a park while holding a frisbee.

Courtesy of Corbis.

dogs. I currently live with five dogs, who have been together for almost 4 years. Although my dog pack has been created ad hoc as I've adopted homeless dogs, the pack dynamics are similar to a wolf pack, which is created based primarily on family relationships: There is a definite pack hierarchy, although, as in wolf packs, the hierarchy is somewhat fluid and has changed over time.

Max is the alpha or dominant dog in the pack, and Scrap is the beta or second in line. Scrap is very confident, and when she first met the other dogs, she welcomed herself by walking through the front door with her head held high and her tail erect. She eyed the pack and immediately started a fight with Max, whom she apparently identified as the alpha dog. (Other dogs I have rescued have reacted differently when introduced to the pack. For example, when Zoie was introduced, she communicated her willingness to submit by tucking her tail and folding back her ears, and she frequently solicited Max's favor by licking his lips.) Scrap and Max had a very tense relationship for several months, which frequently erupted into fights. Although dog fights can appear to be ferocious and involve a great deal of growling, showing of teeth, and mounting, injuries are extremely rare: Most fighting or aggression is ritualized and serves to establish dominance hierarchies or pack order, which in turn enables the pack to cooperate and makes real fights (as opposed to ritual fights) unnecessary. Willie and Sunny are the middle ranking dogs, and Fin is the indisputable omega or low-ranking dog.

The pack hierarchy is reflected in how the dogs interact with each other. Max and Scrap never play together, presumably because of their uncertain status relative to each other. Although Max and Scrap have a more or less amiable relationship, Scrap continues to challenge Max and on any given day may act more like the alpha dog in all situations except those involving food. When it comes to food, Max is always dominant. Willie and Sunny play together constantly and like to chase and wrestle. Fin never plays with the other dogs and, as the omega dog, he's often the target of aggression from the other pack members. When there are altercations between any of the dogs, one of the dogs involved will frequently attack Fin as a way of releasing tension. This also happens in wolf packs, where the omega wolf is the scapegoat. Whenever Fin is attacked, he rolls over, shows his belly, tucks his tail between his legs, yelps loudly and, on very rare occasions, he urinates. These are all signals dogs and wolves use to indicate submission. After the attack, however, Fin will stand up and deliver several high-pitched barks in succession. This is the only time Fin displays this behavior, and I believe he is expressing his outrage or indignation at the unfair treatment he receives.

If you live with multiple dogs, it's important to understand the role you play in pack dynamics because as a pack leader your actions can either reinforce the dominance hierarchies dogs naturally establish between themselves, or your actions can destabilize the pack structure, leading to ongoing and increasingly aggressive conflicts. For example, in his case studies, Nicholas Dodman, a veterinarian and professor of behavioral pharmacology at Tufts University, has documented pack aggression that resulted in fatal injuries to the omega dog, largely because the dog's guardian continually destabilized the pack order by reinforcing the omega dog, instead of reinforcing the leadership role of the alpha dog.

By closely observing the dogs who share your home or frequent your neighborhood park, you can not only learn about canid behavior, but can also deepen the relationship you have with your own dog or dogs. As pack animals, dogs naturally see us as members of their pack and by reading their body language and understanding their social structures, we can be better pack leaders, more effective guardians, and more supportive and understanding friends to the dogs who share our lives.

See also Applied Animal Behavior—*Social Dynamics and Aggression in Dogs*
Communication—Vocal—*Social Communication in Dogs: The Subtleties of Silent Language*
Play—*Dog Minds and Dog Play*
Play—*Social Play Behavior and Social Morality*

Further Resources

Abrantes, Roger. 1997. *Dog Language: An Encyclopedia of Canine Behaviour*. Naperville, IL: Wakan Tanka Publishers.

Dodman, Nicholas. 1996. *The Dog Who Loved Too Much: Tales, Treatments, and the Psychology of Dogs*. New York: Bantam Books.

Fogle, Bruce. 1992. *Know Your Dog: An Owner's Guide to Dog Behavior*. New York: Dorling Kindersley, Inc.

Hare, Brian & Tomasello, Michael. 1999. *Domestic dogs* (Canis familiaris) *use human and conspecific social cues to locate hidden food*. Journal of Comparative Psychology, 113, 173–177.

Hare, B., Brown, M., Williamson, C. & Tomasello, M. 2002. *The domestication of social cognition in dogs*. Science 298, 1634–1636.

Janette Nystrom

■ Social Organization
Socioecology of Marmots

Marmots are the largest ground-dwelling squirrels. The best-known marmot is the woodchuck, which each year on February 2 predicts whether winter will be short or prolonged. Of the 14 species of marmots, the woodchuck is the earliest to emerge from hibernation in the spring and the only marmot species that is not social. This relationship raises the question of whether marmot sociality is related to hibernation. Because marmots are the largest true hibernators, could body size play a role in the evolution of sociality? To answer these questions, we must first consider the evolutionary history of marmots.

The first mammals clearly recognized as marmots lived in North America about 8 million years ago. Marmots reached Eurasia about 2 million years ago, and all existing species have evolved since that time. Marmots inhabited cool moist habitats. In general, the environment was characterized by cold winters and short, warm summers in a grassy landscape. With the general warming trend since the last glaciation, all but two species of marmots disappeared from warm, low-elevation habitats, and came to occupy cool, high-elevation meadows, often in association with rocky outcrops and talus. These environments are relatively harsh; summers are typically short, the spring season is often cold and stormy,

The Species of Marmots (Genus *Marmota*)

Eurasia		North America	
Black-capped	*M. camtschatica*	Woodchuck	*M. monax*
Steppe	*M. bobak*	Yellow-bellied	*M. flaviventris*
Tarbagan	*M. siberica*	Hoary	*M. caligata*
Menzbier's	*M. menzbieri*	Vancouver Island	*M. vancouverensis*
Long-tailed	*M. caudata*	Olympic	*M. olympus*
Gray	*M. baibacina*	Arctic	*M. broweri*
Alpine	*M. marmota*		

and the landscape may be snow covered from September until early July. Thus, the winter season, when food is unavailable, averages 7.2 months in length for all species of marmots. Marmots are too small to migrate to lower elevations for the winter or to remain active and survive on stored fat. They are too large to store food or to seek food under the snow. The alternative adopted by marmots is to accumulate fat during the summer growing season and to hibernate. Because hibernation greatly reduces metabolism, body fat supplies all the energy required to survive the long winter. For at least six marmot species, the body fat must not only support hibernation, it must also support reproduction, which occurs in the burrow before marmots

A hoary marmot forages in a rocky meadow in Alaska where vegetation can be scarce.
Courtesy of Kenneth Armitage.

emerge above ground and before they can forage. This activity is so energy demanding that most marmot species skip one or more years between successful reproductions in order to accumulate sufficient energy for both survival and reproduction. We know very little about how marmots allocate energy to basic maintenance and reproduction or the mechanism that determines whether marmots will initiate reproduction.

These high-energy demands raise the question of why marmots are large. Large animals require more total energy than small animals; could marmots reduce their energy demands if they were smaller? The answer is yes, but large size is highly advantageous in the places where marmots live. A larger body size has a larger digestive tract relative to metabolic rate. Large animals, such as marmots, can process the fibrous diets available in their grassy environments by retaining food in their digestive tracts for longer periods of time. As long as food plants are sufficiently abundant, marmots can support their large size. The home range that a marmot utilizes for foraging varies widely among species; the largest home ranges may be at least 15 times larger than the smallest home ranges. Presumably these differences in home range reflect the abundance of food available. But marmots need not only energy, but also essential nutrients, such as protein and certain fats necessary for successful hibernation. Little is known about marmot nutrition and the quality of available food. Furthermore, how does a large home range affect foraging and antipredator behavior? Are marmots with large home ranges more vulnerable to predation? Or are they more wary? Does parental care of the vulnerable young differ among species of marmots in relation to the area over which they forage? Or do wide-ranging species dig more escape burrows so that they are never far from a refuge?

Large body size has another important advantage. Large animals accumulate fat at the same rate as small animals, but use the fat relatively more slowly. Thus, if food resources are sufficient, large animals such as marmots can accumulate large amounts of

An adult marmot ventures outside its burrow in mid-June on Vancouver Island.
Courtesy of Kenneth Armitage.

fat. Because of its efficient use, marmots have sufficient energy for hibernation, for initiating reproduction, and for coping with unfavorable weather conditions until vegetation becomes available.

But large body size has a major consequence. The short growing season does not provide sufficient time for the young to reach maturity before their first hibernation. There is one exception, the woodchuck. The active season of the woodchuck is about 7.5 months, which is 2–4 months longer than that of any other marmot. The woodchuck disperses away from its natal burrow, hibernates alone, and may reproduce as a yearling (1-year-old). Because of large size and a short growing season, the young of all other species of marmots remain at home for one or more years. This retention of young in their natal environment forms the basic social unit of marmots.

Social behavior is tightly related to social organization. Two major types of social behavior occur; amicable behavior consists of greeting, when one marmot sniffs at the head area of another or two marmots simultaneously sniff at each other's cheeks, and grooming, when one marmot chews at the head, neck, shoulders, or flank of another. Agonistic or conflict behavior is characterized by one marmot chasing another or by one marmot fleeing or avoiding another. Amicable behavior is cohesive; it apparently enables marmots to form social bonds, which are expressed in group sharing of burrows and foraging areas. Agonistic behavior is dispersive, nongroup marmots are excluded from the group and prevented from using the burrow and food resources of the group. Thus, marmot social groups are territorial.

Two young yellow-bellied marmots greet each other above their burrow in Colorado.
Courtesy of Kenneth Armitage.

The woodchuck is the only asocial marmot species. Agonistic behavior characterizes social interactions; in effect, an adult forms a group of one and defends its territory against other woodchucks. The adult male defends a territory that overlaps the territories of several females. Only during mating do adult woodchucks associate together.

In the yellow-bellied marmot, all male yearlings disperse, but about half of the female yearlings remain in their natal area to form mother–daughter social groups. These groups may persist for several generations as matrilines. An adult, immigrant male associates with one or more matrilines and defends them against other male intruders. The matriline defends its territory against female immigrants. Female immigrants do not join an existing matriline, but become successful residents only when some part of the habitat patch is unoccupied and open for settlement. Social behavior within matrilines is primarily amicable, but between marmots from different matrilines, the behavior is almost entirely agonistic.

All other species of marmots live in family groups. Within groups, behavior is primarily amicable. Intruders encounter agonistic behavior from family members of the same sex. In species such as the alpine marmot of the European Alps and the Arctic marmot of the Brooks Range in Alaska, the extended family consists of a territorial, reproductive pair, nonreproductive adults of both sexes, yearlings, and young. In the hoary, Olympic, and Vancouver Island marmots of western North America, the limited family group also consists of a territorial pair, yearlings, and young. The nonreproductive adults are not present in the social group of these species. The difference in social structure is related to the age of dispersal. Dispersal occurs at the age of 2 years in the hoary, Olympic, and Vancouver Island marmots, but may be delayed to the age of 4 years in the alpine and arctic marmots. Although all marmot species are capable of reproduction by the age of 3 years, marmots living in family groups delay reproduction to an older age and those living in extended family groups actually delay dispersal to an age older than that at which they could reproduce. Why do these marmots delay reproduction and dispersal? This question is critical because evolutionary success requires producing offspring that live to reproduce; otherwise the genealogical line becomes extinct. The implication is that the benefits from living in a group are so great that it is worth delaying or possibly forgoing reproduction. If the benefits of group living are so great, why are woodchucks not social and why do yellow-bellied marmots not form family groups? To answer this question, we must first ask what are the benefits of group living?

The major benefit to group living is survivorship. The longer a marmot remains in its social group, the higher the probability that it will live to maturity. However, this benefit does not include young; the known mortality rate of young marmots is about 50%. But the probability of surviving thereafter is greater in those species that live in family groups than it is in yellow-bellied marmots. The difference in survivorship seems to be related to the age of dispersal. The mortality rate of dispersers is greater than that of nondispersers; thus more yearling yellow-bellied marmots die than yearlings of other species because about 75% of yearling yellow-bellied marmots disperse, whereas none of the yearlings disperse in the species living in family groups. Unfortunately, we do not know the mortality rate of woodchucks that disperse as young, but we would expect it to be higher than that of other marmot species. We can now note that there is a trade-off in group living; increased survivorship comes at the cost of delayed reproduction. All things being equal, an individual will have greater reproductive success and contribute more of its genes to the next generation (the criterion for evolutionary success) the earlier it reproduces. Yellow-bellied marmots are under strong natural selection to reproduce at the age of 2 years, the earliest age at which reproduction is possible, and they disperse as yearlings. Woodchucks reproduce as yearlings, and disperse as young. Thus, woodchucks and yellow-bellied marmots disperse at an early age and risk predation in order to initiate reproduction at the earliest age possible. Although marmots that form restricted family groups could disperse as yearlings, these species delay dispersal for another year and disperse at the age of 2 years. These species are larger than the yellow-bellied marmot and may require an additional year of growth to reach reproductive maturity at the age of 3 years.

Thus, we note a tendency for a marmot species to disperse the summer before it is capable of reproduction, but we also noted that marmots living in extended family groups do not disperse until one or more years after they are capable of reproduction. We have come full circle and again ask the question, why do these species lose reproductive opportunities by remaining in their family instead of dispersing at a younger age? The answer apparently lies in the probability that a disperser can survive its first hibernation. A young woodchuck can either find an old burrow to occupy or readily dig a new one. A dispersing yearling yellow-bellied marmot can find many places where unoccupied burrows exist, and they survive hibernation as well as those yearlings that remain in their natal colony. But the situation differs drastically for such marmots as the alpine and steppe marmots. Their environments appear to be saturated with existing families. An alpine marmot has almost no chance either of finding an unoccupied burrow or of digging a new one for hibernation.

The only chance for survival is for the disperser to find a family in which the same sex territorial adult is no longer present and replace the missing adult or to successfully invade a family and drive out the current resident. The loser of such brief, intense conflict will almost certainly die. A small marmot would have almost no chance of winning a conflict; thus alpine marmots remain at home until they are large and strong enough to have a chance of winning the conflict for a territorial position. Also, by remaining at home they have some chance of replacing the same-sex territorial adult of their family. Whatever the nonreproductive adults eventually do, their only chance for eventual reproduction is to remain at home, forego reproduction, and gain size and strength for the eventual conflict over who will reproduce. The conflict over who will reproduce characterizes marmot societies and will be discussed more fully later. For now, let us consider how living in groups increases survivorship.

A gray marmot in a vigilant posture outside its burrow in Kazakhstan.
Courtesy of Kenneth Armitage.

In yellow-bellied marmots, survivorship increases as group size increases. Marmots are vigilant and they commonly sit up on their haunches and scan the environment. When foraging, they look up or stand up while chewing and scan their surroundings for predators or agonistic marmots. When a predator is sighted, the marmot may give an alarm call and run to a nearby burrow or may run to the burrow and then call. The high-pitched, whistle-like cry alerts other marmots to the danger and perhaps informs the predator that it has been detected and might as well go elsewhere. The more marmots that are present in the social group, the more likely one of them will detect the approaching predator.

Survivorship may also be increased by group hibernation. Two or more marmots huddling together in the same hibernation nest can share heat and reduce the heat loss of each individual. Reducing heat loss reduces the use of fat, which extends the time that marmots can hibernate or makes more fat available for mating and reproduction after emerging from hibernation. Although all species of marmots except woodchucks could benefit by the social group hibernating together, very little is known about whether marmots benefit from group hibernation. Only the alpine marmot has been studied extensively. In this species, the survivorship of young is greater when the family has several nonreproductive adults who are closely related to the young. The 3- and 4-year-old adults provide heat to the young, which reduces the use of fat by the young. In effect, these adults provide a form of parental care for their younger brothers and sisters. The adults incur a cost; they lose more body mass when hibernating with related young than when related young are not present.

In contrast to those species that form family groups in which only the territorial, dominant female reproduces, in yellow-bellied marmot matrilines, commonly more than one female reproduces. Reproductive success increases as matriline size increases up to a size of three, then decreases in larger matrilines of four or five. Reproduction is enhanced as the density of the social group increases. This relationship implies some kind of social facilitation of reproduction, but nothing is known about a possible mechanism. Reproductive enhancement is most likely when a female is living with same-age kin or with younger kin (i.e., sisters or daughters). Thus, we could predict that adult females should recruit their daughters to become members of their matriline, and that yearling females should attempt to become residents. But only about 50% of the yearling females remain in their natal matriline; the others disperse. As is so often true in biology, recruitment is complex.

First, a yearling female is very unlikely to join a matriline if her mother is not present. In the mother's absence, older, nonlittermate sisters chase their younger sisters from the colony. Marmots prefer to recruit daughters rather than other kin; females that chase their younger sisters one year may recruit their daughters the next year. Second, recruitment is strongly related to the social environment. If most of the social behavior in the group is amicable, the yearling female is more likely to remain. But if agonistic behavior is common, the yearling female is more likely to disperse. Third, yearling recruitment is markedly affected by the "personality" or behavioral phenotype of the adults and yearlings. One way to measure a behavioral phenotype is by mirror image stimulation. A marmot in an arena is exposed to a mirror and its behavior with the mirror image is recorded. For example, a marmot may approach the image, attempt to greet or groom it or may remain in the back of the arena, chirp at the image, adopt a threat posture, or even attack the image. Each marmot receives an individual score, which represents its behavioral phenotype. Marmots that attempt to contact the image may be categorized as "more sociable," and those that avoid or threaten the image may be categorized as "more agonistic." The "more sociable" adult females are more likely to recruit yearling females, whereas the "more agonistic" yearling females are more likely than the "more sociable" yearlings to be recruited. This process produces an adult population in which individuals vary in their degree of sociability. Interestingly, once a yellow-bellied marmot becomes a resident, its behavioral phenotype minimally affects its reproductive success.

The major reason that the behavioral phenotype does not affect reproductive success is that reproductive suppression of younger adult females by older, dominant females is widespread. For example, a 2-year-old female, regardless of behavioral phenotype, has only about a 20% probability of reproducing if an older adult female is present, but that probability increases to about 48% when the no older adult is present. The major reason that reproductive success decreases in large matrilines is that they consist of a high percentage of young adults that are reproductively suppressed. The mechanism of reproductive suppression is unknown, but available evidence suggests that agonistic behavior disrupts the normal endocrine processes of reproductive physiology.

Although yellow-bellied marmots are reproductively mature at age 2, reproductive suppression increases the average age of first reproduction to 3 years. Similar patterns of reproductive suppression occur in all social marmots. For example, the dominant, territorial female alpine marmot reproduces every other year. Even in the year in which she does not reproduce, she suppresses the reproduction of the other adult females in the family. In gray marmots, only 3% of the 2-year-old females breed, but 70% of females aged 5 years or older breed. In the Vancouver Island marmot, most females do not breed until age 4 or older. But what do males do? Male yellow-bellied marmots exclude other males from their territories, and the territorial males do all the breeding. In those marmots, such as the alpine marmot, where subordinate, mature males are present, some of these males may also mate with the territorial female. The territorial male attempts to suppress the reproduction of the subordinate males, especially when these males are not his sons. Because his sons are critical for warming the young during hibernation, their reproduction may not be suppressed. The sons are more likely to warm the young if there is some likelihood that they fathered some of the young. In fact, the sons of the dominant male father about 13% of the young in alpine marmot families. We may conclude that one characteristic of marmot societies is competition for reproduction.

In conclusion, the physiological adaptations associated with hibernation enabled the large-bodied marmots to live in harsh environments with a short growing season. The physiological adaptations in turn required behavioral and social adaptations leading to the retention of young with their parents to form social groups characterized by both competition and cooperation. Cooperation occurs primarily in the detection of predators and defense against marmot intruders. Competition occurs primarily in determining who will reproduce. In general, the older, dominant males and females reproduce, and reproduction by subordinates is suppressed. Fundamentally, the competition is a contest to determine which individuals achieve evolutionary success by producing offspring. The alternative to remaining in the social group and being reproductively suppressed is dispersal at a body size that has almost no chance of surviving hibernation alone or gaining access to an established territory. Only those species where unoccupied habitat is available have a reasonable probability of successful dispersal. Also, the marmots that remain at home have the possibility of reproducing (not all reproductive suppression is successful), of inheriting the territory and becoming a dominant, reproductive individual, of budding off a new adjoining territory and establishing a new family, or of gaining residency in an adjoining territory either by displacing a territorial resident or by filling a vacancy resulting from the death of a resident. Each individual adjusts its behavior (i.e., to be amicable or agonistic) to its local situation. Ultimately such behavior may lead to reproductive and evolutionary success.

See also Lemurs—*Behavioral Ecology of Lemurs*
       Social Organization—*Dispersal*
       Social Organization—*Monkey Families and Human*
          Families
       Social Organization—*Social Dynamics in the Plateau*
          Pika
       Social Organization—*Territoriality*

Further Resources

Armitage, K. B. 1999. *Evolution of sociality in marmots.* Journal of Mammalogy, 80, 1–10.
Armitage, K. B. 2000. *The evolution, ecology, and systematics of marmots.* Oecologia Montana, 9, 1–18.
Armitage, K. B. 2003. *Marmots (Marmota monax and Allies).* In: *Wild Mammals of North America. Biology, Management, and Conservation.* 2nd edn. (Ed. by G. A. Feldhamer, B. C. Thmopson & J. A. Chapman), pp. 188–210. Baltimore, MA: The Johns Hopkins University Press.
Barash, D. P. 1989. *Marmots. Social Behavior and Ecology.* Stanford, CA: Stanford University Press.
Bibikow, D. I. 1996. *Die Murmeltiere der Welt.* Magdeburg, Germany: Westarp Wissenschaften.

Kenneth B. Armitage

■ Social Organization
Territoriality

Territory is one of many terms in animal behavior which has been borrowed from common usage, and as such, it is easy for us to assume that we know more about the territorial behavior of animals than is actually the case. According to Webster, territory is defined as "the land or waters under the jurisdiction of a nation, state, or ruler." This definition captures one critical element of animal territoriality, namely, the ownership of a particular region of space by an individual or a group of individuals. However, it misses another key component of animal territoriality: the behavior patterns that animals use to defend space against other individuals. Much of the interesting work on territoriality over the last few years has focused on the behavioral processes that animals use to establish and maintain exclusive ownership of particular bits of real estate.

Actually, one of the most useful general definitions of animal territoriality was developed in the middle of the last century, when G. K. Noble defined a territory as "any defended area." Since then, this definition has been refined and elaborated, but not by much. Currently, most scientists focus on "territory" as space that is defended for a long period of time relative to an animal's life span, in order to distinguish the defense of a territory from the defense of more ephemeral resources, such as individual food items or basking sites. Hence, a critical component of animal territoriality is that an individual must use a particular area for an extended period of time, a pattern of behavior which is called *site fidelity*. Animals, like crossbills (a species of bird), that wander nomadically across the landscape cannot be territorial, regardless of the sorts of behavior they use when interacting socially with other members of their species (conspecifics). Conversely, animals which remain in the same area for days to months, as is the case of many dragonflies, have the potential to be territorial, depending on whether or not they defend these areas from other individuals.

The second component of Noble's definition focuses on defense. Territory defense includes a wide range of aggressive behavior patterns that tend to discourage other individuals from remaining in, or returning to, a particular area; these include chases, fights, and the use of jaws, claws, horns or other weaponry to inflict injuries on other individuals. However, many territorial animals are able to discourage other individuals from entering space by using behavior patterns which do not have such an obvious and immediate effect on the recipients. Thus, experimental studies using tape recordings have shown that the song of a territory owner reduces the chances that other members of the species will enter the territory, relative to their entry rates when control sounds are played from the same location. One of the puzzling questions in the literature on territoriality is why the signals produced by territory owners have this effect on other members of the species; that is, why would potential intruders into a territory "respect" the signals produced by territory owners? Below, is one possible answer to this question. For now, the important point is that territory owners often defend their territories using "advertisement signals," which tend to discour-

A domestic male cat sprays urine to mark his territory.
© David Wrobel / Visuals Unlimited.

age other individuals from entering the territory. Familiar examples of territorial advertisement include the songs of birds or crickets, visual signals such as the pushup displays of lizards, or the scent marks deposited around the territory by wolves or rabbits.

Noble's definition is silent with respect to the types of individuals against which a territory is defended. After years of study, it has become apparent that territory owners often make fine discriminations among different categories of intruders when defending their territories.

For instance, male house mice defend exclusive territories against other dominant males, but permit adult females, subordinate males, and juveniles to share their territory. In many species, territory owners behave differently when interacting with individuals with whom they share a common border (neighbors) than when interacting with otherwise comparable strangers. In some species, owners are less aggressive to neighbors than to strangers (the so-called "dear enemy" effect), whereas in other species, the reverse is true, and owners react more aggressively to intrusions by neighbors than to intrusions by strangers. In some animals, discrimination among different categories of intruders may even extend to selected members of other species, as in algae-eating reef fish that defend their territories against other species with similar feeding habits, but ignore species with different feeding habitats.

Territorial Diversity

Early students of territorial behavior in animals were impressed by the variety and diversity of territorial behavior within and among different types of animals. To date, territorial behavior has been described in a bewildering array of taxonomic groups, including cnidarians (sea anemones), mollusks (limpets, octopods), many arthropods (insects, crustaceans, spiders), and of course, many vertebrates (fish, frogs, salamanders, lizards, mammals, and birds). Considerable variation often exists even within taxonomic groups with respect to territory size and the types of behavioral activities that occur within the confines of the defended area. For example, many songbirds defend "all purpose" territories, in which an adult male and an adult female raise offspring, forage, and survive the onslaughts of inclement weather, predators, and disease. However, other types of territories are also common in birds. Blue-footed boobies raise their young within small breeding territories on islands but forage in the oceanic waters surrounding the breeding colony, Rufous hummingbirds defend areas in meadows containing flowers during short stopovers in the middle of the autumn migration; and male sage grouse defend small territories in clusters called leks, which females visit to choose a mating partner before leaving to raise the offspring on their own. There is also considerable variation within and among taxa in the age, sex, and number of owner–defenders per territory. In species with extensive parental care, dependent offspring often share their parent's territory, as in ants, squirrels, and songbirds. Juvenile territoriality has been reported in a variety of insects, fish, lizards, and other groups in which individuals are self-sufficient from the time that they hatch out of their eggs. Territory defense by breeding males has received the most attention from field workers, but female territoriality is also common, as is territory defense by nonbreeding adults of both sexes. Finally, there are a surprising number of species in which two or more owners share and defend the same territory. For instance, shared defense by a male–female pair is common not only in birds, but also in other socially monogamous territorial species such as gibbons, butterfly fish, and prairie voles. Group territories defended by

Juvenile female sheephead fish in a territory dispute off the coast of Baja, California.
Visuals Unlimited.

multiple owners have been observed in a variety of species, including wolves, parrotfish, Florida scrub jays, and ants.

Economic Approaches to Territoriality

Early descriptive studies of territoriality emphasized variation in territorial behavior among different types of animals, and led to the development of extensive lists detailing the different types of territories observed in animals. However, this emphasis on territorial diversity understandably discouraged workers from trying to formulate global models that might apply to a wide range of species. A swing back toward generality was made in the 1960s and 1970s with the development of economic models of territorial behavior and habitat selection. Although originally designed with birds in mind, the economic approach was subsequently applied with some success to territorial animals from other taxonomic groups.

Economic models of territory defense assume that the defense of space is, in fact, a surrogate for the defense of resources located within that space, where *resource* can be loosely defined as any environmental feature that enhances growth, survival, or reproduction and that is in short supply relative to the number of potential users. At the simplest level, animals may defend resources required for their own use, for example, a food supply or a refuge from potential predators. Many animals also defend resources required for the growth and survival of their offspring, as is the case of the typical songbird territory, which includes resources required for nestlings and fledglings, not just for their parents. More circuitous relationships between territoriality and access to important resources have been observed in other species. For example, male dragonflies may defend patches of vegetation that females prefer for laying their eggs; as a result of defending resources preferred by females, males gain access to mates that they would not otherwise enjoy.

The economic approach to territoriality assumes that individuals defend territories when the benefits of territory defense exceed the costs, where both benefits and costs are measured in terms of lifetime reproductive success. Lifetime reproductive success depends on a number of factors, including the chances that an individual will survive long enough to reach maturity, the number of offspring an individual is likely to produce per unit time after it begins to reproduce, and the likelihood that the individual will die at various ages after reaching maturity. For most animals, it is difficult to impossible to measure lifetime reproductive success under field conditions, so although this is the best way to estimate the benefits and costs of territoriality, most field workers take shortcuts, and rely instead on indirect indices of the benefits and costs of territorial behavior. For instance, the benefits of a feeding territory might be estimated by food density or quality, or the benefits of a breeding territory by the number of independent young produced in a single breeding season. Along the same lines, estimates of the cost of territory defense might rely on measurements of the number or intensity of attacks directed at intruders or neighbors, or the rate of production of territorial advertisement signals by territory owners.

Short-term, "snapshot" estimates of territorial costs and benefits are fine as a first approximation, but they can be misleading if field workers forget that individuals have both histories and futures, and that costs and benefits measured at one instant in time may not necessarily reflect costs and benefits across a lifetime. Thus, in many seasonally territorial species, the period of territory establishment is characterized by high rates of aggressive interactions, elevated risks of injury or death as a result of those aggressive interactions, and little time available to enjoy the benefits of territory ownership. Many of the aggressive interactions observed during this period are used to establish boundaries between the territories of neighboring territory owners.

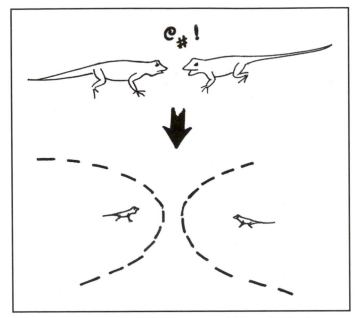

Animals use aggressive interactions to establish and defend territories. As shown here, juvenile lizards use one or more fights to establish a boundary between their adjacent territories. After an initial series of aggressive interactions, neighbors rarely fight again or venture into each other's territory. Both individuals vigorously attack any others that intrude into their territories.

Courtesy of Judy Stamps.

In contrast, once animals have settled into their territories, established stable boundaries with their neighbors, and become familiar with their territory, their neighborhood, and its inhabitants, rates of aggressive interactions decline, and residents begin to reap the benefits from resources contained in their territory. In this type of seasonally territorial species, the costs of territoriality tend to be "frontloaded" during the settlement period, whereas many of the benefits of territory ownership occur after owners have lived on their territory for a period of time. In this situation, any reasonable estimates of the costs and benefits of territorial behavior need to consider both the settlement and the residency periods, rather than exclusively focusing on the period after owners have established their territories (as is often the case).

The economic approach to territoriality has been most successful in predicting territorial behavior in simple systems, in which important resources can be easily quantified and in which owners rapidly adjust their behavior in response to short-term changes in resource levels or defense costs. Many classic studies of territorial behavior focused on nectivorous birds during the nonbreeding season, in which flowers and their energy contents could be measured readily and in which territory owners were prepared to change their territory sizes on a daily basis. In such species, settlement and territory establishment occurs so rapidly that one can capture most of the salient costs and benefits within a relatively short observation period. However, comparably simple tests of economic models have been less successful in predicting the behavior of animals that defend "all purpose" territories for extended periods of time. Such animals frequently do not alter their territory sizes in response to short-term changes in resource levels, suggesting that the owners of these territories may be responding to different types of costs and benefits than apply to the short-term defense of a single (usually food) resource.

Mechanisms for Territory Acquisition and Maintenance

At the same time as the economic approach was being applied to questions about the function and evolution of territorial behavior, behavioral biologists were also beginning to study the proximate behavioral mechanisms involved in territory acquisition and maintenance. Descriptive and experimental studies of the behavior of territorial animals has revealed remarkable convergence in patterns of territorial behavior across a wide range of taxonomic groups. Thus, when individuals who are both new to an area compete for indivisible space (i.e., space that can not be divided or shared), differences in strength, weaponry, experience, or other factors related to fighting ability typically determine which of the opponents will eventually gain ownership of the contested space. The general rule that larger (older, stronger, etc.) animals tend to win space when competing for novel, indivisible space has been validated for territorial animals belonging to virtually every taxon in which territorial behavior has been studied in the laboratory or field. On the other hand, while it seems intuitively obvious that differences in fighting ability should be related to success in territory acquisition, other rules of territorial behavior are less intuitive and more difficult to explain.

One of the most striking of these general phenomena is the "prior residency advantage" which refers to the fact that the owner virtually always wins when competing with strangers for possession of space (see sidebar p. 1036). Another surprising generalization is that when animals compete for divisible space (i.e., space which can be divided among neighboring territory owners), individuals may win space even if they lose every aggressive interaction with a competitor. In this situation, individuals can win space by "nagging" their opponent, when "nagging" in a territorial context involves persistently returning to an area after being chased away from that area by the opponent, repeatedly eliciting chases by the opponent, repeatedly challenging the opponent to engage in an exchange of aggressive signals, and so forth. These and other types of repeated social interactions are costly to both parties, so if an individual can ensure that its opponent must engage in chases, fights, displays, or other sorts of aggressive behavior every time it ventures into an area, that individual may eventually be able to discourage its opponent from returning to the contested area, instead encouraging its opponent to shift to adjacent areas that it can use in peace. Winning space by nagging is a way that smaller, younger, or otherwise less dominant individuals are able to successfully compete for space when interacting with larger, older, or more competitive individuals.

As suggested earlier, when studying the economics of territorial behavior, it is important to work within an appropriate time frame: Short-term, "snapshot" methods of estimating costs and benefits are unlikely to make much sense if most costs are incurred during the settlement period, and benefits accumulate after animals have lived in their territory for an extended period of time. An appropriate time frame is also essential for making sense of the behavior patterns that animals use to establish and maintain their territories.

First, recall that territorial animals are site-faithful: In the absence of interference by others, each individual tends to remain in a localized area for an extended period of time. In turn, site fidelity makes sense in situations in which an individual's ability to use a particular area efficiently improves over time as a result of the individual's familiarity with that area. Years of study by experimental psychologists suggest that the value of familiar space to a resident is likely to increase as a result of several related learning processes: Residents learn which types of resources (e.g., types of food, nest sites, shelter sites) are available in different portions of an area, learn the location of important features of the habitat relative to one another, and learn motor patterns that allow them to use the resources and the area efficiently and quickly. By way of analogy, a person who has just moved into a new apartment in a new city doesn't know the location of stores, restaurants, or laundromats, doesn't know where these sites are located relative to her apartment or place of work, and can't daydream while driving quickly and efficiently to and from work, because she hasn't yet established a familiar route for her commute. After a year, the same person is much more accomplished at using the same area efficiently and effectively. In this example, as is the case for many species with site fidelity, the area hasn't changed, but the individual's ability to use that area effectively has improved as a result of time invested in becoming familiar with that area.

A second reason that time is relevant to territoriality is that aggressive behavior directed at opponents affects their behavior over the long term as well as over the short term. It is clear that aggressive behavior often has immediate effects on the space use of intruders: Animals are

The Prior Residency Advantage

Judy Stamps

When a territory owner and an intruder compete for space, typically the owner wins the space, a phenomenon called the *prior residency advantage*. The prior residency advantage is one of the most puzzling aspects of territorial behavior, because it is not immediately obvious why the amount of time an individual spends in an area would affect its chances of retaining ownership of that area following a series of aggressive interactions with an opponent. One early hypothesis for the prior residency advantage suggested that owners are generally larger, older, or otherwise more competitive than intruders; if the most competitive individuals got the territories in the first place, then it wouldn't be surprising if superior territory owners were able to continue to win contests with inferior intruders. However, a lengthy series of carefully controlled field and laboratory studies put this hypothesis to rest. Even if owners and intruders were carefully matched beforehand with respect to body size, age, reproductive condition, and any other factors that might conceivably affect fighting ability, individuals who were allowed to become familiar with an area for a period of time (i.e., owners) routinely won contests with opponents venturing into the same area for the first time (i.e., intruders). Indeed, in some species, the prior residency advantage is sufficiently strong that residents with low fighting ability are able to retain control of space when competing with strangers with higher fighting ability.

A study of territory establishment in juvenile lizards in their natural habitat in the West Indies illustrates the importance of prior residency when individuals compete for space. When opponents were the same size, prior residents won 86% of their first fights with newcomers, and in virtually all of these cases, the prior resident eventually ended up in possession of the contested space. As one might expect, when one opponent was much larger than the other, the larger individual was more likely to win the first fight, but even in this situation, smaller lizards won more first fights when they were residents and their larger opponents were unfamiliar with the area (38%) than when the reverse was true (0%). Even more surprisingly, many smaller residents did not give up after losing one or more fights to a larger newcomer. Instead, they persisted in returning to the contested area, and continued to engage in aggressive interactions with the intruder in that area. Eventually, many smaller residents "lost the battles" but "won the war," and ended up in possession of contested space after the larger intruder abandoned it. Hence, in this species, the prior residency advantage was an outcome of two processes: residents were more likely to win aggressive interactions with newcomers, and residents were more likely to persist in using the contested area, even after losing multiple aggressive interactions to newcomers in that area.

likely to respond to attacks by leaving the area where they were attacked. What may be less obvious is that in nature, an individual has the option of repeatedly intruding into the same territory, so even if it leaves immediately following a single attack, there is no guarantee that it won't return at a later time. This is where a second important effect of territorial behavior comes in: Attacks, chases, and other types of aggressive behavior can act as "punishment," reducing the chances that the recipient of the behavior will enter that area in the future. According to this view, defensive behavior not only discourages individuals from using a particular area over the short term, but it also helps "train" opponents not to return to that area in the future.

Hence, a temporal perspective is useful for understanding the behavior of territorial animals because much of the behavior of territorial animals involves learning. On the one hand, learning processes which improve the value of familiar space mean that an individual who has lived in a area for an extended period of time is likely to value that area more than an individual who has just set foot in that same area. On the other hand, the fact that aggressive behavior can act as punishment means that attacks directed at opponents at one point in time can influence the space use of those opponents at later points in time. In combination, these two phenomena can explain many of the more puzzling aspects of the behavior of territorial animals.

For instance, let's consider the prior residency effect. The *prior residency effect* says that individuals familiar with space (e.g., owners or residents) win contests with individuals unfamiliar with that same space (e.g., intruders or strangers). A learning-based perspective suggests that the prior residency effect occurs because, as a general rule, a given area is more valuable to an individual who is familiar with that area than to an individual who is not. As a result, owners win contests because they fight more vigorously than intruders, or because they are more persistent than intruders (e.g., they continue to return to the contested area, even after repeatedly losing interactions with the intruder in that area).

Indeed, in species with a strong prior residency advantage, intruders may not even try to compete for space with well-entrenched owners, but simply avoid them, after detecting signals indicating that the owner has lived in that territory for an long period of time. For instance, dominant male mice deposit scent marks around their territories, and these scent marks provide information about the age, vigor, status, and other attributes of the territory owner. An intruding male that encounters scent marks deposited by a vigorous dominant male over an extended period of time would be well-advised to respond by by-passing that area, given the low likelihood that he would ever be successful in prying the owner out of his territory. Thus, a learning-based perspective not only explains the prior residency advantage, but also helps explain why newcomers might avoid areas in which they detect advertisement signals produced by well-established territory owners.

Northern elephant seals fighting near San Simeon, California.

Marc Moritsch / National Geographic Image Collection.

Other patterns of behavior that have been described for a wide range of territorial animals also begin to make sense if approached from a perspective which considers the effects of learning on territorial behavior. For example, if a scientist wants to generate extremely high rates of aggression in territorial animals, the best way to do this is to allow two individuals to become familiar with the same area before they first encounter one another in that area. One can play this trick on animals a number of ways, for example, by allowing two damselflies to establish territories on pads of vegetation located far apart from one another, and then gradually moving these pads closer together, until both damselflies are trying to defend the same place. As a general rule, fights between two owners are among the most vicious observed in the animal kingdom because both opponents fight vigorously and persistently, and both are prepared to remain in, or repeatedly return to, their familiar space, even after sustaining high costs of fighting with the opponent. This pattern of behavior is consistent with the view that the value of space increases as a function of familiarity with that space, so that in situations in which two opponents are familiar with the same space, both are willing to sustain high levels of punishment before either is prepared to abandon that space to the other party.

Practical Implications

Territoriality is a type of animal behavior with direct relevance to many human endeavors. Many species of concern to conservation biologists or pest managers are territorial, and many of the species domesticated by humans are derived from ancestors with sophisticated territorial behavior; examples include dogs, mice, rats or rabbits. As a result, all of these species come equipped with the ability to respond to punishment by avoiding the area in which that punishment occurred, and conversely, are likely to increase their level of aggressiveness toward intruders after becoming familiar with any area suitable for long-term residency. Thus, a basic knowledge of territorial principles can be useful for managing the behavior of these and any other animals with an evolutionary history of territoriality. For instance, if one were interested in training a puppy to avoid a particular area, an effective way to do this would be to deliver swift, consistent punishment the first few times it ventured into that area. In contrast, if the puppy were allowed free access to the area for an extended period before the punishment began, the same level of punishment would have less, or perhaps no, effect on its subsequent use of space. In nature, of course, other members of the same species deliver the punishment that individuals experience when they cross territorial boundaries, and newcomers quickly learn to avoid areas where they are quickly and consistently attacked.

Another reason why it is important to understand territorial behavior is that humans exhibit this behavior: Humans defend group territories against both strangers and neighboring groups. Behavioral biologists can offer new perspectives on some of the patterns observed in territorial conflicts (wars) in humans. For instance, fighting ability is one factor that clearly affects the outcome of territorial interactions in humans, but prior residency is another: Typically, individuals fighting on their home turf fight more persistently and are willing to incur higher costs prior to surrendering than individuals fighting in an strange location far from home. Along the same lines, a behavioral perspective suggests that when two human groups have both become familiar with the same area, contests between them are likely to be especially protracted and costly for both sides. As we look at the history of human warfare, many of the most persistent and bloody conflicts do seem to occur in situations in which two peoples have come to consider themselves to be owners of the same area. Of course, human territorial behavior is more complicated than that of other animals, but it is striking how many of the rules of territorial behavior generated from observations of animals also seem to apply to human territorial behavior. Hence, an appreciation of this branch of animal behavior may prove useful for managing our own future.

See also Social Organization—*Dispersal*
          Social Organization—*Herd Dynamics: Why Aggregations Vary in Size and Complexity*
          Social Organization—*Home Range*

Further Resources

Adams, E. S. 2001. *Approaches to the study of territory size and shape*. Annual Review of Ecology & Systematics, 32, 277–303.

Stamps, J. A. 1994. *Territorial behavior: Testing the assumptions*. Advances in the Study of Behavior, 23, 173–232.

Stamps, J. A. & Krishnan, V. V. 2001. *How territorial animals compete for divisible space: A learning-based model with unequal competitors*. American Naturalist, 157, 154–169.

Temeles, E. J. 1994. *The role of neighbours in territorial systems—when are they dear enemies*. Animal Behaviour, 47, 339–350.

Judy Stamps

■ |Sociobiology

A flock of starlings swerves away from a diving falcon, every bird moving in the same direction in the same instant. A male frog walks across the tropical forest floor with a batch of tadpoles clinging to his back. A peacock spreads and shimmers his tail as a peahen walks past. A man risks his life to save a drowning toddler. Two masses of black ants face off against one another along a boundary of their territories. These are just a few of the social interactions that interest *sociobiologists*, the biologists who study the evolutionary foundations of social behavior.

The first sociobiologist was Charles Darwin, who wrote about such things as the evolution of ant societies and how male beetles used their horns on one another in their aggressive social contests. Sociobiology did not, however, receive its current label until 1975 when Edward O. Wilson published *Sociobiology, The New Synthesis*, a massive summary of what evolutionists had learned about social behavior up to that time. For Wilson, Darwin, and all other sociobiologists (who are sometimes also known as *behavioral ecologists*), the central question has been, and still is, what is the evolved function, the purpose, of a given social ability? Why do starlings have the capacity to behave as one in a flock of hundreds? Why do males of some beetles fight furiously with other males of their species? What do social animals gain by interacting with others?

A swarm of ladybugs covers a sunflower.
Courtesy of John Alcock.

Sociobiologists use Darwin's theory of natural selection to develop answers to evolutionary questions about social behavior. Natural selection is a process that takes place whenever individuals differ in ways that affect their capacity to leave copies of their genes to the next generation. Individuals usually pass on their genes by having surviving offspring, which inherit some of their parents' genes and characteristics. Parents that have relatively many offspring reshape their species genetically in their own image. On the other hand, individuals that fail to reproduce generally fail to donate distinctive genes to their species' evolving gene pool. Over time, therefore, species tend to become composed of individuals who have what it takes to achieve *genetic success* (which can be measured by the number of copies of one's genes that an individual contributes to the next generation). The logic of the selective process is such that when sociobiologists see animals cooperating or fighting or courting in a particular way, they ask, why might these kinds of social tendencies contribute to the genetic success of individuals?

Note that there is nothing about sociobiological (evolutionary) theory that suggests that animal social behavior will be good for the species as a whole. Although the idea that selection helps species avoid extinction has appealed to many, the logic of Darwinian theory tells us that what is good for the species can only become more common over time if the trait first and foremost helps individuals achieve higher genetic success than others in their species. Therefore, any benefits to the entire species from a selected trait will usually be an incidental side effect of an attribute that became more common simply because it happened to help individuals pass on their genes. As a result, sociobiologists today almost never invoke group benefits as an explanation for the evolution of a trait of interest. Instead, they try to figure out how a social ability could help *individuals* achieve genetic success.

Sociobiologists have used this theoretical approach to identify social traits that are especially puzzling, namely, those that appear to *reduce* an individual's reproductive, and thus genetic, output. In this regard, the central challenge to sociobiologists has been to explain *altruism* (self-sacrificing behavior). An altruist foregoes reproduction in order to help others survive and reproduce. But since altruists lower their production of offspring, sometimes to zero, one would think that the hereditary basis of their self-sacrificing actions would quickly disappear from a population. In reality, however, altruism is not all that uncommon in nature. Everyone who has been stung by a honeybee has interacted with an extreme altruist, because the worker dies when she pulls away from the person she stung, leaving behind her still functioning stinger and some of her vital internal organs. Worker bees have evolved a readiness to sacrifice all in order to deter enemies, such as raccoons and bears, from breaking into their hives to steal honey and eat the bee larvae within.

Even worker bees that do not commit suicide on behalf of their hivemates are evolved altruists. They almost never lay eggs, but instead devote themselves entirely to rearing the offspring of their queen mother and collecting the food that the queen and her brood require. Similar altruism is rampant in the colonies of the social ants, wasps, and termites with their sterile worker and soldier castes. Altruism also occurs in many social birds and mammals, with "helpers at the nest" giving up a chance to reproduce for 1–2 years in order to help rear the offspring of other adults. Likewise, in the bizarre naked mole-rat, most members of the underground colony never reproduce at all, but instead defend and feed the few reproducing adults.

A successful explanation for the evolution of altruism was not devised until 1964. In that year, William D. Hamilton III showed how altruism could be adaptive if it enabled individuals to increase their genetic success *indirectly* by making personal sacrifices that raised the reproductive success of their close relatives. Because relatives share genes in common with one another, having inherited them from a recent common ancestor, selective altruism can have the effect of increasing the representation of one's family genes in the next generation. Adaptive altruism can persist because the genes "for" self-sacrifice are expressed only in some members of a family lineage, not all.

Hamilton's theory has been tested by checking whether self-sacrificing helpers do distribute their aid selectively. (An altruist that helps a nonrelative will not share family genes with this other individual and so can gain nothing genetically from its actions.) In fact, altruists of most species from ants to naked mole rats *only* assist closely related individuals, thereby indirectly propagating the genes they share with their brothers and sisters, nephews and nieces. So, for example, when a honeybee worker dies in defense of her colony, she helps her mother survive to produce more eggs that may give rise to future queens and drones, which are siblings of the dead worker. When a queen reproduces successfully, she passes on those genes that she shares with her worker sisters, dead or alive.

In species like buffalo, larger herds form when the vegetation is evenly distributed, due to their ability to use even the lowest quality items and the lack of competition for food.
Courtesy of John Alcock.

Sociobiologists employ the gene counting approach no matter what kind of social behavior they study. One can categorize the interactions between two individuals in terms of the reproductive and genetic consequences they have for the two participants. In an altruistic encounter, the helpful individual gives up some reproductive chances in order to advance the reproductive success of another. Under certain circumstances, altruism can be adaptive in the sense of increasing the genetic success of the helpful altruist.

The Effects of Social Interactions Between Two Individuals on Their Reproductive Success

Individual A	Companion B	Label
+	+	**Cooperation** (or **Reciprocity**, if B repays A later), which raises both A and B's reproductive and genetic success.
−	+	**Altruism** (which can be **Adaptive Altruism** if A's lowered personal reproductive success is more than matched by the increase in A's genetic success that comes from helping B, a close relative with shared genes, to survive).
+	−	**Exploitation** (with A taking reproductive and genetic success from B, as occurs in cases of deceit).
−	−	**Spite** (since both A and B lose reproductive and genetic success as a result of A's behavior, this category of social behavior is expected to be very rare, and it is).

In nature one also sees many cases of nonaltruistic cooperation in which those individuals interacting together all gain reproductively (and genetically) from their social behavior. Why, for example, do lionesses often live in groups, called prides? One possibility is that pride members can kill more prey by hunting together, generating reproductive gains that are sufficiently great to overcome the reproductive disadvantages that come from living side-by-side (increased pressure on the game resources in area, increased risk of contagious diseases, and the like). The result will be increased representation of the genes of competent cooperators in the pride.

Yet another possibility, however, is that lionesses living together sometimes hunt to help their companions who are in special need of a meal, thereby setting the stage for repayment later when the helpers are themselves especially hungry. Cases of this sort involve reciprocity or *reciprocal altruism*, a special form of cooperation, which requires that individuals repay their helpers after having received aid from them. Animals can gain via repayment if that act establishes a long-lasting partnership in which individuals go back and forth in assisting one another.

Consider another kind of social interaction among lionesses, this one between a mother and her relatively inexperienced daughter. Here the mother may do the bulk of the hard work of bringing down an antelope in order to give her offspring more experience with hunting as well as the lion's share of the meal they secure for themselves. This parental interaction carries reproductive costs for the mother because she expends time and energy that could be spent in other ways that might improve her chances of reproducing in the future. Moreover, every hunt carries with it the risk of life-shortening injury. These genetic costs, however, may be more than matched by an increase in the likelihood that the currently existing daughter may reach the age of reproduction and so pass on some of the genes received from her caring mother.

Note that once again an evolutionary understanding of a social interaction requires genetic calculations. If a parental action saves an existing offspring, the parent has in effect saved a carrier of half its genes (since eggs and sperm contain half the parental genotype, one copy of each gene in a maternal or paternal genome that has two copies of each gene). If behaving parentally in a given way substantially helps an offspring survive to pass on the genes received from a parental mother or father, then the action can lead to a net gain in genetic success, provided that the cost in terms of lost opportunities for future reproduction are small enough. Of course, some kinds of parental care of offspring could be so costly as to lower genetic success, in which case these kinds of parental behaviors would become less and less frequent over time, defining the limits of parental helpfulness.

The sociobiological approach has been applied not to just to altruism, reciprocity, parental behavior, and other kinds of helpful actions, but also to the social behaviors that appear in animal conflicts. Fighting is a kind of social interaction that carries with it the possibility of genetic gains (and losses) for the fighters. Sociobiological analyses have shown that the aggressive tendencies of animals appear to have been shaped by natural selection just as much as their helpful tendencies. So, for example, the fact that most potential competitors resolve their fights through noncontact ritualistic displays, rather than all-out grappling, biting, or clawing, has been shown to help fighting animals reduce the costs of their aggression. If an individual can determine through the display behaviors of a rival that the rival would win an all-out fight, then it is to the advantage of the loser to step aside (for the time being) rather than be thrashed, injured, or even killed.

Likewise, sexual interactions, which contain elements of both cooperation and conflict, can be and have been extensively analyzed by sociobiologists. In most species, males could potentially fertilize many more eggs than any one female could produce, a fact that in theory should favor males that attempted to mate with a great many females. In most animals, male behavior does indeed appear to be focused on maximizing sexual contacts, the better to advance the male's genetic success. In contrast, since female genetic success is limited by the relatively small number of eggs she produces (eggs being much larger than sperm), female sexual behavior is expected to be, and often is, focused on acquiring a mate of high quality rather than securing a large retinue of sexual partners. Much recent sociobiological research has revolved around the question of how females manage to choose high quality mates from among the large pool of potential partners available to them. In those species in which males transfer only sperm to their partners while providing no other resources or services of value, female mate choice appears to be heavily influenced by the nature of the often outlandish courtship displays performed by sexually eager males. In the birds of paradise, the sage grouse, assorted pheasants, and many others, these displays involve use of elaborate plumes and other brilliantly colored feather ornaments. In such species, females generally permit only a very small minority of displaying individuals to mate with them, apparently picking certain males on the basis of their display performance. In some cases, the offspring of males preferred by females have better survival rates than the offspring of less attractive males, suggesting that females of these species generally do choose the best males in terms of their genetic quality.

Sociobiologists have also become increasingly active in the study of human social behavior, although these researchers often prefer the label *evolutionary psychologist*. Whatever their title, these persons ask the standard sociobiological question of our own social behavior: What is the evolved purpose or adaptive value of the social tendencies of human beings? This is a reasonable question given the ample evidence that human social behavior depends on the brain, a structure whose development and operation are heavily influenced by our genes. The evolution of the brain is therefore subject to natural selection, such that people living today can be assumed to carry the adaptive hereditary predispositions for sociality exhibited by predecessors who had relatively high genetic success.

If this view is correct, then (for example) we can predict that human altruism will tend to be directed toward relatives. And it is true, for example, that persons in preindustrial societies usually adopt children that are their nephews and nieces in preference to complete genetic strangers. Moreover, stepparents who bring some of their own genetic offspring into a mixed family in North America are more likely to withhold investments from the children of their spouse, directing their care preferentially to their own children.

The imprint of past selection has also been detected in the human sexual psyche. Men around the world are less discriminating about sexual partners than are women, a fact that reflects the evolutionary consequences of the differences between males and females in the minimum investment required for the production of a child for the two sexes.

All aspects of the sexual, cooperative, and aggressive behavior of humans are potentially available for evolutionary analysis, and it seems likely that in time our species will join the honeybee, the lion, the naked mole-rat and a host of other animals in the growing catalog of species whose behavior has been thoroughly examined from the sociobiological perspective.

See also Levels of Analysis in Animal Behavior
Behavioral Phylogeny—*The Evolutionary Origins of Behavior*
Hamilton, William D. III
Recognition—*Kin Recognition*

Further Resources

Alcock, J. 2001. *The Triumph of Sociobiology*. New York: Oxford University Press.

Alcock, J. 2001. *Animal Behavior, An Evolutionary Approach*. Sunderland, MA: Sinauer Associates.

Dawkins, R. 1989. *The Selfish Gene*. 2nd edn. New York: Oxford University Press.

Gaulin, S. J. H. & McBurney, D. H. 2001. *Psychology, An Evolutionary Approach*. Upper Saddle River, NJ: Prentice Hall.

Williams, G. C. 1966. *Adaptation and Natural Selection*. Princeton, NJ: Princeton University Press.

Wilson, E. O. 2000. *Sociobiology, The New Synthesis, Twenty-fifth Anniversary Edition*. Cambridge, MA: Belknap Press.

John Alcock

■|Telepathy and Animal Behavior

For many years, animal trainers, pet owners and naturalists have reported various kinds of perceptiveness in domesticated and wild animals that suggest the existence of unexplained powers. Surprisingly little research has been done on these phenomena. Biologists have been inhibited by the taboo against "the paranormal," and psychic researchers and parapsychologists have, with few exceptions, confined their attention to human beings.

For the last 10 years, with the help of hundreds of animal trainers, shepherds, blind people with guide dogs, veterinarians, pet owners, and naturalists, I have been investigating some of these unexplained powers of animals. There are three major categories of seemingly mysterious perceptiveness: namely, telepathy, the sense of direction, and premonition. (For a detailed discussion see Sheldrake 1999.)

Animal Telepathy

According to recent random household surveys in England and the United States, many pet owners believe their animals are sometimes telepathic with them. An average of 48% of dog owners and 33% of cat owners said that their pets responded to their thoughts or silent commands (Sheldrake, Lawlor & Turney 1998). Many horse trainers and riders believe that their horses can pick up their intentions telepathically.

The most common kinds of seemingly telepathic response by companion animals include:

- the anticipation by dogs and cats of their owners coming home
- the anticipation of owners going away
- the anticipation of being fed
- cats disappearing when their owners intend to take them to the veterinarian
- dogs knowing when their owners are planning to take them for a walk
- cats and dogs that get excited when their owner is on the telephone, even before the telephone has been answered. For example, when the telephone rings in the household of a noted professor at the University of California at Berkeley, his wife says she knows when her husband is on the other end of the line because Whiskins, their silver tabby cat, rushes to the telephone and paws at the receiver. "Many times he succeeds in taking it off the hook and makes appreciative meows that are clearly audible to my husband at the other end," she says. "If someone else telephones, Whiskins takes no notice." The cat responds even when he telephones home from field trips in Africa or South America.

As skeptics rightly point out, some of these responses could be explained in terms of routine expectations, subtle sensory cues, chance coincidence, and selective memory, or put down to the imaginations of doting pet owners. These are reasonable hypotheses, but they should not be accepted in the absence of any evidence. To test these possibilities, it is necessary to do experiments.

My colleagues and I have investigated in great detail the alleged ability of some animals to know when their owners are coming home, often 10 minutes or more in advance. The pets typically wait at a door, window, or gate. In random household surveys in Britain and America, an average of 51% of dog owners and 30% of cat owners said they had noticed such anticipatory behavior (Sheldrake 1999). Some parrot and parakeet owners have also observed that their birds become excited before a member of the family returns.

The return-anticipating animal that I have studied most is a terrier called Jaytee, who belongs to Pam Smart in Ramsbottom, Greater Manchester, England. Pam adopted Jaytee from Manchester Dogs' Home in 1989 when he was still a puppy and soon formed a close bond with him.

In 1991, when Pam was working as a secretary at a school in Manchester, she left Jaytee with her parents, who noticed that the dog went to the French window almost every weekday at about 4:30 p.m., around the time she left work. He waited there until she arrived home some 45 minutes later. She worked routine office hours, so the family assumed that Jaytee's behavior depended on some kind of time sense.

Pam was laid off in 1993 and was subsequently unemployed, no longer tied to any regular pattern of activity. Her parents did not usually know when she would be coming home, but Jaytee still anticipated her return.

In 1994, Pam read an article about my research and volunteered to take part. To check that Jaytee was not reacting to the sound of Pam's car or other familiar vehicles, we investigated whether he still anticipated her arrival when she traveled by unusual means: by bicycle, by train, and by taxi. He did (Sheldrake & Smart 1998). Then on more than 100 occasions during Pam's absences, we videotaped the area by the window where Jaytee waited, providing a continuous, time-coded record of his behavior which was scored "blind" by a third party who did not know the details of the experiments (Sheldrake & Smart 2000a).

We carried out tests in which Pam traveled at least 5 miles from her home and started her return at times selected at random after she had left home, communicated to her by means of a telephone pager. In these experiments, Jaytee started waiting at the window around the time Pam started for home, even though no one at home knew when she would be coming. The odds against this being a chance effect were more than 100,000 to one. Jaytee behaved in a very similar way when he was tested repeatedly by skeptics anxious to debunk his abilities (Sheldrake & Smart 2000a).

The evidence indicates that Jaytee was reacting to Pam's intention to come home even when she was many miles away. Telepathy seems the only hypothesis that can account for the facts.

We have subsequently obtained very similar positive results, highly significant statistically, in trials with a Rhodesian ridgeback (Sheldrake & Smart 2000b).

I have also been investigating other kinds of animal telepathy experimentally; for example, the apparent ability of dogs to know when they are going to be taken for walks. In these tests the dogs were kept in a separate outbuilding and videotaped continuously. Meanwhile their owner, at a randomly selected time, thought about taking them for a walk and then 5 minutes later did so. On the videotapes, the dogs showed obvious excitement when their owner was thinking about taking them out, and by the time she entered the door were sitting around it in eager anticipation. They could not have known when she was planning to take them out by normal sensory means. They did not manifest such excitement at other times (Sheldrake 1999).

The most remarkable case of apparent telepathy between a person and a companion animal that I have come across concerns an African Grey parrot, N'kisi, who lives with Aimée Morgana in Manhattan, New York (Sheldrake 2003).

Reproductive Behavior

Peacocks exhibit promiscuous mating. Females prefer and choose males with elaborate tails, although an elaborate tail can be cumbersome (and a handicap) to the male. Mating in peacocks is an example of sexual selection.
© Fritz Polking / Visuals Unlimited.

Reproductive Behavior

Sockeye salmon spawn in a river in British Columbia. External fertilization creates many more opportunities for sperm competition than does internal fertilization.
© Brandon Cole / Visuals Unlimited.

A male sage grouse performs a strut display to females on a lek (a mating arena). Males who display the most vigorously mate with the most females.
© Tom Walker / Visuals Unlimited

A great egret (Casmerodius albus) *displays his breeding plumage while sitting on his nest.* © Theo Allofs / Visuals Unlimited.

Reproductive Behavior

An adaptation for dealing with sperm competition involves lengthening the time spent copulating. This has been widely observed in arthropods, such as these millipedes.
Courtesy of Getty Images / Digital Vision.

Anemonefish, such as this spine-cheek, are an interesting example of a species in which males are smaller than females. Males are able to undergo a sex change and become female.
© Marty Snyderman / Visuals Unlimited.

In some species, such as this longsnout seahorse, males, as well as females, can become pregnant.
© Brandon Cole / Visuals Unlimited.

Sharks

A male nurse shark prepares to haul away a cooperative female for mating.
© *Nick Caloyianis / National Geographic Image Collection.*

Social Facilitation

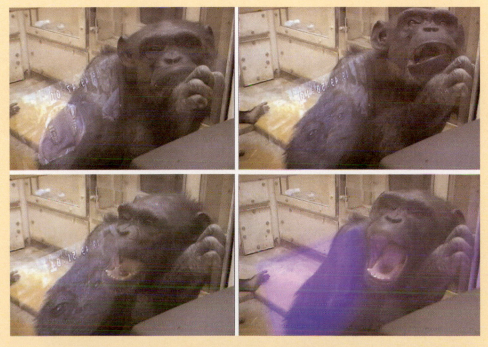

This combination of pictures released by the Primate Research Institute at Kyoto University in Japan shows a chimpanzee yawning after seeing videos of other chimps yawning.
© *AFP / Getty Images.*

Social Organization

A blue-footed booby attempts to defend its territory against an intruding male.
© Gerald and Buff Corsi / Visuals Unlimited.

Olive baboons groom each other to remove parasites and pests. Grooming also serves an important function for reconciling after an aggressive encounter.
© Joe McDonald / Visuals Unlimited.

Tools

In the Mahale Mountains National Park, Tanzania, a chimpanzee fishes for ants in a tree using a twig. This is a common form of tool use observed in chimpanzees.
Michael Nichols / National Geographic Image Collection

Wellfare, Well-Being, and Pain

Like all waterfowl, ducks take good care of their feathers. A duck's very life depends on maintaining the healthy condition of its feathers, which provide buoyancy, waterproofing, and protection from extremes of cold or heat. Excessive feather pecking has been shown to be a sign of distress in captive birds, and this abnormal behavior can harm the condition of the feathers.
Courtesy of Anthony F. Chiffolo.

Wellfare, Well-Being, and Pain

Two young falcon chicks take food from a falcon puppet which serves as a foster parent. If they are released into the wild, these falcons will be less likely to associate food with humans.
© Kennan Ward / Corbis.

A recovery team member force feeds an orphaned baby Florida manatee. It is important to take precautions to ensure that the animal does not bond or become too familiar with people.
© Kike Calvo / V&W / The Image Works.

N'kisi the Parrot

Inspired by the work of Irene Pepperberg with African grey parrots, in 1997 Aimée began training a young male, N'kisi, (pronounced "in-key-see") in the use of language. She did so by teaching him as if he were a human child, starting when he was 5 months old. She used two teaching techniques known as "sentence frames" and "cognitive mapping". In *sentence frames*, words were taught by repeating them in various sentences such as, "Want some water? Look, I have some water." *Cognitive mapping* reinforced meanings that might not yet be fully understood. For example, if N'kisi said "water," Aimée would show him a glass of water.

By the time he was 5 years old, N'kisi had a contextual vocabulary of more than 700 words. He apparently understood the meanings of words and used his language skills to make relevant comments. He ordinarily spoke in grammatical sentences, and by January 2002, Aimée had recorded more than 7,000 original sentences.

Although Aimée's primary interest was in N'kisi's use of language, she soon noticed that he said things that seemed to refer to her own thoughts and intentions. He did the same with her husband, Hana. After reading about my research on telepathy in animals (Sheldrake 1999), in January 2000 she contacted me, summarizing some of her observations. At the same time, she began keeping a detailed log of seemingly telepathic incidents and has continued to do so. By January 2002, there were 630 such incidents on record. Here are some examples:

"I was thinking of calling Rob, and picked up the phone to do so, and N'kisi said, 'Hi, Rob,' as I had the phone in my hand and was moving toward the Rolodex to look up his number."

"I read the phrase, 'The blacker the berry the sweeter the juice.' He said, 'That's called black,' at the same instant."

"I was in a room on a different floor, but I could hear him. I was looking at a deck of cards with individual pictures, and stopped at an image of a purple car. I was thinking it was an amazing shade of purple. Upstairs he said at that instant, 'Oh wow, look at the pretty purple.'"

Of all the various incidents, perhaps the most remarkable occurred when N'kisi interrupted Aimée's dreams. (He usually slept by her bed.) For example: "I was dreaming that I was working with the audio tape deck. N'kisi, sleeping by my head, said out loud, 'You gotta push the button,' as I was doing exactly that in my dream. His speech woke me up." On another occasion, "I was on the couch napping, and I dreamed I was in the bathroom holding a brown dropper medicine bottle. N'kisi woke me up by saying, 'See, that's a bottle.'"

Together we designed and set up a series of trials to test whether this apparent telepathic ability would be expressed in formal tests. Aimée and the parrot were in different rooms, on different floors, under conditions in which the parrot could receive no sensory information from Aimée or from anyone else. During these trials Aimée and the parrot were both videotaped continuously. At the beginning of each trial, Aimée opened a numbered sealed envelope containing a photograph, and then looked at it for 2 minutes. These photographs corresponded to a prespecified list of key words in N'kisi's vocabulary, and were selected and randomized in advance by a third party. The recordings of N'kisi during these trials were transcribed blind by three independent transcribers. Their transcripts were generally in good agreement.

Using a majority scoring method, in which at least two of the three transcribers were in agreement, N'kisi said one or more of the key words in 71 trials. He scored 23 hits: The key words he said corresponded to the target pictures. In a Randomized Permutation Analysis (RPA), there were as many or more hits than N'kisi actually scored in only 5 out of 20,000 random permutations, giving a p value of 5/20,000 or 0.00025. These findings were consistent with the hypothesis that N'kisi was reacting telepathically to Aimée's mental activity (Sheldrake & Morgana 2003).

There is much potential for further research on animal telepathy. And if domestic animals are telepathic with their human owners, then it seems very likely that animals are telepathic with each other, and that this may be important in the wild. Some naturalists have already suggested that the coordination of flocks of birds and herds of animals may involve something like telepathy, as may communication among members of packs of wolves (Sheldrake 1999).

The Sense of Direction

Homing pigeons can find their way back to their loft over hundreds of miles of unfamiliar terrain. Migrating European swallows travel thousands of miles to their feeding grounds in Africa and in the spring return to their native place, even to the very same building, where they nested before. Some dogs, cats, horses, and other domesticated animals also have a good sense of direction and can make their way home from unfamiliar places many miles away.

Most research on animal navigation has been carried out with homing pigeons, and this research over many decades has served only to deepen the problem of understanding their direction-finding ability. Navigation is goal-directed and implies that the animals know where their home is even when they are in an unfamiliar place and have to cross unfamiliar terrain.

Pigeons do not know their way home by remembering the twists and turns of the outward journey, because birds taken in closed vans by devious routes find their way home perfectly well, as do birds that have been anesthetized on the outward journey or transported in rotating drums. They do not navigate by the sun, because pigeons can home on cloudy days and can even be trained to navigate at night. However, they may use the sun as a simple compass to keep their bearings. Although they use landmarks in familiar terrain, they can home from unfamiliar places hundreds of miles from their home, where no familiar landmarks are visible. They cannot smell their home from hundreds of miles away, especially when it is downwind, although smell may play a part in their homing ability when they are close to familiar territory. Pigeons deprived of their sense of smell by researchers were still able to find their homes (Sheldrake 2002).

Some biologists hope that the homing of pigeons might turn out to be explicable in terms of a magnetic sense. But even if pigeons have a compass sense (which is not proven), this could not explain their ability to navigate. If you were taken blindfolded to an unknown destination and given a compass, you would know where north was, but not the direction of your home.

The failure of conventional attempts to explain pigeon homing and many other kinds of animal navigation implies the existence of a sense of direction as yet unrecognized by institutional science. This could have major implications for the understanding of animal migrations and would shed light on the human sense of direction, much better developed in traditional peoples, such as the bushmen of the Kalahari or Polynesian navigators, than in modern urban people (Sheldrake 2002).

Premonitions of Earthquakes

There is a large body of evidence for unusual animal behavior before earthquakes, including historical accounts from ancient Greece. I have collected more detailed accounts from recent earthquakes in California, including the Loma Prieta earthquake in 1987 and Northridge in 1994, from the 1995 Kobe earthquake in Japan, and the 1997 earthquake in

Assisi, Italy (Sheldrake 1999), the 1999 earthquakes in Turkey and Greece, and the 2001 quake near Seattle, WA (Sheldrake 2003). In all those cases there were many reports of both wild and domesticated animals behaving in fearful, anxious, or unusual ways hours or even days before the earthquakes struck.

With very few exceptions, Western seismologists have ignored such stories of unusual animal behavior, dismissing them as anecdotal or even superstitious. By contrast, in China since the 1970s, the authorities have encouraged people to report unusual animal behavior, and they have an impressive track record in predicting earthquakes, in some cases evacuating cities hours before devastating earthquakes struck, saving tens of thousands of lives.

By paying attention to unusual animal behavior, as the Chinese do, rather than ignoring it, earthquake warning systems might be feasible in California, Greece, Turkey, Japan, and elsewhere by enlisting the help of thousands or even millions of volunteers, using modern communications technologies like telephones and the Internet.

No one knows how some animals seem to sense earthquakes coming. The three main theories are that they somehow pick up subtle sounds, vibrations or movements of the earth; that they respond to subterranean gases released prior to earthquakes; or that they respond to changes in the Earth's electrical field preceding earthquakes. As well, or instead, animals may somehow "sense" in advance what is about to happen in a way that lies beyond current scientific understanding. In other words they may be presentient or precognitive, that is, having a feeling that something is about to happen. This hypothesis would be unnecessary if all the facts could be explained satisfactorily by more conventional theories. At present the electrical theory seems sufficiently promising to justify ignoring this more radical possibility. But several other kinds of animal premonition cannot be explained electrically.

Other Kinds of Premonition

On February 23, 1999, an avalanche devastated the Austrian village of Galtur in the Tyrol, killing dozens of people. It was the worst avalanche disaster in Austria since 1954. The previous day, the chamois (small goat-like antelopes) came down from the mountains into the valleys, something they usually never do (Sheldrake 2003).

Through surveys in villages in the Austrian and Swiss Alps, kindly conducted on my behalf by Theodore Itten, I have found that the animals most often said to anticipate avalanches are chamois and ibexes, and also dogs. Some dogs are said to have barked persistently for no apparent reason for hours before an avalanche struck; and some refused to go outside (Sheldrake 2003).

As in the case of earthquakes, it is not clear how these animals anticipated the coming disasters. Perhaps they reacted to electrical or other physical changes. But if so, no one knows what these changes are. Or perhaps they have a more mysterious presentiment of danger.

However they do it, an ability to anticipate avalanches would be of obvious survival value in mountain animals, and would be favored by natural selection. But many animals also anticipate man-made catastrophes that would not have occurred in the natural world, such as air raids.

During the World War II, many families in Britain and Germany relied on their pets' behavior to warn them of impending air raids before any official warnings were given. These warnings occurred when enemy planes were still hundreds of miles away, long before the animals could have heard them coming. Some dogs in London even anticipated the explosion of German V-2 rockets, even though these missiles were supersonic and hence could not have been heard in advance (Sheldrake 1999). If animals were not anticipating air raids by hearing approaching bombers or rockets, how did they sense the attacks were coming? Precognition or presentiment seems the only possibility.

Very little research has been done on animal premonitions, even in the case of earthquakes where such warnings could prove very useful.

Conclusions

The three types of unexplained perceptiveness discussed above—telepathy, the sense of direction, and premonitions—seem better developed in nonhuman species like dogs than they are in people. Nevertheless they occur in the human realm too, but they seem to be stronger in traditional cultures than in the modern industrial world. Maybe we have lost some of these abilities because we no longer need them: Telephones and television have superseded telepathy; maps and global positioning systems have replaced the sense of direction. And perceptiveness is not cultivated in our educational system. Indeed the existence of unexplained powers is often denied and even ridiculed.

Nevertheless, human "sixth senses" have not gone away. They look more natural, more biological, when they are seen in the light of animal behavior. Much that appears "paranormal" at present looks normal when we expand our ideas of normality. But we need to expand our view of physics as well as of biology if these phenomena are to be explained at a more fundamental level.

Telepathy from people to animals usually occurs only when there are close emotional bonds. This may well be an important factor in human telepathy too. My own hypothesis is that these bonds depend on fields that link together members of a social group, called social fields. These are members of a more general class of fields called morphic fields. These bonds continue to link members of the social group together even when they are far apart, beyond the range of sensory communication, and can serve as a medium through which telepathic communications can pass (Sheldrake 2003).

Morphic fields may also underlie a sense of direction. Animals are not only linked to members of their social group by morphic fields, but also to significant places, such as their home. These fields continue to connect them to their home even when they are far away, rather like invisible elastic bands. These bonds can consequently give directional information, "pulling" the animal in a homeward direction (Sheldrake 1999).

We have much to learn from our companion animals about animal nature, and about our own.

Further Resources

Sheldrake, R. 1999. *Dogs that Know When Their Owners Are Coming Home, and Other Unexplained Powers of Animals*. New York: Crown.

Sheldrake, R. 2002. *Seven Experiments that Could Change the World*. 2nd edn. Rochester, VT: Park Street Press.

Sheldrake, R. 2003. *The Sense of Being Stared at, and Other Aspects of the Extended Mind*. New York: Crown.

Sheldrake, R., Lawlor, C. & Turney, J. 1998. *Perceptive pets: A survey in London*. Biology Forum, 91, 57–74.

Sheldrake, R. & Morgana, A. 2003. *Testing a language-using parrot for telepathy*. Journal of Scientific Exploration, 17, 601–615.

Sheldrake, R. & Smart, P. 1998. *A dog that seems to know when his owner is returning: Preliminary investigations*. Journal of the Society for Psychical Research, 62, 220–232.

Sheldrake, R. & Smart, P. 2000a. *A dog that seems to know when his owner is coming home: Videotaped experiments and observations*. Journal of Scientific Exploration, 14, 233–255.

Sheldrake, R. & Smart, P. 2000b. *Testing a return-anticipating dog, Kane*. Anthrozoös, 13, 203–212.

Rupert Sheldrake

■ Tinbergen, Nikolaas
(1907–1988)

Nikolaas Tinbergen, born in the Hague, Netherlands, in 1907, grew up loving the outdoors, and explored the coastal flatlands and animal inhabitants of his native country. Tinbergen had a genius for conducting simple, yet insightful, experiments in the field; he was the archetypical "curious naturalist." In 1949, he moved to England in an effort to enhance the spread of ethology in the English-speaking world. In 1973, Tinbergen won the Nobel Prize in Medicine or Physiology, along with Karl von Frisch and Konrad Lorenz. Tinbergen was a humane and modest man. He is well-known for developing the four questions he believed should be asked of any animal behavior, which are related to function, causation, development, and evolutionary history. He is also especially famous for his research on sea gulls, which led to important findings on courtship and mating behavior. He became a British citizen in 1955, taught at Oxford University, and died in 1988.

Donald A. Dewsbury

■ Tinbergen, Nikolaas
Ethology and Stress Diseases

Nobel Lecture, December 12, 1973

Many of us have been surprised at the unconventional decision of the Nobel Foundation to award this year's prize 'for Physiology or Medicine' to three men who had until recently been regarded as 'mere animal watchers.' Since at least Konrad Lorenz and I could not really be described as physiologists, we must conclude that our *scientia amabilis* is now being acknowledged as an integral part of the eminently practical field of Medicine. It is for this reason that I have decided to discuss today two concrete examples of how the old method (1) of 'watching and wondering' about behaviour (which incidentally we revived rather than invented) can indeed contribute to the relief of human suffering—in particular of suffering caused by stress. It seems to me fitting to do this in a city already renowned for important work on psychosocial stress and psychosomatic diseases (2).

Note: The illustrations in the original paper have not been reproduced here. To view them go to
http://nobelprize.org/medicine/laureates/1973/tinbergen-lecture.html

My first example concerns some new facts and views on the nature of what is now widely called Early Childhood Autism. This is a set of behavioural aberrations which Leo Kanner first described in 1943 (3). To us, i.e. my wife Elisabeth and me, it looks as if it is actually on the increase in a number of western and westernised societies. From the description of autistic behaviour—or Kanner's syndrome (4)—it is clear, even to those who have not themselves seen these unfortunate children, how crippling this affliction is. In various degrees of severity, it involves, among other things: a total withdrawal from the environment; a failure to acquire, or a regression of overt speech, and a serious lagging behind in the acquisition of numerous other skills; obsessive preoccupation with a limited number of objects; the performance of seemingly senseless and stereotyped movements; and an EEG pattern that indicates high overall arousal. A number of autists recover (some of them 'spontaneously') but many others end up in mental hospitals, where they are then often diagnosed, and treated, as schizophrenics.

In spite of a growing volume of research on the subject (5), opinions of medical experts on how to recognise autism, on its causation, and therefore on the best treatment vary widely. Let me consider this briefly, point by point.

1. There is disagreement already at the level of diagnosis and labeling. For instance, for 445 children Rimland compared the diagnosis given by the doctor who was consulted first, with a 'second opinion' (6). If the art of diagnosis had any objective basis, there should be a positive correlation between first and second opinions. In fact, as Rimland points out, there is not a trace of such a correlation—the diagnoses are practically random. What these doctors have been saying to the parents is little more than: 'You are quite right; there is something wrong with your child'.

And yet, if we use the term autism in the descriptive sense of 'Kanner's syndrome', it does name a relatively well-defined cluster of aberrations.

2. The disagreement about the causation of autism is no less striking. It expresses itself at two levels. First, there is the usual 'nature–nurture' controversy. The majority of experts who have written on autism hold that it is due either to a genetic defect, or to equally irreparable 'organic' abnormalities—for instance brain damage such as can be incurred during a difficult delivery. Some of the specialists are certainly emphatic in their assertion that autism is 'not caused by the personalities of the parents, nor by their child-rearing practices' (7). If this were true, the outlook for a real cure for such children would of course be bleak, for the best one could hope for would be an amelioration of their suffering. But there are also a few experts who are inclined to ascribe at least some cases of autism to damaging environmental causes—either traumatising events in early childhood, or a sustained failure in the parent–infant interaction (8). If this were even partially correct, the prospect for a real cure would be brighter of course.

The confusion about causation is also evident in the disagreement about the question what is 'primary' in the overall syndrome—what is 'at the root of the trouble'—and what are mere symptoms. Some authors hold that autism is primarily either a cognitive, or (often mentioned in one breath) a speech defect (9). Others consider the hyperarousal as primary (10). Those who subscribe to the environmental hypothesis think either in terms of too much overall input (11), or in terms of failures in the processes of affiliation, and of subsequent socialisation (8).

3. In view of all this it is no wonder that therapies, which are often based on views concerning causation, also differ very widely. Nor is it easy to judge the success rates of

A Visit with Niko and Lies Tinbergen

Donald A. Dewsbury

One of my most memorable experiences was a Sunday afternoon I spent with Niko and Lies Tinbergen at their home in Oxford in August 1988, four months before he died. This always-modest laureate reflected that it was difficult to accept that people put him on a pedestal and treated him differently—as if he were a superior person—just because he had won a Nobel Prize. He had written:

> My own reaction was a mixture of personal pride, of embarrassment at having been placed on this kind of pedestal..., and of a not wholly pleasant awareness that too may people suddenly began to look up to me as if I were a kind of superior know-all, understand-all, and generally a *wise* man... If *only* people realize that a Nobel Prize, just as any other distinction, is no more than a pat-on-the-back, a "well done" from other, far from infallible, human beings (Tinbergen 1985, p. 456).

The couple expressed great concern about the social issues of the day such as population growth, pollution, nuclear waste, the arms race, world politics, and prospects for peace, issues that also had been of great concern to Konrad Lorenz. Niko, having recovered from a stroke, was a most engaging and modest host.

any of these therapies, for the numbers of children treated by any individual therapist or institution are small; also, the descriptions of the treatments are inevitably incomplete and often vague. Unless one observes the therapist in action it is not really possible to judge what he has actually been doing.

In short, as O'Gorman put it not long ago (4. p. 124) '. . . our efforts in the past have been largely empirical, and largely ineffectual'.

In view of all this uncertainty any assistance from outside the field of Psychiatry could be of value. And it is such assistance that my wife and I have recently tried to offer (12). Very soon our work led us to conclusions which went against the majority opinion, and we formulated proposals about therapies which, with few exceptions, had not so far been tried out. And I can already say that, where these treatments have been applied, they are leading to highly promising results, and we feel that we begin to see a glimmer of hope.

Before giving my arguments for this optimistic prognosis, let me describe how and why we became involved.

Our interest in autistic children, aroused initially by what little we had seen of the work that was being done in the Park hospital in Oxford, remained dormant for a long time. But when, in 1970, we read the statement by Drs. John and Corinne Hutt that '. . . apart from gaze aversion of the face, all other components of the social encounters of these autistic children are those shown by normal non-autistic children'. (13, p. 147), we suddenly sat up, because we knew from many years of child watching that normal children quite often show all the elements of Kanner's syndrome.

Thinking this over we remembered the commonsense but sound warning of Medawar, namely that 'it is not informative to study variations of behaviour unless we know beforehand the norm from which the variants depart' (14, p. 109), and we realised that these words had not really been heeded by psychiatrists. In their literature we had found very little about normal children that could serve as a basis for comparison.

We also realised that, since so many autists do not speak (and are often quite wrongly considered not to understand speech either) a better insight into their illness would have to be based on the study of their non-verbal behaviour. And it is just in this sphere that we could apply some of the methods that had already proved their value in studies of animal behaviour (15).

Therefore we began to compare our knowledge of the non-verbal behaviour that normal children show only occasionally, with that of true autists, which we had not only found described in the literature but also began to observe more closely at first hand.

The types of behaviour to which we soon turned our attention included such things as: the child keeping its distance from a strange person or situation; details of its facial expressions; its bodily stance; its consistent avoidance of making eye contact etc.—an extremely rich set of expressions that are all correlated with overt avoidance. The work of professional child ethologists is beginning to show us how immensely rich and subtle the repertoire is of such non-verbal expressions (16).

But, apart from observing these behaviours themselves, we also collected evidence about the circumstances which made normal children revert to bouts of autistic behaviour.

What emerged from this dual approach was quite clear: Such passing attacks of autistic behaviour appear in a normal child when it finds itself in a situation that creates a conflict between two incompatible motivations. On the one hand, the situation evokes fear (a tendency to withdraw, physically and mentally) but on the other hand it also elicits social, and often exploratory behaviour—but the fear prevents the child from venturing out into the world. And, not unexpectedly, it is 'naturally' timid children (by nature or by nurture, or both) that show this conflict behaviour more readily than more resilient, confident children. But my point is that they all respond to the environment.

Once we had arrived at this interpretation, we tested it in some simple experiments. In fact, we realised that in our years of interaction with children we had already been experimenting a great deal. Such experiments had not been aiming at the elicitation of autistic behaviour, but rather at its opposite: its elimination. As we have written before, each of these experiments was in reality a subtly modulated series of experiments. For a description of what we actually did, I quote from our original publication. We wrote (12, pp. 29–30):

'What we invariably do when visiting, or being visited by a family with young children is, after a very brief friendly glance, ignoring the child(ren) completely, at the same time eliciting, during our early conversations, friendly responses from the parent(s). One can see a great deal of the behaviour of the child out of the corner of one's eye, and can monitor a surprising amount of the behaviour that reveals the child's state. Usually such a child will start by simply looking intently at the stranger, studying him guardedly. One may already at this stage judge it safe to now and then look briefly at the child and assess more accurately the state it is in. If, on doing so, one sees the child avert its glance, eye contact must at once be broken off. Very soon the child will stop studying one. It will approach gingerly, and it will soon reveal its strong bonding tendency by touching one—for instance by putting its hand tentatively on one's knee. This is often a crucial moment: one must not respond by looking at the child (which may set it back considerably) but by cautiously touching the child's hand with one's own. Again, playing this 'game' by, if necessary, stopping, or going one step back in the process, according to the child's response, one can soon give a mildly reassuring signal by *touch*, for instance by gently pressing its hand, or by touching it quickly, and withdrawing again. If, as is often the case, the child laughs at this, one can laugh oneself, but still without looking at the child. Soon it will become more daring, and the continuation of contact, by touch and by indirect vocalisation, will begin to cement a bond. One can then switch to the first, tentative eye contact. This again must be done with caution, and step by step; certainly with a smile, and for brief moments at first. We find that first covering one's face with one's hands, then turning towards the child (perhaps saying 'where's Andrew?' or whatever the child's name) and then briefly showing one's eyes and covering them up at once, is very likely to elicit a smile, or even a laugh. For this, incidentally, a child often takes the initiative (see, e.g. Stroh and Buick (11)). Very soon the child will then begin to solicit this; it will rapidly tolerate increasingly long periods of direct eye contact and join one. If this is played further, with continuous awareness of and adjustments to slight reverses to a more negative attitude, one will soon find the child literally clamouring for intense play contact. Throughout this process the vast variety of expressions of the child must be *understood* in order to monitor it correctly, and one must oneself *apply* an equally large repertoire in order to give, at any moment, the best signal. The 'bag of tricks' one has to have at one's disposal must be used to the full, and the 'trick' selected must whenever possible be adjusted to the child's individual tastes. Once established, the bond can be maintained by surprisingly slight signals; a child coming to show proudly a drawing it has made is often completely happy with just a 'how nice dear' and will then return to its own play. Even simpler vocal contacts can work; analogous to the vocal contact calls of birds (which the famous Swedish writer Selma Lagerloef correctly described in 'Nils Holgersson' as, 'I am here, where are you?') many children develop an individual contact call, to which one has merely to answer in the same language.

'The results of this procedure have been found to be surprisingly rapid, and also consistent *if one adjusts oneself to the monitoring results*. Different children may require different starting levels, and different tempos of stepping-up. One may even have to start by staying away from the child's favourite room. It is also of great significance how familiar to the child the physical environment is. Many children take more than one day; with such it is important to remember that one has to start at a lower level in the morning than where one left off the previous evening. We have the impression that the process is on the whole completed sooner if one continually holds back until one senses the child longing for a more intense contact.'

With all these experiences with normal children in mind, we began to reconsider the evidence about permanently autistic children—again using our own observations as well as the reports we found in the literature. And two things became clear almost at once: Neither for genetic abnormalities nor for gross brain damage was there any convincing, direct evidence; all we found were inferences, or arguments that do not hold water.

The main argument for a genetic abnormality is the statement (and one hears it time and again) 'these children have been odd from birth'. And we also found that, for various reasons, neither the specialists nor the parents are very willing to consider environmental influences. But in view of what we know about the effects of non-genetic agents that act in *utero*—of which the new indications about the effects of rubella contracted by pregnant women is only one (17)—the 'odd-from-birth' argument is of course irrelevant. And at least two cases are known of identical twins of whom only one developed Kanner's syndrome (18).

Equally unconvincing are the arguments in favour of gross brain damage; this idea too is based mainly on inference.

On the other hand, the body of positive evidence that points to environmental causes is growing. For instance, many workers report that the incidence of autism is not random. Relatively many autists are first-born children (19). There is also a pretty widespread conviction that the parents of autists are somehow different—for instance many of them are very serious people, or people who are themselves under some sort of strain. And to a trained observer it is also very obvious that autists respond to conditions, which to them are frightening or intrusive, by an intensification of all their symptoms. Conversely, we have tried out our 'taming procedure' as described for normal children on some severely autistic children, and succeeded in 'drawing them out of their shells', and making them snuggle up to us, and even in making them join us in, for instance, 'touch games'. I cannot possibly go into all the evidence, but there are several good indications, firstly, that many autists are potentially normal children, whose affiliation and subsequent socialisation processes have gone wrong in one way or another, and secondly: this can often be traced back to something in the early environment—on occasion a frightening accident, but most often something in the behaviour of the parents, in particular the mothers. Let me hasten to add that in saying this we are not *blaming* these unfortunate parents. Very often they seem to have been either simply inexperienced (hence perhaps the high incidence among first-borns); or over-apprehensive; or over-efficient and intrusive; or—perhaps most often—they are people who are themselves under stress. For this, and many other reasons, the parents of autists deserve as much compassion, and may be as much in need of help, as the autists themselves.

Now if we are only partially right in assuming that at least a large proportion of autists are victims of some kind of environmental stress, whose basic trouble is of an emotional nature, then one would expect that those therapies that aim at reducing anxiety—by *allowing* spontaneous socialisation and exploration whenever it occurs—would be more successful than those that aim at the teaching of specific skills. Unfortunately (as I have already said), it is hardly possible to judge from published reports what treatment has actually been applied. For instance one speech therapist may behave rather intrusively, and turn a child into a mere 'trained monkey,' leaving all the other symptoms as they were, or even making them worse. Another speech therapist may have success simply by having proceeded in a very gentle, motherly way. One has to go by those instances where one has either been involved oneself or where one knows pretty precisely how the therapist has in fact proceeded. It is with this in mind that I will now mention briefly three examples of treatments that seem to hold great promise.

Firstly, even before we published our first paper, the Australian therapist Helen Clancy had been treating autistic children and their families along lines that are very similar to, in fact are more sophisticated than those recommended by us in 1972.

The gist of Clancy's method is as follows (8): firstly, since she considers the restoration of affiliation as the first goal of treatment of autism, she treats both mother and child, and the family as well. She does this by provoking in the mother an increase in maternal, protective behaviour. Secondly, she uses a form of operant conditioning for speeding up the child's response to this change in the mother. In other words, she tries to elicit a mutual emotional bond between mother and child, and refrains, at least at first, from the piecemeal teaching of particular skills.

With those mothers who were willing to cooperate, Clancy has achieved highly encouraging success—although of course a few families (4 out of appr. 50 treated over a period of 14 years) have failed to benefit.

Secondly, after the first public discussion of our work, my wife received invitations to visit some schools for autists, and to observe what was being done. She found that in one of them, a small day school which already had an impressive record of recoveries, the treatment was likewise aimed at the restoration of emotional security, and teaching as such, including some gentle speech therapy, was never started until a child had reached a socially positive attitude. Much to our dismay, this school has since been incorporated into a school for maladjusted children—the experiment has been discontinued.

Thirdly, a regional psychiatrist invited us a year ago to act as advisors in a fascinating experiment which she too had begun well before she had heard of our work. Three boys, who are now 9, 9 1/2 and 11 1/2 years of age, and who had all been professionally diagnosed as severely autistic, are now being gently integrated into a normal primary school. This involves a part-time hometutor for each boy, a sympathetic headmaster, and willingness of the parents to cooperate. The results are already little short of spectacular. In fact, a specialist on autistic children who visited the school recently said to us: 'Had the records not shown that these three children were still severely autistic a couple of years ago, I would not now believe it'. This experiment, which is also run along lines that are consistent with our ideas, is being carefully documented. It is this type of evidence, together with that provided by a number of already published case histories (20), that has by now convinced us that many autists can attain full recovery, if only we act on the assumption that they have been traumatised rather than genetically or organically damaged.

I cannot go into further details here, but I can sum up in a few sentences the gist of what the ethological approach to Early Childhood Autism has produced so far:

1. There are strong indications that many autists suffer primarily from an emotional disturbance, from a form of anxiety neurosis, which prevents or retards normal affiliation and subsequent socialisation, and this in its turn hampers or suppresses the development of overt speech, of reading, of exploration, and of other learning processes based on these three behaviours.

2. More often than has so far been assumed these aberrations are not due to either genetic abnormalities or to gross brain damage, but to early environmental influences. The majority of autists—as well as their parents—seem to be genuine victims of environmental stress. And our work on normal children has convinced us not only that this type of stress disease is actually on the increase in western and westernised countries, but also that very many children must be regarded as semi-autistic, and even more as being seriously at risk.

3. Those therapies that aim at the reduction of anxiety and at a re-starting of proper socialisation seem to be far more effective than for instance speech therapy per se and enforced social instruction, which seem to be at best symptom treatments, and to have only limited success. Time and again treatment at the emotional level has produced an explosive emergence of speech and other skills.

If I now try to assess the implications of what I have said, I feel at the same time alarmed and hopeful.

We are alarmed because we found this corner of Psychiatry in a state of disarray, and because we discovered that many of the established experts—doctors, teachers and therapists—are so little open to new ideas and even facts. Another cause for alarm is our conviction that the officially recognised autists are only a fraction of a much larger number of children who obviously suffer to some degree from this form of social stress.

We feel hopeful because attempts at curing such children at the emotional level, while still in the experimental stage, are already leading to positive results. And another encouraging sign is that, among the young psychiatrists, we have found many who are sympathetic to our views, or even share them, and begin to act on them.

In the interest of these thousands of unfortunate children we appeal to all concerned to give the 'stress view' of autism at least the benefit of the doubt, and to try out the forms of therapy that I have mentioned.

My second example of the usefulness of an ethological approach to Medicine has quite a different history. It concerns the work of a very remarkable man, the late F. M. Alexander (21). His research started some fifty years before the revival of Ethology for which we are now being honoured, yet his procedure was very similar to modern observational methods, and we believe that his achievements and those of his pupils deserve close attention.

Alexander, who was born in 1869 in Tasmania, became at an early age a 'reciter of dramatic and humorous pieces'. Very soon he developed serious vocal trouble and he came very near to losing his voice altogether. When no doctor could help him, he took matters into his own hands. He began to observe himself in front of a mirror, and then he noticed that his voice was at its worst when he adopted the stances which to him felt appropriate and 'right' for what he was reciting. Without any outside help he worked out, during a series of agonising years, how to improve what is now called the 'use' of his body musculature in all his postures and movements. And, the remarkable outcome was that he regained control of his voice. This story, of perceptiveness, of intelligence, and of persistence, shown by a man without medical training, is one of the true epics of medical research and practice (22).

Once Alexander had become aware of the mis-use of his own body, he began to observe his fellow men, and he found that, at least in modern western society, the majority of people stand, sit and move in an equally defective manner.

Encouraged by a doctor in Sydney, he now became a kind of missionary. He set out to teach—first actors, then a variety of people—how to restore the proper use of their musculature. Gradually he discovered that he could in this way alleviate an astonishing variety of somatic and mental illnesses. He also wrote extensively on the subject. And finally he taught a number of his pupils to become teachers in their turn, and to achieve the same results with their patients. Whereas it had taken him years to work out the technique and to apply it to his own body, a successful course became a matter of months—with occasional 'refresher' sessions afterwards. Admittedly, the training of a good Alexander teacher takes a few years.

For scores of years a small but dedicated number of pupils have continued his work. Their combined successes have recently been described by Barlow (23). I must admit that his physiological explanations of how the treatment could be supposed to work (and also a touch of hero worship in his book) made me initially a little doubtful, and even sceptical. But the claims made, first by Alexander, and reiterated and extended by Barlow sounded so extraordinary that I felt I ought to give the method at least the benefit of the doubt. And so, arguing that medical practice often goes by the sound empirical principle of 'the proof of the pudding is in the eating', my wife, one of our daughters and I decided to undergo treatment ourselves, and also to use the opportunity for observing its effects as critically as we could. For obvious reasons, each of us went to a different Alexander teacher.

We discovered that the therapy is based on exceptionally sophisticated observation, not only by means of vision but also to a surprising extent by using the sense of touch. It consists in essence of no more than a very gentle, first exploratory, and then corrective manipulation of the entire muscular system.

This starts with the head and neck, then very soon the shoulders and chest are involved, and finally the pelvis, legs and feet, until the whole body is under scrutiny and treatment. As in our own observations of children, the therapist is continuously monitoring the body, and adjusting his procedure all the time. What is actually done varies from one patient to another, depending on what kind of mis-use the diagnostic exploration reveals. And naturally, it affects different people in different ways. But between the three of us, we already notice, with growing amazement, very striking improvements in such diverse things as high blood pressure, breathing, depth of sleep, overall cheerfulness and mental alertness, resilience against outside pressures, and also in such a refined skill as playing a stringed instrument.

So from personal experience we can already confirm some of the seemingly fantastic claims made by Alexander and his followers, namely that many types of under-performance and even ailments, both mental and physical, can be alleviated, sometimes to a surprising extent, by teaching the body musculature to function differently. And although we have by no means finished our course, the evidence given and documented by Alexander and Barlow, of beneficial effects on a variety of vital functions no longer sounds so astonishing to us. Their long list includes first of all what Barlow calls the 'rag bag' of rheumatism, including various forms of arthritis; but also respiratory troubles, even potentially lethal asthma; following in their wake, circulation defects, which may lead to high blood pressure and also to some dangerous heart conditions; gastro-intestinal disorders of many types; various gynaecological conditions; sexual failures; migraines and depressive states that often lead to suicide—in short a very wide spectrum of diseases, both 'somatic' and 'mental', that are not caused by identifiable parasites.

Although no one would claim that the Alexander treatment is a cure-all in every case, there can be no doubt that it often does have profound and beneficial effects—and, I repeat once more, both in the 'mental' and 'somatic' sphere.

The importance of the treatment has been stressed by many prominent people, for instance John Dewey (24), Aldous Huxley (25), and—perhaps more convincing to us—by scientists of renown, such as Coghill (26), Raymond Dart (27), and the great neurophysiologist Sherrington (28). Yet, with few exceptions, the medical profession has largely ignored Alexander—perhaps under the impression that he was the centre of some kind of 'cult', and also because the effects seemed difficult to explain. And this brings me to my next point.

Once one knows that an empirically developed therapy has demonstrable effects, one likes to know how it could work; what its physiological explanation could be. And here some recent discoveries in the borderline field between neurophysiology and ethology can make some aspects of the Alexander therapy more understandable and more plausible than they could have been in Sherrington's time.

One of these new discoveries concerns the key-concept of 're-afference' (29). There are many strong indications that, at various levels of integration, from single muscle units up to complex behaviour, the correct performance of many movements is continuously checked by the brain. It does this by comparing a *feedback report*, that says 'orders carried out', with the feedback *expectation* for which, with the initiation of each movement, the brain has been alerted. Only when the expected feedback and the actual feedback match does the brain stop sending out commands for corrective action. Already the discoverers of this principle, von Holst and Mittelstaedt, knew that the functioning of this complex mechanism could vary from moment to moment with the internal state of the subject—the 'target value' or *Sollwert* of the expected feedback changes with the motor commands that are given. But what Alexander has discovered beyond this is that a lifelong mis-use of the body muscles (such as caused by, for instance, too much sitting and too little walking) can make the entire system go wrong. As a consequence, reports that 'all is correct' are received by the brain (or perhaps interpreted as correct) when in fact all is very wrong. A person can feel 'at ease' e.g. when slouching in front of a television set, when in fact he is grossly abusing his body. I can show you only a few examples, but they will be familiar to all of you.

It is still an open question exactly where in this complex mechanism the matching procedure goes wrong under the influence of consistent mis-use. But the modern ethologist feels inclined, with Alexander and Barlow, to blame phenotypic rather than genetic causes for mis-use. It is highly unlikely that in their very long evolutionary history of walking upright, the Hominids have not had time to evolve the correct mechanisms for bipedal locomotion. This conclusion receives support from the surprising, but indubitable fact that even after forty to fifty years of obvious mis-use one's body can (one might say) 'snap' back into proper, and in many respects more healthy use as a result of a short series of half-hourly sessions. Proper stance and movement are obviously genetically old, environment-resistant behaviours (30). Mis-use, with all its psychosomatic or rather, somato-psychic consequences, must therefore be considered a result of modern living conditions—of a culturally determined stress. I might add here that I am not merely thinking of too much sitting, but just as much of the 'cowed' posture that one assumes when one feels that one is not quite up to one's work—when one feels insecure.

Secondly, it need not cause surprise that a mere gentle handling of body muscles can have such profound effects on both body and mind. The more that is being discovered about psychosomatic diseases, and in general about the extremely complex two-way traffic between the brain and the rest of the body, the more obvious it has become that too rigid a distinction between 'mind' and 'body' is of only limited use to medical science—in fact can be a hindrance to its advance.

A third biologically interesting aspect of the Alexander therapy is that every session clearly demonstrates that the innumerable muscles of the body are continuously operating as an intricately linked web. Whenever a gentle pressure is used to make a slight change in leg posture, the neck muscles react immediately. Conversely, when the therapist helps one to 'release' the neck muscles, it is amazing to see quite pronounced movements for instance of the toes, even when one is lying on a couch.

In this short sketch, I can do no more than characterise, and recommend, the Alexander treatment as an extremely sophisticated form of rehabilitation, or rather of re-deployment, of the entire muscular equipment, and through that of many other organs. Compared with this, many types of physiotherapy which are now in general use look surprisingly crude and restricted in their effect—and sometimes even harmful to the rest of the body.

What then is the upshot of these few brief remarks about Early Childhood Autism and about the Alexander treatment? What have these two examples in common? First of all they stress the importance for medical science of openminded observation—of 'watching and wondering'. This basic scientific method is still too often looked down on by those blinded by the glamour of apparatus, by the prestige of 'tests' and by the temptation to turn to drugs. But it is by using this old method of observation that both autism and general mis-use of the body can be seen in a new light: to a much larger extent than is now realised both could very well be due to modern stressful conditions.

But beyond this I feel that my two excursions into the field of medical research have much wider implications. Medical science and practice meet with a growing sense of unease and of lack of confidence from the side of the general public. The causes of this are complex, but at least in one respect the situation could be improved: a little more open-mindedness (31), a little more collaboration with other biological sciences, a little more attention to the body as a whole, and to the unity of body and mind, could substantially enrich the field of medical research. I therefore appeal to our medical colleagues to recognise that the study of animals—in particular 'plain' observation—can make useful contributions to human biology not only in the field of somatic malfunctioning, but also in that of behavioural disturbances, and ultimately help us understand what psychosocial stress is doing to us, It is stress in the widest sense, the inadequacy of our adjustability, that will become perhaps the most important disruptive influence in our society.

If I have today stressed the applicability of animal behaviour research I do not want to be misunderstood. As in all sciences, applications come in the wake of research motivated by sheer intellectual curiosity. What this occasion enables me to emphasise is that biologically oriented research into animal behaviour, which has been done so far with very modest budgets, deserves encouragement—whatever the motivation and whatever the ultimate aims of the researcher. And we ethologists must be prepared to respond to the challenge if and when it comes.

Notes and References

1. I call the method old because it must already have been highly developed by our ancestral hunter–gatherers, as it still is in non-westernised hunting–gathering tribes such as the Bushmen, the Eskimo and the Australian Aborigines. As a scientific method applied to Man it could be said to have been revived first by Charles Darwin in 1872 in 'The Expression of the Emotions in Man and the Animals.' London, John Murray.
2. Levi, L. (ed.) 1971 Society, Stress, and Disease. Vol. 1.: The Psychosocial Environment and Psychosomatic Diseases. London, Oxford University Press.
3. Kanner, L. 1943 Autistic disturbances of affective contact. Nerv. Child 2, 217–250. Recently, Kanner has published a selection of his papers in Childhood Psychosis 1973 Washington D.C., Winston & Son (distributed by John Wiley & Sons).
4. When I speak of Kanner's syndrome, I refer to the largely descriptive list of symptoms given by O'Gorman, G. 1970 The Nature of Childhood Autism. London, Butterworth. This is a slightly modified version of the description given in Creak, M. 1961 (Chairman) The schizophrenic syndrome in childhood. Progress report of a working party. Brit. Med. J. 2, 889–890. Many other definitions of autism in its various forms are mixtures of observed behavioural deviations and interpretations. For a discussion of the confusion surrounding the word 'autism' see Tinbergen & Tinbergen (see below, note 12) pp. 45–46.
5. For the purpose of finding one's way in this literature we can refer to Rutter, M. 1965 Infantile Autism. London, Methuen; and to the quarterly, started in 1971: Journal of Autism and Childhood Schizophrenia, published by Winston & Son, Washington, D.C., which prints original articles as well as reviews. The most recent and most exhaustive review is: Ornitz, E. M. 1973 Childhood Autism. (Medical Progress). Calif. Med. 118, 21–47.

   Throughout the literature (not only on autism but on many other psychiatric issues as well) one finds one fundamental error in scientific reasoning. Time and again we receive the comment that we overlook the 'hard' evidence of internal malfunctioning, in autists as well as in other categories of the mentally ill. I assure my readers that we do not *overlook* such evidence (such as that on blood platelets, on lead contents, on EEG patterns, etc.). The erroneous assumption underlying most of the arguments in which such facts are used for the purpose of throwing light on the causation of the behavioural deviation is almost invariably due to the *confusion between correlations and cause–effect relations*. With some exceptions (such as the deleterious effect of lead) the physiological or biochemical evidence is considered, without any ground whatsoever, to indicate causes, whereas the correlations found could just as well point to consequences or side-effects. It is just as *nonsensical to say that retarded bone growth, or abnormalities in the blood platelet picture (or for that matter speech defects, or high overall arousal) are 'causes' of autism as it is to say that a high temperature is the 'cause' of typhoid, or pneumonia.* Unless there is evidence, clinical and ultimately experimental, indicating what is cause and what is effect the opinions based on 'hard' evidence are in fact worthless. Our experimental evidence discussed on pp. 10 and 11 is hard, whereas evidence on correlations—however impressive the techniques might be by which they are found—are scientifically useless until an attempt is made to place it into cause-effect context. This is what I mean in my final paragraphs by 'the glamour of apparatus'—the idolisation of techniques, coupled with the failure to think about the meaning of evidence, is a serious disease of medical research.
6. Rimland, B. 1971 The differentiation of childhood psychoses. J. Aut. Childh. Schizophr. 1, 161–175.
7. Wing, L. 1970 The syndrome of early childhood autism. Brit. J. Hosp. Med. 381–392 (p. 381). See also Tinbergen and Tinbergen (see below, note 12) p. 51.

8. One of the most prominent exponents of this view is: Bettelheim, B. 1967 The Empty Fortress: Infantile Austism and the Birth of Self. London, Collier-Macmillan. See also: Clancy, H. and G. McBride 1969 The autistic process and its treatment. J. Child Psychol Psychiatry 10, 233–244.

9. Rutter, M. et. al. 1971 Autism-a central disorder of cognition and language? In: M. Rutter (ed.) Infantile Autism: Concepts, Characteristics, and Treatment. London, Churchill.

10. Hutt, C., Hutt, S. J., Lee, D. and Ounsted, C., 1964 Arousal and childhood autism. Nature 204, 908–909.
Hutt, S. J. and Hutt, C. (eds.) 1970 Behaviour Studies in Psychiatry. Oxford, Pergamon Press.

11. Stroh, G. Buick, D. 1970 The effect of relative sensory isolation on the behaviour of two autistic children. In: S. J. Hutt and C. Hutt (eds.) Behaviour Studies in Psychiatry. Oxford, Pergamon Press. pp. 161–174.

12. Tinbergen, E. A. and Tinbergen, N. 1972 Early Childhood Autism-an Ethological Approach. Advances in Ethology 10, l–53. Berlin, Paul Parey.

13. Hutt, S. J. and Hutt, C., 1970 Direct Observation and Measurement of Behaviour. Springfield, Ill., Charles C. Thomas.

14. Medawar, P. B. 1967 The Art of the Soluble. London, Methuen.

15. For a recent review about the analysis of non-verbal signs of mixed motivation, or motivational conflicts, see e.g. Manning, A. 1972 An Introduction to Animal Behaviour. London, Arnold. (Chapter 5) and Hinde, R. A. 1970 Animal Behavior. New York, McGraw-Hill. (Chapter 17). Both books give further references.

16. See e.g. Blurton Jones, N. G. (ed.) Ethological Studies of Infant Behaviour. London, Cambridge University Press.

17. Chess, S. 1971 Autism in children with congenital rubella. J. Aut. Childh. Schizophr. I, 33–47. The point I want to make with this brief reference is that, while one should call rubella an early environmental influence and therefore not 'congenital' in the sense of 'genetic', it might well be correct to call it 'organic', even though rubella could well create a state of anxiety already during pregnancy in mothers who have heard about other damaging effects of the disease. And this in itself could well cause a complex psychosomatic state.

18. Kamp, N. L. J. 1964 Autistic syndrome in one of a pair of monozygotic twins. Psychiatr., Neurol., Neurochir. 67, 143–147; and Vaillant, G. E. 1963 Twins discordant for early infantile autism. Arch. Gen. Psychiatry 9, 163–167. While I do not of course intend to underrate the possibility of genetic predisposition, the hypothesis of a purely genetic deviation conflicts with this type of observation. At the same time we know that even when twins grow up in the same family, their experiences can never be identical.

19. See, e.g. Wing, L. 1971 Autistic Children. London, Constable. p. 8.

20. Although not all the authors of the following books label their subject as 'autistic', I mention them because the descriptions of the initial behaviour conform in whole or in part to Kanner's syndrome; and, as I have said, I consider such descriptions the only acceptable starting points.
d'Ambrosio, R. 1971 No Language but a Cry. London, Cassell.
Axline, V. 1964 Dibs—in Search of Self. (since 1971 available as a Pelican Book, Harmondsworth, Penguin Books Ltd.)
Copeland, J. and J. Hodges 1973 For the Love of Ann. London, Arrow Original.
Hundley, J. M. 1973 (first published in 1971) The Small Outsider. Sydney, Angus and Robertson.
Park, C. C. 1972 (first published in 1967) The Siege. Pelican Book, Harmondsworth, Penguin Books Ltd.
Wexler, S. S. 1971 (first published in 1955) The Story of Sandy. New York, A Signet Book.
Thieme, G. 1971 Leben mit unserem autistischen Kind. Lüdenscheid, Hilfe für das autistische Kind, e.V.
No two of these seven children received the same treatment, but on the whole one can say that those who were treated primarily at the emotional level rather than at the level of specific skills showed the most striking improvement.

21. The clearest introduction is: Alexander, F. M. 1932 The Use of Self. London Chaterston; but a great deal of interest can also be found in Alexander, F. M. 1910 Man's Supreme Inheritance. London, Chaterston, and Alexander, F. M. 1942 The Universal Constant in Living. London, Chaterston.

22. The history of medical science is full of such examples of 'breakthroughs' due to a re-orientation of attention. Compare e.g. Jenner's discovery that milkmaids did not contract smallpox; Goldberger's observation that the staff of a 'lunatic asylum' did not develop pellagra; Fleming's wondering about empty areas round the *Pennicillium* in his cultures.

23. Barlow, W. 1973 The Alexander Principle. London, Gollancz.

24. Dewey, K. See e.g. 1932 Introduction to Alexander, F. M. The Use of Self. London, Chaterston.

25. Huxley, A. 1937 Ends and Means. London: Chatto and Windus. Huxley, A. 1965 End-gaining and means whereby. Alexander J. 4, 19.

26. Coghill, G. E. 1941 Appreciation: the educational methods of F. Matthias Alexander. In: Alexander, F. M. 1941 The Universal Constant in Living. New York, Dutton.

27. Dart, R. A. 1947 The attainment of poise. South Afr. Med. J. 21, 74–91. Dart, R. A. 1970 An anatomist's tribute to F. M. Alexander, London, the Shelldrake Press.

28. Sherrington, C. S. 1946 The Endeavour of Jean Fernel. London, Cambridge University Press. Sherrington, C. S. 1951 Man on his Nature. London, Cambridge University Press.

29. von Holst, E. and Mittelstaedt, H. 1950 Das Reafferenzprinzip. Naturwiss. 37, 464–476.

30. Tinbergen, N. 1973 Functional ethology and the human sciences. Proc. Roy. Soc. Lond. B. 182, 385–410.

31. This plea is nowadays heard more often: see e.g. Kanner, L. 1971 Approaches: retrospect and prospect. J. Aut. Childh. Schizophr. 1, 453–459 (see p. 457).

See also Frisch, Karl Von (1886–1982)

History—*History of Animal Behavior Studies*
History—Niko Tinbergen and the "Four Questions" of Ethology
Lorenz, Konrad (1903–1989)
History—*Some Leaders in Animal Behavior Studies*
Nobel Prize—*The 1973 Nobel Prize for Physiology or Medicine*

■ Tools
Tool Manufacture by a Wild Bonnet Macaque

Many wild and captive primates, belonging to virtually all families of nonhuman primates, except marmosets, tamarins and the primitive prosimians, are known to *use* tools—an example of a behavioral pattern among animals that has interesting implications for the evolution of material culture by human beings. Authentic examples of the *manufacture* of tools by primates, on the other hand, appear to be extremely rare, restricted primarily to chimpanzees and orangutans among the apes and capuchins among monkeys; the latter modify tools readily to obtain hidden foods, but only in captivity! Of the African and Asian monkeys, only lion-tailed macaques in captive social groups were observed to manufacture tools by detaching sticks from larger branches to extract food through narrow openings of enclosed containers. Such observations of primate tool use and modification—both in the wild and in the laboratory—have led to the view that tools are primarily used by primates to facilitate the acquisition and processing of food.

Some primates may, however, rarely use tools in very different contexts. In fact, the only example of elaborate tool manufacture by any monkey in the wild is that by a female bonnet macaque (*Macaca radiata*), approximately 15 years of age, that lived in a large troop of about 45 individuals in semi-arid scrublands on the outskirts of the city of Bangalore in southern India. She was observed to repeatedly insert a dry stick, stiff leaf or grass blade, or a leaf

*A bonnet macaque (*Macaca radiata*) washes a mango in a forest stream in the deciduous forest of Tamil Nadu in southern India. This rare behavior, displayed by two particular individuals, may mark the beginning of a social tradition in which a tool (water) is used to clean the surface of the fruit before it is consumed.*
Courtesy of Anindya Sinha.

midrib into her vagina and scratch vigorously, in response to some irritation that appeared to bother her persistently over months. During some 21 hours of intensive observation between March 1993 and September 1994, this female scratched her genitalia on at least 18 occasions. Her dependence on these objects is shown by the fact that on 15 of these occasions she used a tool a total of 34 times, while only on 3 occasions did she use her forefingers alone.

What was more remarkable, however, was that on 8 out of the 15 occasions on which she used objects, she actively manufactured or modified her tools. She did this by usually removing the leaf blade of dry *Eucalyptus globosus* or *Acacia auriculiformis* leaves with her fingers or teeth, breaking the midrib into several pieces and using only a single short piece. Occasionally, she also detached short sticks from branched twigs, broke them into several pieces, and then used one of these shortened sticks.

The remarkable ability of this individual to use and appropriately modify different objects to achieve the same goal suggests that she could perhaps comprehend the function of detached objects in mediating changes in another out-of-reach object through systematic control. However, does her versatility reflect an insightful use of these tools? Did she have a mental model of a tool to which she could repeatedly refer? The use of different leaf midribs after removal of the blade strongly suggests that she could have indeed recognized a tool-like pattern (stick) within an apparently dissimilar object (leaf) through an appropriate mental representation of her ideal tool. If both these inferences are true, these cognitive mechanisms would correspond to the two highest levels of development in Piaget's series of (human) sensorimotor intelligence, a model that has been invoked in nonhuman primates as well.

Sophisticated tool-using abilities are believed to have evolved independently in different primate groups primarily as adaptation for the retrieval and processing of embedded foods. This example, however, clearly shows that some species may indeed possess the potential to use and make tools under very different contexts. This observation, coupled with the fact that all laboratory studies on primate tool use have so far used the paradigm of food acquisition, suggests that it is vitally important to document the occurrence of rare behavioral events and patterns, and also their contexts, in wild primate groups; similarly, alternative experimental protocols may have to be designed in order to conduct controlled studies of tool use technology under captive conditions.

See also Tools—*Tool Use*
                Tools—*Tool Use and Manufacture by Birds*
                Tools—*Tool Use by Dolphins*
                Tools—*Tool Use by Elephants*

Further Resources

Beck, B. 1980. *Animal Tool Behavior: The Use and Manufacture of Tools by Animals*. New York: Garland STPM Press.

Berthelet, A. & Chavaillon, J. (Eds). 1993. *The Use of Tools by Human and Non-human Primates*. Oxford: Clarendon Press.

Chevalier-Skolnikoff, S. 1989. *Spontaneous tool use and sensorimotor intelligence in Cebus compared with other monkeys and apes*. Behavioral and Brain Sciences, 12, 561–627.

Tomasello, M. & Call, J. 1997. *Primate Cognition*. New York: Oxford University Press.

van Schaik, C. P., Deaner, R. O. & Merrill, M. Y. 1999. *The conditions for tool use in primates: Implications for the evolution of material culture*. Journal of Human Evolution, 36, 719–741.

Anindya Sinha

■ Tools
Tool Use

Only a few centuries ago, most people believed that humans were the only species intelligent enough to use tools. Tool use, so it was believed, was one of the few capacities that separated our species from other animals. Today, however, scientists realize that humans are not alone in their skilled use of objects. In fact, many different animals—from crabs to crows to chimpanzees—use tools in complex, functional ways.

In an influential book, Benjamin Beck (1980) provided a comprehensive and much cited definition of animal tool use. To be a tool user, according to Beck, an organism must carry, modify, or manipulate an external object and then use it to effect a change on another object in the environment. This definition suggests that in order to qualify as tool-using, an animal must utilize an object outside its own body to achieve a particular goal; simply using one's own body part (e.g., a strong beak, prehensile tail, long claw) in a functional way does not count as tool use. Similarly, Beck's definition requires organisms to demonstrate some action on the object (e.g., orienting it correctly, carrying it, modifying it) prior to use. As such, merely using an external object functionally (e.g., pulling a vine to get an attached piece of fruit) does not qualify. To date, scientists have observed many examples of animal behavior that qualify as tool use under Beck's definition (see Beck 1980 for review). Since a comprehensive review of animal tool use is beyond the focus of this entry, we instead focus on a few of the most common examples of tool use in wild nonhuman animals.

Many classic examples of nonhuman tool use are seen in the domain of foraging. A number of species use tools to carry food during foraging: Ants (Genus *Aphaenogaster*) create sponges from leaves and wood and use these sponges to carry fruity liquids back to their colony (Fellers & Fellers 1976). Similarly, a number of researchers have observed wild chimpanzees using leaf sponges to carry water from streams and holes in trees (Goodall 1986).

Other animals use tools to open hard objects that contain food. Sea otters, for example, use stone tools both as levers to pry clams from their fixed underwater location and as anvils placed on their chest to use while hammering the clams (see Griffin 1992). Similarly, Egyptian vultures (*Neophron percnopterus*) find hard stones to throw at vulture eggs they desire to break open (Goodall & van Lawick 1966). In perhaps the most famous example, wild chimpanzees (*Pan troglodytes*) in the Tai forest crack hard nuts using stone hammers and anvils (see McGrew 1992 for review). Moreover, Boesch & Boesch (1983) observed that chimpanzees carry heavier stones longer distances, suggesting that they know to invest in a particularly effective hammering stone.

Animals also use tools to extract foods from hard-to-reach places. In addition to their skills with stone hammers, wild chimpanzees are famous for their extraction tools; chimpanzees use twigs to extract foods as diverse as termites, ants, insect larvae, honey, and bone marrow from hard to reach places (Goodall 1986). Likewise, woodpecker finches (*Cactospiza pallida*) from the Galapagos islands hold cactus spines in their beaks and use these to pry insects from inside their burrows (Grant 1986). The New Caledonian crow (*Corvus moneduloides*) also forages for insects and larvae that live in the tiny cavities of trees. Hunt (1996, 2000) observed that wild crows extract these foods using at least two different types of extraction tools, each designed for the task at hand. The first, a long thin hook-shaped twig is used to hook insects hiding in narrow holes. The other, a rough horizontal tool made from leaves, is designed for extracting foods from less narrow kinds of holes.

Animals also use objects to protect themselves and their young from predation and other unwanted social situations. A number of primate species break branches and use them during defensive displays against predators or conspecifics (e.g., Beck 1980; Boinski 1988; Goodall 1986). Female sand wasps (*Ammophilia campestris*) protect their buried eggs by using stone tools to pack down the soil and sand above their burrows (Beck 1980). Moreover, certain crab species pick up stinging anemones and hold them with their claws to ward off potential predators (see Griffin 1992). Other animals use tools to protect themselves from uncomfortable physical situations; chimpanzees, for example, break off pieces of bark to use as primitive slippers when climbing up prickly trees to get fruit (Alp 1997).

Although many animals use tools, only a few species actually modify objects and make new tools. One such species is the chimpanzee, which modifies a number of different objects (see Tomasello & Call 1997 for review). Chimpanzees will chew leaves to make their sponges more absorbent, bite branches to make effectively-sized probing tools, and alter the height of a stone anvil using smaller pebbles. Chimpanzees can also use tools in combination to solve problems (Sugiyama 1997). Chimpanzees have been observed to use small twigs to push their leaf sponges inside the cavities of trees and to employ two different tools in succession to perforate and then dip into termite mounds. But chimpanzees are not the only species that modifies tools. Recent research suggests that New Caledonian crows also modify leaves and twigs to create better tools. In order to learn what these birds understood about creating effective extraction tools, Weir, Chappell, & Kacelnik (2002) presented captive crows with a problem in which a bucket containing food was lodged in a narrow transparent tube. To retrieve the bucket, birds were allowed to use straight pieces of pliant wire, a material they had no prior experience with. During the study, one female crow found a solution to the extraction problem; she took the wire and bent it to create a curved cane shape that could be used to hook the bucket. These results suggest that crows, like chimpanzees, are able to modify objects flexibly and spontaneously to solve novel foraging tasks.

See also Tools—*Tool Manufacture by a Wild Bonnet Macaque*
Tools—*Tool Use and Manufacture by Birds*
Tools—*Tool Use by Dolphins*
Tools—*Tool Use by Elephants*

Further Resources

Alp, R. 1997. *Stepping-sticks and seat-sticks: New types of tools used by wild chimpanzees* (Pan troglodytes) *in Sierra Leone*. American Journal of Primatology, 41, 45–52.

Beck, B. B. 1980. *Animal Tool Behavior: The Use and Manufacture of Tools*. New York: Garland Press.

Boinski, S. 1988. *Use of a club by a wild white-faced capuchin* (Cebus capucinus) *to attack a venomous snake* (Bothrops asper). American Journal of Primatology, 14, 177–179.

Fellers, J. & Fellers, G. 1976. *Tool use in a social insect and its implications for competitive interactions*. Science, 192, 70–72.

Goodall, J. 1986. *The Chimpanzees of Gombe: Patterns of Behavior*. Cambridge, MA: Harvard University Press.

Goodall, J. & van Lawick, H. 1966. *Use of tools by the Egyptian vulture Neophron percnopterus*. Nature, 212, 1468–69.

Grant, P. R. 1986. *Ecology and Evolution of Darwin's Finches*. Princeton, NJ: Princeton University Press.

Griffin, D. R. 1992. *Animal Minds*. Chicago: University of Chicago Press.

Hunt, G. R. 1996. *Manufacture and use of hook-tools by New Caledonian crows*. Nature, 379, 249–251.

McGrew, W. 1992. *Chimpanzee Material Culture: Implications for Human Evolution*. Cambridge, UK: Cambridge University Press.

Hunt, G. R. 2000. *Human-like, population-level specialization in the manufacture of pandanus tools by New Caledonian crows* Corvus moneduloides. Proceedings of the Royal Society London B, 267, 403–413.

McGrew, W. C. 1992. *Chimpanzee Material Culture: Implications for Human Evolution*. Cambridge: Cambridge University Press.

Sugiyama, Y. 1997. *Social tradition and the use of tool-composites by wild chimpanzees*. Evolutionary Anthropology, 6, 23–27.

Tomasello, M. & Call, J. 1997. *Primate Cognition*. Oxford: Oxford University Press.

Weir, A. A. S., Chappell, J., & Kacelnik, A. 2002. *Shaping of hooks in New Caledonian crows*. Science, 297, 981.

Laurie R. Santos

■ Tools
Tool Use and Manufacture by Birds

Of course, we all know that only humans make and use tools, right? Well, OK, some primates might be able to, but certainly no other animals.

This was the traditional view of tool use, but is, in fact, very wrong. Tool manufacture and use is surprisingly widespread among animals (see Griffin 2001 or Hauser 2001 for examples), including several birds. Most common is the use of twigs or pieces of bark to extract insects from crevices, but there are also birds that drop stones onto eggs to break them (or onto intruders to scare them off), birds that attract fish with bait, and even birds that use herbs to control nest parasites such as mites!

Perhaps the most impressive avian tool users, however, are New Caledonian crows (*Corvus moneduloides*). These crows are endemic to the island of New Caledonia in the South Pacific. They make at least four types of tools—straight twigs, hooked sticks, barbed vine leaves, and stepped-cut pandanus probes—and use these to extract invertebrates from crevices. The pandanus probes are made from the long, broad leaves of the pandanus tree (or "screw pine"): The crow cuts into the leaf, then tears it longitudinally, repeating this sequence several times until it has fashioned a tapering, stepped tool, with the optimal combination of stiffness (due to its wide base) and usefulness (the narrow tip). Such complex tool manufacture is almost unknown outside of humans.

Intriguingly, experiments with captive New Caledonian crows have shown that their tool use is not highly rigid and stereotyped, but instead seems to be flexible and sophisticated. For example, they can choose sticks of the correct length and diameter to retrieve pieces of meat hidden inside a tube, and if a suitable one is not available, they will make one from a branch. These tasks are similar to situations they face in the wild—but they can also create completely novel tools for novel problems.

Betty removes a bucket containing meat from a vertical tube by using a wire she has bent into a hook.
Courtesy of Alex Kacelnik.

In one study, they were given the task of retrieving a small container of meat from the bottom of a vertical pipe. This was something they were very proficient at, having been given sticks with hooks on the ends in previous experiments. However, on one occasion when they were offered pieces of wire instead of sticks, the male subject stole the hooked wire, leaving the female with the straight one. After trying for a while with this, she suddenly took it to a corner of the apparatus and bent it into a hook, which she immediately used to retrieve the bucket. In subsequent trials in which she was only given straight wire, she repeated the behavior almost every time.

The reason that observations of tool use are so interesting is that they might provide a window into the mind of the animal using the tool. The sort of flexible, innovative tool use found in New Caledonian crows isn't even seen in our closest relatives, chimpanzees, and might help shed light on how our own minds evolved. Tools, unlike behavior, "fossilize" and leave behind a tangible record of their manufacture and use. This can tell us something about how cultures develop, and about the social transfer of information within a species—and there is tantalizing evidence that wild New Caledonian crow tool use has indeed "evolved" culturally (see Hunt & Gray 2003). Tool use also throws up intriguing questions about the general intelligence of tool users: Is their brain specialized for using tools, or are they all-round problem solvers? These are all questions that remain to be answered.

See also Corvids—*The Crow Family*
    Tools—*Tool Manufacture by a Wild Bonnet*
        Macaque
    Tools—*Tool Use*
    Tools—*Tool Use by Dolphins*
    Tools—*Tool Use by Elephants*

Further Resources

Chappell, J. & Kacelnik, A. 2002. *Tool selectivity in a non-mammal, the New Caledonian crow* (Corvus moneduloides). Animal Cognition, 5, 71–78.

Griffin, D. R. 2001. *Animal Minds: Beyond Cognition to Consciousness.* Chicago: University of Chicago Press.

Hauser, M. D. 2001. *Wild minds: What animals really think.* London: Penguin.

Hunt, G. R. 1996. *Manufacture and use of hook-tools by New Caledonian crows*. Nature, 379, 249–251.

Hunt, G. R. 2000. *Human-like, population-level specialization in the manufacture of pandanus tools by New Caledonian crows* Corvus moneduloides. Proceedings of the Royal Society of London B, 267, 403–413.

Hunt, G. R. & Gray, R. D. 2003. *Diversification and cumulative evolution in New Caledonian crow tool manufacture*. Proceedings of the Royal Society of London B, 270, 867–874.

Pain, S. 2002. *Look, no hands!* New Scientist, 175 (no. 2356, 17 August), 44–47.

Weir, A. A. S., Chappell, J. & Kacelnik, A. 2002. *Shaping of hooks in New Caledonian crows*. Science, 297, 981.

Also see our website: http://users.ox.ac.uk/~kgroup/tools/tools_main.html.

Alex Kacelnik, Jackie Chappell, Alex A. S. Weir, & Ben Kenward

■ Tools
Tool Use by Dolphins

Humans are tremendously resourceful tool users. Our ancestors made spears, arrows, traps, hammers, and many other tools from bone, wood, and stone. Now we make and use remote controlled robots, space probes, and guided missile launchers. Until recently, tool use was considered to be the hallmark of our species—yet another indication of our superior intelligence and manipulative skill. In the mid-1960s, however, Jane Goodall shook us up with her startling discovery that chimpanzees select and modify twigs to use as effective tools for capturing termites.

Since then, the study of animal behavior has grown, and more and more diverse creatures, entirely unrelated to humans, have been discovered to make and use tools. Finches probe for grubs under bark using thorns, otters pound open abalone shells with rocks, and caddis fly larvae encase themselves in suits of armor composed of bits of debris glued together. The need for a more rigorous definition of tool use became necessary, and Benjamin Beck (1980) devised one that is now widely accepted: "the external employment of an unattached environmental object to alter more efficiently the form, position or condition of another object, another organism or the user itself, when the user holds or carries the tool just prior to use and is responsible for the proper and effective orientation of the tool." By this definition, the sponge carrying of dolphins in Shark Bay, West Australia, is a form of tool use (Smolker et al. 1997). Sponge carrying was first observed in 1984. A dolphin, later named "Halfluke" because of her damaged tail fin, surfaced near our boat with a large reddish and knobby "growth" on her face. She took several breaths at the surface and then dove, bringing her damaged tail out of the water, an indication that she was diving straight down toward the bottom. After watching for some time it became evident that Halfluke was carrying something which she occasionally dropped.

It was not a "growth," but rather a piece of sponge. Halfluke was one of several dolphins who regularly carried sponges, usually within a particular deep-water channel. These "sponge carriers" were all female (most had offspring, some of whom have matured to carry sponges as well). The sponge carriers have been observed on many occasions during fieldwork carried out almost every year since 1984, and almost invariably carrying sponge. Sponge carrying is their specialty and they have been doing it for many years. Other dolphins living in the same vicinity do not carry sponges, though on very rare occasions, a few other dolphins were observed carrying sponges, but only very briefly.

From the surface, sponge carrying looks very like other forms of foraging—a few breaths followed by a "tail-out" dive and a period of time spent below. The travel direction is often difficult to predict since the dolphins change direction while below the surface. Occasionally a sponge-carrying dolphin will surface without a sponge or will trade in one piece for another, different in shape or size (but the same variety). The sponges (*Echino-dictyum mesenterinium*), are roughly cone-shaped and fit over the dolphin's beak. Some are large and flop back over the dolphin's face, even obstructing vision. Others are small and fit just over the tip of the beak. The shape of the sponge and the backward pressure of the water as the dolphin moves forward apparently keep it in place. Sometimes they do fall off, however, and are quickly retrieved by the dolphin.

After years of observing sponge carrying, we have never seen exactly how the sponges are used, since the dolphins appear to use them while submerged in deep and generally murky water. Given the amount of time that sponge carriers spend with sponges, similar to the amount of time spent foraging by other dolphins, we suspect that the sponges are used as an aid to foraging. Also, we sometimes see the sponge carriers surface rapidly without sponge and snap up prey near the surface. The sponge is likely used during search and then dropped prior to pursuit and capture of prey.

All evidence fits best with the theory that dolphins use the sponges as protective shields, guarding against stingers or spines of noxious organisms encountered while searching for prey along the bottom. The bottom is rocky in some areas and lionfish are not uncommon here. The dolphins may also be searching for prey in sandy areas, poking their beaks into the sand to flush out burrowing prey. The sponges may protect against abrasion.

The use of sponges as shields is not the first example of tool use by dolphins. Tayler & Saayman (1973) reported that a captive dolphin used pieces of tile to rub the tank walls and dislodge bits of seaweed. Brown & Norris (1956) described an incident in which two captive dolphins sought out and captured a lionfish, carried it to a crevice within which an eel was hiding and poked at the eel with the spiny fish. The eel abandoned its hiding place and was captured by the dolphins. These examples are indicative of the resourcefulness and intelligence of dolphins, but they are one-time only behaviors. Sponge carrying is engaged in by several dolphins and has been ongoing for many years. It has not spread throughout the dolphin population (as, for example, sweet potato washing did among Japanese macaques), but rather is restricted to a few individuals. Thus, like termite fishing in chimpanzees, sponge carrying is a well-established tradition of tool use by an animal living in a vastly different ecosystem and bearing little relatedness to humans.

See also Tools—*Tool Manufacture by a Wild Bonnet Macaque*
　　　　　　Tools—*Tool Use*
　　　　　　Tools—*Tool Use and Manufacture by Birds*
　　　　　　Tools—*Tool Use by Elephants*

Further Resources

Beck, B. 1980. *Animal Tool Behavior*. New York: Garland ATPM Press.

Brown, D. H. & Norris, K. S. 1956. *Observations of wild and captive cetaceans*. Behaviour, 37, 311–326.

Smolker, R., Richards, A. F., Connor, R. C., Mann, J. & Berggren, P. 1997. *Sponge-carrying by dolphins (Delphinidae, Tursiops sp.): A foraging specialization involving tool use?* Ethology, 103, 454–465.

Tayler, C. K. & Saayman, G. S. 1973. *Imitative behavior by Indian Ocean bottlenose dolphins (Tursiops aduncus) in captivity*. Behaviour, 44, 286–298.

Rachel Smolker

■ Tools
Tool Use by Elephants

Most people have used fly swatters. In fact, not long ago almost every house had fly swatters, and they were frequently used to swat flies that were constantly landing on the countertops or even on bodies. Although we may not think of it as such, a fly swatter is a tool with a function much like a hammer or a screwdriver. Tools are used often by people and on some occasions by animals. A tool is an object one gets from the environment and manipulates with a purpose or function. If you were sitting around a campfire and were pestered by flies or mosquitoes and wanted a fly swatter, you might grab a leafy branch from the ground to swat or chase away flies. This is an example of using a tool as you found it in nature. A fly swatter made out of metal with a net at the end is a manufactured tool.

People are great at manufacturing tools, and even some animals are known to manufacture tools. Although not always true, the manufacture of tools requires quite a bit of brain power or thinking, so it is animals with large, complex brains, such as chimpanzees, orangutans, and elephants, that primarily use tools. When animals manufacture tools, this may be as simple as breaking a stick for scratching, as some elephants do, or modifying a bushy branch with which to fly switch, much as we might do around a campfire.

Elephants and Fly Switching

Elephants are large targets and easily seen by flies. Of course, they supply a lot of blood, and there are parts of their body, such as the belly or neck, where the skin is thin enough to be pierced by flies that are looking for a good blood meal. When the biting flies are out, elephants, especially the Asian elephants that live in dense jungles, cannot go into a screened porch to get away from flies. Elephants have a hand-like trunk that can easily pick up a branch and manipulate it much as a person can. They also have a large brain, actually twice as large as the human brain, so they have no trouble figuring out how to use, and even modify, a branch for switching off flies. And the elephant's habitat is full of potential fly swatters because branches from bushes and trees abound, and if an animal pulls off a branch, even one that it may eat later, it has a handy fly switch. In the figure, an elephant is using a fly switch in a dense jungle; in this instance the switch is from the wild fig tree. The elephant used it as a switch prior to eating the leaves. In our study of fly switching in Asian elephants, we noticed that those kept in captivity for taking tourists into the forest used branches lying around the camp for switches and even carried the switch along with them so that it was available when needed.

Early naturalists who studied wildlife in Africa were aware of the use of fly switches by elephants. The adventurer Colonel Harris, in talking about his travels in 1838, described

An elephant uses a tree branch to keep insects away from its head.

Courtesy of Benjamin and Lynette Hart.

seeing elephants emerging into an open field bearing in their trunks the branches of trees with which they systematically protected themselves from flies (Harris 1838, p. 169). The modification of branches was noted by the naturalist, S. Peal, in 1879, when he observed a captive Asian elephant stripping down a branch before breaking it off and using it as a switch. Surely a well-known example of this type of tool use by elephants was given to us by the great storyteller Rudyard Kipling in 1902 in the *Just So* story, "The Elephant's Child." The elephant's child, having acquired a new trunk, courtesy of the crocodile, starts home, and we are told, "when flies bit him he broke off the branch of a tree and used it as a fly whisk" (Kipling 1902). Now we know that part of the *Just So* story is really a true one.

We became interested in learning more about tool use in elephants and did our research in the Nagarhole Park in India where we were able to observe wild elephants as well as captive elephants which were used to take tourists on rides. Like the naturalists mentioned above, we too observed wild elephants using branches for fly switches. We concluded that the use of fly switches was the most frequent type of tool use exhibited by elephants living in the wild, at least when flies are around. Of course, if no flies are around, elephants will not engage in fly switching, so whether or not one sees elephants switching depends on the time of year and the presence of biting flies.

We were also interested in whether the use of fly switches by elephants actually repels flies. Is there some actual value to the behavior? We assume that when we use a fly switch and chase off or kill a fly from time to time, that the switch is actually having some function. We found, in fact, that the interest of elephants in picking up a fly switch and using it to whisk off flies varied as a reflection of the number of biting flies that were around. By counting flies on and around elephants and either giving them a branch for switching or offering them no branch, we found that when they were able to engage in fly switching, flies were reduced by almost 50%. Clearly, there is a benefit to the use of fly switches.

In another study of captive elephants kept under a naturalistic system, we presented branches that were too long or bushy to be used effectively as switches. When given 5 minutes to do what they wanted with the long or bushy branch, most of the elephants modified the branch to a smaller one and then switched with that smaller branch. The most common type of modification was holding the stem with the front foot and pulling off a side branch with the trunk.

A baby elephant mimics the behavior of an adult.
Courtesy of Benjamin and Lynette Hart.

An interesting part of our observations was on baby elephants that were kept with their mothers with access to the branches that we were giving their mothers. We noticed that an 18-month-old baby elephant was able to remove a side branch and switch with it as did its mother, but that a 9-month-old baby had not yet developed the coordination to be able to modify or actually switch with the branch and simply tossed it around.

We should note that, if you go to a zoo or other area where captive elephants are kept, you are not likely to see elephants engaging in fly switching. This is because elephants will only use a branch for fly switching when they are in an environment where branches are plentiful. In captivity, branches are not usually given to elephants for forage, and if they have access to some they would simply eat them before giving any thought to using them as a fly switch. In places like India, where elephants are normally fed plenty of branches, they use some of them for switching.

We believe that the modification and use of fly switching by elephants is a reflection of their large brain and capacity for cognitive and thinking behavior. The modification of branches for fly switching places the Asian elephant tool use behavior at the same level as that of the chimpanzees and orangutans as highly intelligent animals that share our planet. We owe these animals all the love and protection of their natural habitat that we can give them.

See also Tools—*Tool Manufacture by a Wild Bonnet Macaque*
　　　　　Tools—*Tool Use*
　　　　　Tools—*Tool Use and Manufacture by Birds*
　　　　　Tools—*Tool Use by Dolphins*

Further Resources

Beck, B. B. 1980. *Animal Tool Behavior: The Use and Manufacture of Tools by Animals.* New York: Garland STPM Press.

Harris, W. C. 1838/1967. *Narrative of an Expedition into Southern Africa.* New York: Abercrombie & Fitch.

Hart, B. L. & Hart, L. A. 1994. *Fly switching by Asian elephants: Tool use to control parasites.* Animal Behaviour, 48, 35–45.

Hart, B. L., Hart, L. A., McCoy, M., & Sarath, C. R. 2001. *Cognitive behaviour in Asian elephants: Use and modification of branches for fly switching.* Animal Behaviour, 62, 839–847.

Kipling, R. 1902. *Just So Stories.* London: Macmillan.

Sukumar, S. 1995. *Elephant Days and Nights: Ten Years with the Indian Elephant.* Delhi: Oxford University Press.

Benjamin Hart & Lynette Hart

■ Turner, Charles Henry
(1867–1923)

Charles Henry Turner (1867–1923) was a pioneer of the comparative psychology/animal behavior movement in America and one of the most influential scientists working on problems of comparative behavior from the late 1890s through the early 1920s. Trained as a zoologist, his research focused on questions about behavior. His contributions include: the development of techniques to measure learning and other types of behavior of several invertebrate species (some of these techniques are still used more then 80 years after their invention), placing an emphasis, which was unique at the time, on how training variables such as intertrial interval, intersession interval, individual differences, and age of subjects influence performance; publishing 70 scientific papers, and initiating the first controlled studies of color vision and pattern vision in honeybees. In addition to his behavioral work, Turner also published extensive anatomical studies of both the avian and crustacean brain, discovered a new species of aquatic invertebrate, and wrote on social and educational issues of the day. Turner was a leader in the civil rights movement in St. Louis, Missouri, and was also instrumental in developing social services for African-Americans in the St. Louis area.

Charles Henry Turner.
Courtesy of Charles I. Abramson.

Dr. Turner was born on February 3, 1867, in Cincinnati, Ohio, 2 years after the Civil War ended. His father was a church custodian and his mother a practical nurse. Highlights of his scholastic life include being selected as valedictorian of his high school class, earning his B.S. degree from the University of Cincinnati in Biology, and becoming the first African-American to be granted a Ph.D. from the University of Chicago.

Upon receiving his doctorate, Dr. Turner could not secure an academic appointment at the University of Chicago or Tuskegee Institute despite having published 29 papers, including the first paper by an African-American in the journal *Science*. As a result, Dr. Turner spent much of his early academic career moving from high school to high school before settling, at the age of 41, in Sumner High School (St. Louis, Missouri). He would remain at Sumner until his retirement in 1922 at age 55.

While at Sumner High School he continued to publish. His publication rate (two articles per year) exceeded most of his white contemporaries working at colleges and universities. His achievements are made even more remarkable when we consider that he worked with unconventional organisms that required mastery of a wide range of rearing techniques. These animals included various species of ants, bees, cockroaches, crustaceans, moths, pigeons, spiders, and wasps. He worked with plants. Moreover, many of his papers included investigations of several species within the same manuscript. And, throughout his career, he had few or no formal laboratory facilities, no easy access to research libraries, no opportunity to train research students at the undergraduate or graduate level, heavy teaching loads, low pay, and restricted laboratory time. Regarding the latter, many of his experiments were run in the summer and/or in the evenings when the opportunity presented itself.

The importance of Turner's work was recognized by leaders in the animal behavior movement in the early twentieth century. John B. Watson, for example, in considering a study on ant behavior, called Turner's method "ingenious." Turner's work was also discussed by such well-known scientists of the day as Margaret Washburn, E. L. Thorndike, T. C. Schneirla, and Eugene-Louis Bouvier. In the animal behavior literature of France, a particular type of ant movement toward the nest is known as "Turner's circling" in honor of its discoverer.

Turner's Research

Morphology and Anatomy

Professor Turner began his research career by continuing a line of research begun by his mentor Charles L. Herrick. Herrick maintained a profound interest in the study of morphology and anatomy and contributed several articles on brain anatomy. Professor Turner contributed original articles on the comparative anatomy of the pigeon brain, directly compared the brains of arthropods and annelids, and studied the mushroom bodies of crayfish. His last morphological paper appeared in 1901 in which he outlined the brain of crayfish.

Experiments on Insect Navigation

A second line of research attempted to answer the question of how insects navigate. At the time of this research, there were four views. Some researchers believed that insects navigate by reflexes and others believed they navigate by instinct. Two other groups of investigators believed that insects use simple or complex learning respectively, to navigate.

Turner constructed an elaborate maze with pupae as the reward and tested 12 species of ants. The experiments were begun at Clark University in Atlanta, GA, and concluded at the University of Chicago. The project took 5 years to complete. His results showed that the movements of ants are not the results of tropisms or a homing instinct. Although ants are influenced by olfactory, visual, kinesthetic, and tactile stimuli, they also possess variability in their behavior, have memory, and exhibit a form of associative learning. In summarizing the results he notes, "Ants are much more than mere reflex machines; they are self-acting creatures guided by memories of past individual (ontogenetic) experience." Professor Turner confirmed previous observations that the direction of light influences the behavior of ants. The experiment also contains observations on the influence of caste membership on performance and sex differences. Professor Turner noted, "The males seem unable to solve even the simplest problems."

This experiment is noteworthy, not only because of the method and the research design, but because he instituted controls that were not considered at the time. For example, he ran many replicates of his observations and used heat filters when using light as a stimulus. He was also concerned with such contemporary factors as sex and age differences on performance.

Professor Turner continued this line of research by extending it to other species, including honeybees, wasps, and caterpillars, and came to the conclusion that, like ants, these invertebrates can learn by experience.

Experiments on Learning

The results of Professor Turner's experiments on homing suggested that associative learning played a large factor in the ability of an insect to return to its nest. The study of invertebrate learning was a main research interest. Professor Turner conducted experiments on learning in a wide variety of insects including honeybees, ants, and cockroaches. In some cases he used learning procedures to investigate other processes, such as determining the existence of color and pattern vision in honeybees.

His earliest paper on learning was mainly observational, although he did introduce some experimental manipulations. This study was entitled "Psychological notes upon the gallery spider—Illustrations of intelligent variations in the construction of the web" (1892). In this study he not only presents a classification of web design, but also intentionally changes the environment to assess the changes in web design.

Another example in which Professor Turner introduced modifications into a "natural" situation is found in the paper, "Do ants form practical judgments" (1907). This paper appeared before his 1907 dissertation manuscript. In this experiment, he intentionally introduced modifications in the nest, such as making a crack in the brood chamber and removing "trash" from the nest. Following these manipulations he would observe the behavior of the workers. The rationale behind the experiment was to determine whether ants could use "instinctive" behavior patterns in novel ways. The answer for Professor Turner was yes.

In addition to his 1907 dissertation paper on the homing of ants, perhaps the best examples of his use of learning procedures to investigate problems in insect behavior are his papers on color vision and pattern vision of honeybees. For both papers, elaborate apparatus was constructed, and controls were implemented that conclusively showed that honeybees can perceive both color and pattern.

The rationale behind the color vision paper (1910) was to provide data under controlled conditions on the ability of bees to see color. Such experiments were theoretically important because of the perceived interactions between honeybees and flowers. Professor Turner begins the paper with a scholarly review of the literature in which the various theories of why bees should see colors are enumerated, followed by a discussion of the limitations of the existing data.

To investigate the problem, he studied honeybees in O'Fallon Park in St. Louis, Missouri. He designed various colored disks, colored boxes, and "cornucopias" into which the bees were trained to fly. Thirty-two experiments were designed, and controls for the influence of odor and brightness were instituted. The results of his experiments showed that bees can see colors and can discriminate among them. It is interesting that, in considering the results of his experiments, he believed that bees may be creating, in his words, "memory pictures" of the environment. The idea of memory pictures is certainly contemporary.

The second paper of the series on honeybee learning concerned pattern vision (1911). The methods were identical to the color vision paper with the exception that various patterns were used, as were colors. The use of patterns and colors is the first use, in my view, of compound conditioning methods so popular in contemporary studies of animal learning. The study contains 19 experiments, and the results showed that honeybees can readily distinguish patterns.

Professor Turner also worked with cockroaches. The rationale for his first experiment was to determine if the reaction of roaches to light can be modified by experience (1912). Consistent with his previous work, he used a wide range of subjects, including adult males and females and larval females. In some cases he studied roaches with amputated antennae. At first he tried to develop a new apparatus; after several designs he decided to use the method developed earlier by Szymanski (1912) because the experiments could be performed faster. A roach was introduced into the illuminated side of a two-compartment chamber. Within moments of being introduced into the apparatus, it would run out of the illuminated chamber and into a dark chamber. Once in the dark chamber, the roach received an electric shock until it returned to the illuminated side. The experiment terminated when the roach made 10 successive "refusals" to enter the darkened compartment.

Professor Turner interpreted the avoidance of a previously preferred place, not as a reversal of a phototropic response, but as a result of learning to associate darkness with punishment. In some cases, memory of the task lasted 21 days. Additional results suggested that male roaches learn faster than female roaches, that there are large individual differences in performance, and that avoidance of a dark chamber survives molting. Of particular interest is a variation of the paradigm in which roaches trained to avoid darkness in one apparatus retained the behavior when transferred to a differently shaped apparatus.

Professor Turner extended this line of investigation in a second paper (1913). In this experiment, a new type of maze, constructed from copper, is described. The reward was a return to the home container, which was constructed from a "jelly jar." If the roach fell off the maze, it landed in water. The results indicated that the roaches could benefit from experience. Interesting aspects of the results include discussions of "jumping activities," "acrobatic feats," and "toilet-making habits." This paper also contains a discussion of the tactile, olfactory, auditory, and visual senses of the roach and how it relates to learning the maze. Anticipating some modern animal cognitivists, Professor Turner suggests that the roaches ". . . act as though experiencing the emotion the psychologists call *will*." The roach experiments are also an excellent example of the type of variables of interest to Professor Turner. These include intertrial and intersession intervals, individual differences, and age and sex related changes in performance. Investigating such variables is now a standard part of animal behavior experiments.

Experiments on "Playing Possum"

One of the more interesting lines of research conducted by Professor Turner is in the area of "death feigning." The technical term is *letisimulation* (from *letum*, death, and *simulare*, to feign) and refers to a death-like posture exhibited by many animals when handled roughly. For Professor Turner's first study on the issue, he investigated the ant lion. This study reports the results of five experiments and investigates such parameters as the relative duration of successive death feints, the influence of temperature, stimulus intensity, and hunger.

The experiment on ant lions is also unique because it provides one of the earliest descriptions of ant lion behavior—descriptions of, for example, the pits constructed to capture its prey, methods of excavation, posture of the ant lion, feeding behavior, locomotion, growth cycle, and the types of stimulation needed to elicit death feigning. Such naturalistic observations of invertebrate behavior are common in Professor Turner's work and form the basis of many of his contributions.

In a 1923 paper, Professor Turner summarized the literature on letisimulation. Consistent with the comparative nature of his research, he extends the discussion of letisimulation from invertebrates to vertebrates. He discusses the psychological implications of death feigning in bulls, foxes, and opossums and concludes that until further experimentation, the underlying behavioral mechanism of "playing possum" is similar in vertebrates and invertebrates.

Naturalistic Observations

Throughout his career, Professor Turner engaged in the observation and classification of invertebrates. Most of his papers, whether or not they were specially designed as observational studies, contain descriptions of behavior. The descriptions are often in such detail that one is left with the impression that his descriptive skills were acquired from his earlier morphological work. Many of his descriptive papers also contain the results of preliminary experiments.

Professor Turner provides the descriptions of micro-organisms of Cincinnati (1892). This paper is unique in that it contains descriptions of a new species which he discovered. He extended this work by providing descriptions of, for instance, the Cladocera of Georgia (1894) and additional comments on Ostracoda (1899). Also available are descriptions, for instance, of wasps (1908), flies (1908), the breeding habits of red ants (1909), and parasitic bees (1911).

Acoustic Experiments

One of Professor Turner's major contributions to the insect behavior literature was providing the first experimental evidence that insects can hear airborne sounds (1914). Consistent with his comparative approach, five species of moth were tested. The procedure involved sounding a "Galton whistle" at various distances from the moth and observing the reactions. Controls were implemented to ensure that the animals were not responding to the sight of the whistle, or air puffs. The results indicated that *Catocala* moths respond to high-pitched notes by either flying away or by making "quivering" movements and do not respond to low frequency sounds. *C. vidua* and *C. neogama* were the least responsive to sounds, and *C. flebilis*, *C. habilis*, and *C. robinsoni* were the most responsive.

Professor Turner suggested that the failure to respond to low frequency sounds does not necessarily mean that the moths do not hear such sounds. The experiment provides evidence that the moths will respond to such sounds if they have "life significance." He describes a brief experiment in which an air puff was paired with a low frequency sound. After one pairing the moth readily responded to a low frequency sound.

For a companion paper, he conducts a laboratory test of the ability of four species of moth to hear sounds. Individual animals were confined beneath "wire dish covers" in such a way that the wings could move but not provide lift. Sounds were produced by an adjustable organ pipe, an adjustable pitch pipe, and a Galton whistle. The dependent variable was wing movement. Psychophysical experiments were performed to determine thresholds. Of the moths tested only *Telea polyphemus* showed little responsiveness to auditory stimuli. Following up on his suggestion in the previous 1914 paper that a sound might need to be made "significant," he performed a classical conditioning experiment in which a tone was paired with "rough handling." This experiment may well represent the first example of classical conditioning in an insect.

Civil Rights Papers

Of 70 papers, 4 were concerned with civil rights issues. The first of these papers appeared in 1897. In this and his other papers on "Black–White" relations, the emphasis is always on education. In reading these papers one is struck by how contemporary they seem.

Dr. Turner died from acute myocarditis—an inflammation of the muscular walls of the heart at the home of his son Darwin Romanes Turner in Chicago, Illinois, on February 14, 1923—11 days after his 55th birthday and shortly after his retirement from Sumner High School. In recognition of his outstanding achievements as a scientist, educator, and humanitarian, the first school for African-American children with disabilities was named in his honor (Charles Henry Turner Open Air School for Crippled Children—founded in 1925).

Over the years several schools in an area of St. Louis known as "The Ville" bears his name. These include Charles Henry Turner Middle Branch (founded 1954) and the new Charles Henry Turner MEGA Magnet Middle School for 6th–8th graders (founded 1999). In 1962, Turner–Tanner Hall (now known as Tanner–Turner) at Clark College in Atlanta, Georgia, was named in his honor.

Following his death there were a number of tributes at Sumner High School. As Dr. A. G. Pohlman wrote:

> It is for you who knew Doctor Turner to satisfy yourselves that here indeed was a great man. It is for you to determine in your own hearts if this man possessed the strength of character, the devotion to work, the faithfulness to ideals, the respect to truth, and the unselfishness in sharing that which he possessed. Was he indeed the humble man of science who might well be taken into the fold of the most highly esteemed?

See also History—*History of Animal Behavior Studies*

Further Resources

Abramson, C. I., Jackson, L. D., & Fuller, C. L. (Eds.) 2003. *Selected Papers and Biography of Charles Henry Turner (1867–1923), Pioneer of Comparative Animal Behavior Studies.* New York: Edwin Mellen Press.
Cadwallader, T. C. 1984. *Neglected aspects of the evolution of American comparative and animal psychology.* In: *Behavioral Evolution and Integrative levels: The T. C. Schneirla Conference Series* (Ed. by G. Greenberg & E. Tobach), pp. 15–48. Hillsdale, NJ: Lawrence Erlbaum Associates.

Szymanski, J. S. 1912. *Modification of the innate behavior of cockroaches*. Journal of Animal Behavior, 2, 81–90.

Web Sites

Charles Henry Turner: Contributions of a Forgotten African-American in Scientific Research. http://psychology.okstate.edu/museum/turner/turnermain.html
Infography of Charles Henry Turner.
http://www.infography.com/content/215287765588.html

Charles I. Abramson

■|Unisexual Vertebrates

What Are Unisexual Vertebrates?

The vast majority of animal species reproduces sexually. Sexual reproduction involves reshuffling of the genes in every generation, typically via a genetic process called meiosis. Although it is extremely prevalent in the animal kingdom, sexual reproduction is not without alternatives. These alternative modes of reproduction—although often differing in important details—are usually lumped into the category "asexual" reproduction. If there are several ways to reproduce, why is sexual reproduction so common? What is the adaptive advantage of sexual reproduction? This question is still among the unsolved problems in evolutionary biology. Theoretically, being asexual provides several major advantages, especially in the short run. For detailed treatments of this problem refer to the books by Maynard-Smith (1978), Bell (1982), or to a recent article by West et al. (1999).

Asexual vertebrates are an exceptional group within the vertebrates. They have also been dubbed unisexual because usually they do not have two genders as most species, but instead typically are all-female. However, there are exceptions, most notably functional males in the European waterfrog (*Rana esculenta*). There are also so called "pseudomales" in a fish, the Amazon molly (*Poecilia formosa*). All unisexual vertebrates apparently are of hybrid origin. They have lost recombination, and most of them are completely clonal, so that mothers and daughters are genetically identical. Currently about nine groups of unisexual fishes and amphibians are recognized. Within these groups, asexuality often evolved multiple times, leading to a dazzling diversity of clonal forms. In addition to these fishes and amphibians, roughly 14 groups of asexual reptiles are known, showing a similar pattern of multiple origins of asexual lineages within these groups. For a detailed review of these species, refer to the article by Dawley (1989). Asexuality does not occur naturally in birds and mammals.

Given their recent sexual ancestry, most asexuals do not differ strongly in most aspects of their behavior from their sexual ancestors. Clearly, *sexual* behavior is the most interesting and intriguing aspect of their behavior and the remainder of this article will focus on that aspect using only a few examples to illustrate the behavior of unisexual vertebrates.

Sperm Dependency of Some Unisexual Vertebrates

Asexual vertebrates essentially come in two types. Some species, despite being all-female, still need to mate. They are sperm dependent. All asexual fishes and amphibians belong into this category. No unisexual reptile is sperm dependent, but even those sometimes show sexual behavior.

In the sperm-dependent species, males from a different species provide sperm for the asexual females. These species are often called host species. It is difficult to understand why males of a host species would mate with an asexual female, because their genes are not being transmitted. The sperm only serves to trigger the development of the egg. In other words, natural selection should favor males that avoid such "wrong" matings. Because obviously such matings do occur in nature, we are trying to understand the mechanisms that lead to such matings and their consequences.

Sperm dependent unisexuals use sperm in two different ways. In some cases, sperm is needed only to trigger the development of the embryo. This is called *gynogenesis*. In other cases, the genetic material of the sperm is included in the growing organism, but later during the formation of the gametes (eggs or sperm), those paternal genes are removed again. This mechanism is called *hybridogenesis*. In the case of gynogenesis the whole genome is transmitted clonally, whereas in the case of hybridogenesis, half of the genome is transmitted clonally. Consequently, this mode of inheritance is also called hemiclonal.

Behavior of Unisexual Fishes and Amphibians

Because of their intriguing sperm dependency, mating behavior of unisexual fish and amphibians is being studied intensively.

In most fish species, fertilization is external and males release sperm into the water. This makes it relatively easy for asexual females to obtain sperm. They can interfere with a mating event and release their eggs together with the eggs of a sexual female. This has been described, for example, for the crucian carp (*Carassius auratus*). However, several asexual forms have internal fertilization. Intuitively, it seems to be more likely that sperm dependent asexuals have external fertilization, but actually several lineages have to interact very closely with males to obtain sperm. Some asexual fish are livebearing and thus have internal fertilization (mollies of the genus *Poecilia* and topminnows of the genus *Poeciliopsis*), and asexual salamanders of the American genus *Ambystoma* have to pick up spermatophores. Frogs of the European genus *Rana* have to get into the typical mating posture, called amplexus, in which the male holds on to the female and both release their gametes in a coordinated way.

The mating behavior of the Amazon molly (*Poecilia formosa*) is being studied as one example of a sperm-dependent species. This species occurs in southeastern Texas and in northeastern Mexico. Here, the unisexual females have to "seduce" males from a different species to actually copulate with them. Amazon mollies mainly use males from two different species as hosts: the Atlantic molly, *Poecilia mexicana*, and sailfin molly, *Poecilia latipinna*. As seems to be typical for asexual vertebrates, these two species are also exactly the two species that were involved in the hybridization event that lead to the Amazon molly. An obvious problem is to understand why males of the host species "waste" their sperm by mating with Amazon mollies. In nature, Amazon mollies and at least one of the host species form mixed groups. Thus, males typically have to choose between females of the Amazon molly and females of their own sexual species. The benefit of distinguishing is clear: With females of their own species, the males have their own sons and daughters, with females of the Amazon molly they have no offspring that is genetically related to them. In this peculiar case—and the Amazon molly is only an example here—male choice is pivotal to understanding why and how such mating complexes can be maintained. If males

*An Amazon molly (*Poecilia formosa*).*
Photo by Ingo Schlupp.

would never mate with the asexual females, those would become extinct. If males would mate randomly with all available females, the asexual females would show enormous population growth and, after a few generations, the asexuals would dominate the population and drive the sexual females to extinction. Being sperm dependent, the asexuals would consequently also go extinct. Apparently, this does not happen, and a key factor to explain this stability is male choice.

How do these heterospecific matings happen? A first explanation could be that males cannot distinguish between the two types of females. In experiments investigating this, it was shown that males clearly can make this distinction, and in most cases, it was found that they show a preference for females of their own species. Male sailfin mollies from populations that live together with the Amazon molly (*sympatry*) generally are better at distinguishing, compared to males from populations without the Amazon molly (*allopatry*). These findings lead to the conclusion that males mate with the "wrong" females despite being able to recognize them. Females may be under selection to make it difficult for males and thus mimic the sexual females. Such sexual mimicry in spot patterns around the female genital opening has been reported in asexual topminnows (*Poeciliopsis*), another well-studied complex of unisexual fish. For more details on this fascinating complex, refer to the article by Vrijenhoek (1994). Another mechanism leading to "wrong" matings is apparently mediated by chemical communication. In livebearing fish, the females have a sexual cycle, and males are extremely attracted to females around the time of ovulation. One typical behavior pattern males show is frequent nipping at the female's genital opening. Apparently males can taste whether a female is in this attractive phase of the cycle. Given a choice between an attractive Amazon molly and a nonattractive sexual female, males strongly prefer the Amazon molly. Thus, under certain circumstances the asexuals may even have a mating advantage.

An alternate—but not mutually exclusive—explanation would be to assume that even the heterospecific matings are beneficial for the males. A mechanism that does allow for this is "mate copying." It has been shown that females of several species include social information in their mating decisions: They simply imitate the mating preferences of other females. For sailfin mollies it has been shown that females are attracted to males they have seen interact with asexual Amazon mollies. Thus, by mating with Amazon mollies, males gain an indirect benefit that may be larger (or equal to) the costs of mating with the "wrong" female. Females of the sailfin molly and the Amazon molly can distinguish between the two species. They prefer to associate with females of their own species, but their tendency to join *a group* of fish overrides this tendency. Larger groups of fish provide better protection against predation, so it is probably adaptive for the females to join even females of a different species and form the mixed groups needed for mate copying to occur.

Another intriguing aspect of behavior in Amazon mollies is to compare the behavior of sexual males to the behavior of masculinized individuals of the Amazon molly. To masculinize females, Amazon mollies were treated with male hormones. After the treatment, they looked like males and behaved almost like males of the sexual species *P. mexicana*. This indicated that the genes coding for many male traits were still present in the all-female fish.

Amphibians provide another important example. In the European genus *Rana*, hybridization of the sexual species *R. ridibunda* and *R. lessonae* led to a hemiclonal hybrid, the edible frog, *R. esculenta*. In this exceptional case *R. esculenta* males exist. Thus, matings of hybrid males and hybrid females are possible, and females have to choose among different males. Offspring resulting from matings with *R. esculenta* males do not usually survive. Therefore, females should avoid these males. Females of both the hemiclonal *R. esculenta* and females of the sexual *R. ridibunda* show preferences for sexual males, based on contact or acoustical information. However, females cannot exercise their preferences freely, because they cannot completely avoid being clasped by the "wrong" *R. esculenta* males. In this case, they exercise "cryptic female choice": They release fewer eggs with this undesired partner and may find a better partner after the "wrong" male released them.

Behavior of Unisexual Reptiles

Unisexual reptiles are not sperm dependent. They reproduce by true parthenogenesis, which is independent of any external trigger. This gives them the potential to replace their sexual ancestors, and it seems likely that this has happened. Despite being independent of sperm, some unisexual whiptail lizards of the *Cnemidophorus* group from the southwestern United States show sexual behavior. Female *C. uniparens* engage in "pseudocopulations" in which one female acts as a "pseudomale." These pseudocopulations strongly resemble actual matings of closely related species. Females that were not allowed to engage in pseudocopulations had fewer offspring than females that did have pseudocopulations. This fascinating behavior is mediated by hormones. The hormonal cycle of these females has two phases, and the state relative to ovulation determines which of the unisexual whiptail lizards acts as male and which one acts as female. Females in the preovulatory phase behave female-like, but postovulatory females behave male-like.

In summary, the behavior of unisexual vertebrates provides an excellent opportunity for using a different angle to increase our understanding of animal behavior.

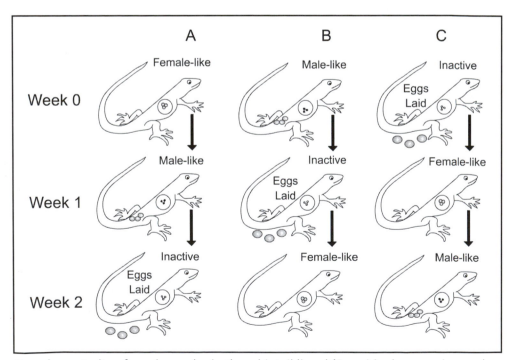

Complementarity of ovarian cycles in the whiptail lizard (Cnemidophorus uniparens) (from: Crews 1987).
Courtesy of Ingo Schlupp.

Further Resources

Bell, G. 1982. *The Masterpiece of Nature*. London: Croom Helm.

Crews, D. 1987. *Courtship in unisexual lizards: A model for brain evolution*. Scientific American, 255, 116–121.

Dawley, R. M. 1989. *An introduction to unisexual vertebrates*. In: *Evolution and Ecology of Unisexual Vertebrates* (Ed. by R. M. Dawley & J. P. Bogart), pp. 1–19.

Hellriegel, B. & Reyer, H.-U. 2000. *Factors influencing the composition of mixed populations of a hemiclonal hybrid and its sexual host*. Journal of Evolutionary Biology, 13, 906–918.

Maynard Smith, J. 1978. *The Evolution of Sex*. Cambridge: Cambridge University Press.

Reyer, H.-U., Frei, G. & Som, C. 1999. *Cryptic female choice: Frogs reduce clutch size when amplexed by undesired males*. Proceedings of the Royal Society Biological Sciences Series B, 266, 2101–2107.

Schlupp, I., Marler, C. & Ryan, M. J. 1994. *Benefit to male sailfin mollies of mating with heterospecific females*. Science, 263, 373–374.

Schlupp, I., Nanda, I., Doebler, M., Lamatsch, D. K., Epplen, J. T., Parzefall, J., Schmid, M. & Schartl, M. 1998. *Dispensable and indispensable genes in an ameiotic fish, the Amazon molly* Poecilia formosa. Cytogenetics & Cell Genetics, 80, 193–198.

Vrijenhoek, R. C. 1994. *Unisexual fish: Model systems for studying ecology and evolution*. Annual Review of Ecology and Systematics, 25, 71–96.

West, S. A., Lively, C. M. & Read, A. F. 1999. *A pluralist approach to sex and recombination*. Journal of Evolutionary Biology, 12, 1003–1012.

Ingo Schlupp

Washburn, Margaret Floy
(1871–1939)

Margaret Floy Washburn was an early pioneer of the psychological study of animal behavior. Perhaps her greatest contribution to the animal behavior literature was the book, *The Animal Mind*. In addition to this book, she published articles on animal behavior, reviewed books describing the life of animals, and served on the editorial board of journals that published articles on animal behavior. Her career lasted over 40 years, and during that time, she received numerous honors including election as president of the American Psychological Association and election to the National Academy of Science.

Margaret Floy Washburn was a member of the first generation of American women psychologists. She was born in the Harlem section of New York City on July 25, 1871, in the same year Charles Darwin published *The Descent of Man*. Her father was the Reverend Francis Washburn, and her mother was Elizabeth Floy Washburn. Margaret was an only child and spent much of her early childhood alone. This first generation was white, middle class, and in general descended from colonial families that originally settled from England. They also came from stable families that supported the quest for advanced education. Opportunities for advanced degrees at the doctorate level (Ph.D., MD) were customarily denied to women of this generation. When a woman obtained an advanced degree, it was often impossible to find a professorship at a research university as opposed to a "teacher's" or "women's college." Moreover, unlike their male counterparts, they were often required to choose between a career and marriage.

One of the more well-known women of this first generation is Dr. Washburn. Her 45-year career stands as a testament to diligence and hard work. Unlike many of her female contemporaries, her career comes closest to fitting the stereotype of an academic psychologist. Following her undergraduate training, for example, she entered graduate school and, upon earning the Ph.D., embarked on a career as a teacher and scholar.

Trained as a psychologist, first with James McKeen Cattell of Columbia University and later under the tutelage of Edward Bradford Titchener (often called the "Dean of American Empirical Psychology") at Cornell University, her work covered the entire range of research topics available to psychologists during the late 1800s until the mid-1900s.

These topics included the development of apparatus such as those used to measure the sense of touch, the perception of color and distance, after images, emotions, verbal learning, and motor learning. In addition to her contributions in these areas, she conducted research on the

Margaret Floy Washburn at a convention in 1926.
© *Bettmann / Corbis.*

affective value of color and of sound, the influence of verbal suggestions on judgments, the association of odors, visual imagery, and how aesthetic judgments are made. Dr. Washburn also dabbled in such applied research as extrasensory perception and intelligence testing. Perhaps her greatest interest was in the area of animal behavior.

Her major contribution in the field of animal behavior was the book, *The Animal Mind*, which appeared in four editions (1908, 1917, 1926, and 1936) and was translated into Japanese. It has the distinction of being the first comprehensive survey of experimental work in animal psychology. Until the publication of this book, students had no readily available source on the animal behavior literature. It not only contained scholarly reviews of the literature, but also interpretations of experiments and suggestions for future research. Reviewing the four editions of the book offers the reader a splendid opportunity to see how the field of animal behavior changed during the early days of American psychology.

In the first edition of the book, Dr. Washburn reviews the existing literature, including work using the anecdotal method. Developed by George Romanes, a colleague of Charles Darwin, the anecdotal method consisted of what might best be described as animal stories. The goal of these particular stories was to provide evidence for the evolution of intelligence. Dr. Washburn critiqued the anecdotal method and suggested that the experimental method is a better method. The book also contains a discussion on the criteria for the existence of mind in animals and the relationship between mind and consciousness.

Later editions of the book continued her analytical reviews of the literature, including the behaviorist position as exemplified by John B. Watson (an early psychologist who forcefully argued that the proper subject matter of psychology should be the study of behavior, not of the mind) and expanded upon her position for the existence of mind in animals. Always interested in the comparison of animals and humans, she continued to make the point that mind exists in higher animals, and that it is a worthwhile effort to discover when mind first appeared in the animal kingdom. In each edition, she advanced the hypothesis that it is possible to explore consciousness of humans by studying the consciousness of animals. She suggested, for example, that social consciousness first evolved from social *behavior* of lower animals. The social behavior of lower animals is characterized by motor behavior leading to the escape or avoidance of predators. There is no "mental state" per se in lower animals. However, this motor behavior evolved into the social consciousness seen in higher animals. Her position predates what is now known as the field of animal cognition.

The final edition of the book published just 3 years before her death contains a discussion of the experiments performed by the Gestalt psychologists. Gestalt psychologists such as Wolfgang Köhler believed that the proper study of animal behavior should be focused on "what the animal is doing" rather than "how the animal is doing it." The Gestalt view was that the behaviorist position could not explain such behavior as learning to solve a problem through insight; learning relationships, not absolutes; and that reward is not necessary for learning. The fourth edition also contains discussions of new experiments that illustrated the importance of emotions, incentives, and drives in animal behavior.

Dr. Washburn had a long and distinguished career. In addition to her many research papers and books, she was an active participant in the administration of professional psychology. She served as editor or consulting editor for a number of psychology journals including the *American Journal of Psychology, Psychological Review, Journal of Comparative Psychology*, and the *Journal of Animal Behavior*. She was also elected president of the American Psychology Association (the APA is one of the governing bodies of professional psychology) in 1921. She was the second female president of the APA (Mary Whiton Calkins was the first, having served in 1905).

Her contributions to the field of psychology were such that she received many honors. Perhaps one of the more unique was the prize from the Edison Phonograph Company for the best research on the effect of music. Her article, "The emotional effects of instrumental music," won the $500 prize. Dr. Washburn also received an honorary degree from Wittenberg College in 1927 and earned election to various national scientific organizations such as the National Academy of Science in 1931 (the second woman so honored—Dr. Florence Sabin was the first), and the United States delegate to the International Congress of Psychology in 1932. In 1927, volume 39 of the *American Journal of Psychology* was dedicated to her in recognition of 33 years of "Distinguished Service to Psychology."

Dr. Washburn died on the afternoon of October 29, 1939, in Poughkeepsie, New York. Her death was caused by a cerebral hemorrhage that had occurred March 18, 1937. As Dr. Karl M. Dallenbach wrote in his obituary, "Her death removes from American psychology one of the most active and honored members of the *Fach*. Psychology will not see her like again."

See also History—*History of Animal Behavior Studies*

Further Resources

Dallenbach, K. 1940. *Margaret Floy Washburn (1871–1939)*. American Journal of Psychology, 53, 1–5.

Kambouropoulou, P. A. 1940. *A bibliography of the writings of Margaret Floy Washburn: 1928–1939*. American Journal of Psychology, 53, 19–20.

Martin, M. F. 1940. *The psychological contributions of Margaret Floy Washburn*. American Journal of Psychology, 53, 7–18.

Mull, H. K. 1927. *A bibliography of the writings of Margaret Floy Washburn, 1894–1927*. American Journal of Psychology, 39, 428–436.

Pillsbury, W. G. 1940. *Margaret Floy Washburn (1871–1939)*. Psychological Review, 47, 99–109.

Scarborough, E., & Furumoto, L. 1987. *Untold Lives: The First Generation of American Women Psychologists*. New York: Columbia University Press.

Washburn, M. F. 1908. *The Animal Mind*. New York: Macmillan.

Washburn, M. F. 1930. *Autobiography*. In: *History of Psychology in Autobiography (vol. 2)*. (Ed. by C. Murchison), pp. 333–358. Worcester, MA: Clark University Press.

Web Site

http://psychology.okstate.edu/museum/women/cover2.html
  (*describes the contributions of women to psychology*)

Charles I. Abramson

■ Welfare, Well-Being, and Pain
Animal Welfare and Reward

Animal welfare is concerned with an individual's quality of life. Although it is important how we conceptualize this "quality of life," what really matters is how animals perceive it. In important respects, animal welfare science is therefore the science of subjective experiences.

Thus far, the study of animal welfare has been preoccupied with poor welfare over good welfare. Thus, many studies focused on the effects on welfare of acute and chronic stressors associated with commercial production environments. However, in addition to understanding the cause of negative subjective experiences, it is equally important to know the cause of positive subjective experiences.

What do animals like? To answer this question, we not only need to know what *kind* of stimuli are being liked (food, shelter, or a social partner), but we also have to know under which conditions such stimuli are *perceived* as pleasurable. To illustrate this point, let us consider the presentation of food to pigs kept in commercial production units. According to popular belief, there is nothing more pleasurable for a pig than getting its food "for free" without having to work for it. Thus, fattening pigs generally are fed ad lib with standard pig pellets, whereas production sows are generally fed twice daily to satiety with pellets that are consumed in not more than 10 minutes. All the necessary nutrients are in the pellets. But do these pigs experience all the pleasurable feelings that may possibly be associated with food? There are several indications that the answer is negative.

First of all, pigs kept in seminatural environments are shown to rely on their natural foraging strategy of rooting and grazing for more than 60% of their active time even when they have access to standard pig pellets that fullfill their nutritional requirements. In other studies, it has been shown that animals may prefer to "work" for their food even when they have free access to food. This phenomenon is referred to as "contrafreeloading." It should be noted however, that contrafreeloading is not observed under all conditions. For instance, if an animal is very hungry, or the efforts required to obtain the food are very high, animals may prefer the "free" food anyway. Finally, there is a growing body of evidence indicating that, under certain conditions, animals may prefer food rewards that are presented in a variable and/or unpredictable ("risky") way rather than food rewards that are presented in a standard and predictable way. It is as if animals may opt for a strategy in the lotto in which small chances for big rewards are promised. Also in this case, such preferences will depend on various factors, including species-specific characteristics or the animal's energy budget. However, studies of animal welfare have thus far paid little attention to the phenomenon that potentially rewarding stimuli may be perceived as more or less pleasurable, depending on the conditions in which these stimuli are presented.

Theories of learning and reinforcement association may help to clarify when stimuli are perceived as pleasurable. By definition, a reward is considered any stimulus for which an animal will work. Thus, pigs will work for food, but if they are fed to satiety, they will rapidly lose their interest in working for more. It is stated that the reward value of food has decreased. Reward may be the presentation of a positive stimulus like food, but it may also be the termination of an aversive stimulus (like the threat of a predator). Some stimuli are unlearned or primary reinforcers. Primary reinforcers will require no learning in order to be effective. Many stimuli that are needed for survival will be primary reinforcers. Thus, food and water, but also eating and foraging behavior may be primary reinforcers for members of most species. Other stimuli may become reinforcing through its association with primary reinforcers. These are called secondary reinforcers. Examples of secondary reinforcers are, for instance, a tone or a bell or the approach of an animal caretaker that signals to the pig that food is coming. Eventually, the behavior an animal shows in anticipation of a food reward, or the behavior it has to perform in order to get the food, may become reinforcing through its association with the primary reinforcer (food). In fact, the latter is considered one of the possible mechanisms underlying the phenomenon of "contrafreeloading."

In relation to animal welfare, reinforcement learning predicts that positive experiences associated with ad lib feeding will be lower than those associated with feeding procedures that facilitate appetitive and anticipatory behaviors in addition to the consummation of food. In an experimental procedure in the rat, anticipatory behaviors were facilitated in a number of studies of Spruijt and collaborators through a classical Pavlovian conditioning procedure. In this procedure, the conditioning stimulus (CS) signals the presentation of a rewarding stimulus (US). When the delay between CS and US is gradually increased, anticipatory behavior develops. Using this paradigm in the rat, it can be shown that facilitation of anticipatory behaviors by repeated announcements of a reward, can prevent the development of stress-related symptoms during a period of chronic stress. More in general, these findings imply that providing opportunities to anticipate rewards may help to compensate for deleterious effects of stressful conditions.

The importance of positive stimuli for animal welfare is further stressed by studies dealing with the ontogeny, or development, of behavior. These studies show that early rearing conditions may influence the way animals perceive their living conditions in adulthood. Thus, stressful events during early ontogeny, like abrupt separation from the mother, may increase the sensitivity to the deleterious effects of stressors in adulthood. Even stress exposure to the gestating mother may affect the stress sensitivity of the offspring in adulthood. Recent evidence now indicates that early rearing conditions may also affect the adult sensitivity to positive stimuli: Thus, play deprivation of 3- to 4-week-old rats is shown to decrease the sensitivity of responding to both social and nonsocial rewards in adulthood. These findings are important because they imply that good animal welfare for animals requires care in order to provide optimal rearing conditions in the first place.

One of the intriguing questions is how the potential reward value of stimuli is evaluated by the brain. In that respect, the vertebrate brain shows remarkable similarities in design. According to this general design, evaluation and interpretation of incoming stimuli is hierarchically organized. Thus, different properties of a stimulus are first separately "decoded" in one structure before the information is transported to a next structure where integration of this information takes place. In a number of steps of increasing complexity, the reward value of a stimulus may be evaluated. As an illustrative example, Rolls (2000), for instance, described the evaluation of the taste of a food stimulus by the primate brain in a few hierarchical steps: Unimodal representation of taste properties (sweet, salt, bitter, etc.) takes place in the primary taste cortex. Information from this primary taste cortex is subsequently projected to the secondary taste cortex (the orbitofrontal cortex) where information about sweet, salt, and bitter is integrated together with information about odorous and visual characteristics of the food. This secondary taste cortex (but not the primary taste cortex) is implicated in a first and simple evaluation of the reward value of taste. Projections from this secondary taste cortex to the amygdala provides the necessary information for analysis at a next level of complexity: The amygdala, in particular, is considered as the area that is critically involved in the integration of information about conditioned stimuli (a tone or a bell or approach of the animal caretaker) that gain emotional value through its association with a primary reward (for instance the taste of food).

Since animals live in a complex environment, the reward value of many different stimuli should be evaluated and compared by the brain in order to make a decision on how to behave in an adequate and efficient way. Such a final evaluation requires a "common currency" in the brain which allows comparison of the reward value of, for instance, access to food or access to a sexual partner. For this final evaluation, two groups of neuromodulators seem of crucial importance. These include the endogenous opiate system and the dopaminergic system. Both systems play a role in drug abuse such as the use of heroin (which acts on the endogenous opiate system) and cocaine or amphetamine (which acts on the dopaminergic system). In relation to the function of these neuromodulators, two separate processes are discriminated, for example, that of "liking" and that of "wanting." "Liking," which is supposed to be mediated by the opioid system, refers to the evaluation of the potential reward value of primary and secondary reinforces, whereas "wanting," which is supposed to be mediated by the dopaminergic system, refers to the motivation to behave in order to get the stimulus that is "liked." In rats, for instance, play behavior facilitates the release of opioid peptides in the brain. In addition, play deprivation, which is shown to decrease adult sensitivity to social and nonsocial rewards, is also shown to modulate the sensitivity of the adult opioid system.

Reward evaluating mechanisms in the brain should be conceptually dissociated from those that generate "pleasurable feelings." For instance, although the amygdala appears to be critically involved in the evaluation of the reward value of conditioned stimuli, it seems unlikely that the amygdala also *cause* pleasurable feelings. Although theories about the neurophysiological processes responsible for subjective experiences are only emerging, the generation of such pleasurable feelings most likely involves the complex integration of contextual cues, complex learning, and information about the bodily state.

Knowledge about the causes of positive subjective experiences is essential in order to find effective strategies to improve animal welfare. Many of the animals kept by humans are typically kept under impoverished environments in which the primary needs (like hunger, thirst, escape from life threats) are met, while opportunities for positive subjective experiences associated with the fulfilling of these primary needs are limited. When we try to design welfare-friendly housing systems for animals, a more challenging living environment can be achieved by increasing the rewarding effects of primary reinforcers (such as food) through their association with various secondary reinforcers, because it is both the primary reinforcer and the appetitive and anticipatory behaviors preceding the presentation of primary reinforcers that may activate the endogenous opiate system associated with the emergence of pleasurable subjective experiences.

See also Welfare, Well-Being, and Pain—*Psychological*
    Well-Being
   Welfare, Well-Being, and Pain—*Feather Pecking*
    in Birds

Further Resources

Appleby, M. C. & Hughes, B. O. (Eds.). 1997. *Animal Welfare*. Wallingford, UK: CAB International.

Bateson, M. 2002. *Recent advances in our understanding of risk-sensitive foraging preferences*. Proceedings of the Nutrition Society, 61(4), 509–516.

Berridge, K. C. 1996. *Food reward: brain substrates of wanting and liking*. Neuroscience and Biobehavioral Reviews, 20(1), 1–25.

De Jonge, F. H. 1997. *Animal Welfare? An ethological contribution to the understanding of emotions in pigs*. In: *Animals and Consciousness and Animal Ethics. Series Animals in Philosophy and Science* (Ed. by M. Dol, S. Kasamoentalib, S. Lijmbach, E. Rivas, R. van den Bos), pp. 101–113. The Netherlands: Van Gorcum Publishers.

Inglis, I. R, Forkman, B., & Lazarus, J. 1997. *Free food or earned food? A review and fuzzy model of contrafreeloading*. Animal Behaviour, 53(6), 1171–1191.

Jensen, P. 2002. *The Ethology of Domestic Animals: An Introductory Text*. Wallingford, UK: CABI.

Rolls, E. T. 2000. *Memory systems in the brain*. Annual Review of Psychology, 51, 599–630.

Spruijt, B. M. 2001. *How the hierarchical organization of the brain and increasing cognitive abilities may result in consciousness*. Animal Welfare, 10, S77–87.

Spruijt, B. M., Bos, R. van den, Pijlman, F. T. A. 2001. *A concept of welfare based on reward evaluating mechanisms in the brain: Anticipatory behaviour as an indicator for the state of reward systems*. Applied Animal Behaviour Science, 72(2), 145–171.

Francien H. de Jonge & B.M. Spruijt

Welfare, Well-Being, and Pain
Back Scratching and Enrichment in Sea Turtles

Environmental enrichment provides captive animals with opportunities to display a range of species-specific behaviors, and provides individuals with choices designed to give them some degree of control over their lives. Most research on the effects of enrichment centers around mammals and birds, whereas very little has been done on other species of animals. The animal group containing sea turtles has had very little research involving enrichment. Sea turtles are highly migratory reptiles that spend almost their entire lives in salt water, moving between feeding and nesting grounds, living in the open ocean and around coral reefs. The only time a sea turtle leaves the water is when the female lays her eggs on a sandy beach, after which she immediately returns to the water. Sea turtles are vulnerable to many dangers, which include being hit by boats, drowning in fishing nets, marine pollution, and loss of beach habitat. All of the eight species of sea turtles are either endangered or threatened with extinction.

Although not much is known about sea turtle behavior in the wild, research shows that green turtles, *Chelonia mydas*, remain submerged most of the time they are in the open ocean, neither basking at the surface nor drifting passively. Green turtles remain submerged about 20 minutes, which may mean that they are comparatively inactive during this time. Their dive periods are longer during the night than during the day. During the nesting season, green turtles often stay within a short distance of the beach. In the wild, sea turtles often hide under rocks on the ocean's floor for protection. They also may scratch their backs on rocks or other rough materials such as coral reefs under the water to free themselves of barnacles or other debris that may be attached to their backs, which may create drag and disturb a turtle's hydrodynamics.

If an injured or sick sea turtle is found on the beach or in the water, it is often rescued and taken to a rehabilitation facility. Once healed, the turtle is released back into the ocean. However, some sea turtles are too sick or physically damaged to be released into the wild and are kept in captivity for the remainder of their lives, which may be decades. Because they spend most of their lives swimming in the open ocean, information on their natural behavior in the wild is extremely sketchy. As a result, it is hard to create natural-type habitats for captive sea turtles, and sea turtles are housed in simple, round tanks (1 m high x 3 m in diameter). Unlike sea turtles in the wild, sea turtles in captivity do not have any rocks, materials, or other devices on which to scratch their backs.

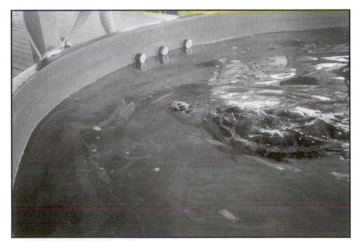

Tank containing a green sea turtle; note the lack of any type of environmental stimulation (the pipes in the tank's walls are overflow pipes).
Courtesy of Heidi Marcum.

Their tanks are kept purposefully barren to decrease the time it takes to clean the tanks and to decrease the chances for some form of contamination entering the water. However, the major reason the tanks are kept barren is because, up to now, people have felt that turtles are not capable of experiencing any desires beyond feeding and swimming, and that they exhibit no intentional behaviors. Because they are reptiles, people do not think that turtles need any form of enrichment or environmental stimulation. As a result of their sterile environment, the turtles spend the vast majority of their time swimming around the perimeter of the tank, stopping only occasionally to switch directions or take a breath of air. Captive sea turtles have little opportunities for exploration, finding food, or locating mates.

Although very little is known about how to improve the environment for captive sea turtles, one opportunity for enrichment is to provide them with a device on which to scratch their backs and stomachs. Students and faculty at Baylor University in Waco, Texas, have created a back scratching device for captive sea turtles. This device is made of several PVC pipe pieces glued together in a form that slips over the side of a captive sea turtle's tank. A lower portion of the device hangs free in the water in the shape of a large open square that the turtles can swim through, climb on, rest under, or scratch and rub their bodies on. The device has been tested for the past 3 years at a captive sea turtle facility called Sea Turtle Inc. on South Padre Island, Texas. This facility houses a variety of species of sea turtles including the green sea turtle, which has been used to evaluate the effectiveness of this back scratching device as a captive enrichment tool. These captive turtles spend almost 100% of their time during the day swimming around the perimeter of the tank when the device is not in the water. However, when the device is placed into their tanks, their activities change considerably. The time that they spend swimming declines from 100% to 24–34%. The rest of the time that they would have originally spent swimming without the device is now divided into several behaviors, depending on how the turtles use the device. The turtles spend much of their time rubbing against the device, mostly rubbing their backs, heads, and bottom shell. The turtles spend 5% of their time resting on the device, and another 5% of their time swimming in and out of the device. The sea turtles interact with the device so much that when a person tries to move it or take it out of the tank, it is often difficult to get the turtle off of the device or away from it. After over 2 years, the turtles still use the device whenever it is placed in the tank.

The same sea turtle as in the previous photograph, rubbing his back on the enrichment device.
Courtesy of Heidi Marcum.

Even though these captive sea turtles do not have barnacles attached to their backs, their backs and stomachs have numerous sensory nerves that are stimulated by rubbing and scratching on rough material, which in some way gives these turtles a pleasurable sensation. Thus, by placing a back scratching device in the tanks of these usually unenriched captive sea turtles, people have given captive sea turtles a more stimulating environment that presumably gives them some amount of pleasure in captivity that they would have not otherwise had.

See also Welfare, Well-Being, and Pain—*Enrichment for Chickens*
   Welfare, Well-Being, and Pain—*Enrichment for Monkeys*

Further Resources

Bjorndal, K. A. 1995. *Biology and Conservation of Sea Turtles*. Washington, D.C.: Smithsonian Institution Press.
Bjorndal, K. A. 1995. *World Conference on Sea Turtle Conservation*. Washington, D.C: Smithsonian Institution Press.
Klemens, M. (Ed.) 2000. *Turtle Conservation*. Washington, D.C.: Smithsonian Institution Press.
Sizemore, E. 2002. *The Turtle Lady: Ila Fox Loetscher of South Padre*. Plano, TX: Republic of Texas Press.

Heidi Marcum & Chantel Dawson

■ Welfare, Well-Being, and Pain
Behavior and Animal Suffering

Suffering

Suffering can be defined as

> A negative emotional state, which may derive from various adverse physical or physio-
> logical or psychological circumstances, and which is determined by the cognitive capac-
> ity of the species and the individual being, as well as its life's experience.

This proposed definition addresses the mental distress that may be caused in some animals through their perception of their external environment, particularly through senses such as smell, sight and sound, as well as their internalized individual predicament through feelings such as pain, or an instinct to carry out certain behaviors (e.g., migration when caged). This mental distress will also be affected by an animal's experiences in life and the ability to recall them and to recognize contextual similarities. Let me give some examples to amplify this definition, although first I wish to define animal.

Only animals that have the neurological development and capability to experience (feel) adverse states are the subjects of concern here. More primitive forms such as amoeba, simple multicellular organisms without a complexly organized nervous system are unlikely to feel, although they may well respond and even have simple programming mechanisms. However, they are unable to interpret novel circumstances. Here we will consider animals at a level of consciousness that includes the ability to assimilate new information and to apply general learned principles to novel circumstances. As far as we are aware, this level of awareness is generally found in vertebrates, but not exclusively (e.g., octopuses seem to have an ability to recall adverse experiences and take avoiding action). Moreover, the ability to feel pain and other adverse states varies among different phyla and, not surprisingly, progressively develops throughout gestation or incubation, and so at some stage would apply to the fetus. In humans, this seems to be somewhere between 18 and 26 weeks (probably later rather than earlier). The development of the nervous system in some mammalian species has been shown to continue after birth. For example, the descending inhibitory pathways that control the passage of nociceptive impulses up to the brain (so-called "gating" because it is a hurdle to the continuing passage of impulses up to the brain) continue to develop for several weeks after birth. This has led to the speculation that neonatal and young animals feel more pain than when their nervous system has fully matured and the gating mechanism fully developed. Finally, the development of self-awareness and therefore the ability to reflect on one's own circumstances (plight?), could add another dimension to any experience of suffering, and appears to develop at around 2 years of age in humans, but there is little data on this in animals, other than that it may be present or absent in adults. In this essay, I am concerned with those beings (animals) that are sentient, that is, capable of feelings such as pain, fear, frustration, boredom, and possibly other "human" feelings such as happiness, pleasure, grief, guilt.

So having laid out some preconditions as to what sort of animals I am considering, I now wish to elaborate on this definition of suffering.

"A negative emotional state. . ."

Animals that are sentient can feel positive and negative emotions, and suffering may occur when these feelings are overwhelmingly negative. In some situations, there may be a mixture of positive and negative states, and obtaining food at the price of an electric shock may still be an overall positive experience, and an animal may return to such a situation to maintain its homeostasis (e.g., to meet its feeling of hunger). It is obvious that animals can experience a range of emotions from those indicating pleasure and happiness in some way (dogs wagging tails, cows eager to get out to grass even though they have ample food before them, cats purring) to the other end of the spectrum where animals may deliberately avoid situations that they have found unpleasant (e.g., taking a puppy back to the veterinary clinic where it had previously experienced pain, sheep avoiding a shed where something they have found aversive in the past such as [electro] immobilization and a surgical procedure such as foot trimming when lame). Such negative experiences can be recalled by an animal from an earlier event, and animals may take avoiding action when given a chance. Not all negative experience require prior exposure (e.g., thirst, the desire to migrate, to mate, to play).

". . . which may derive from various adverse physical or physiological or psychological circumstances, . . ."

Examples of adverse physical states include environments which induce abnormal behaviors, or where abnormal lengths of time are spent carrying out normal behaviors. Examples of barren environments are cages or pens where animals have little opportunity to carry out instinctive behaviors such as dust bathing in chickens, digging in rabbits and gerbils, and nest building for mice and sows at the time of farrowing. This seems to set up an internal conflict for an animal (akin to a feeling of frustration) when they are unable to satisfy their instincts and leads them to subconsciously carry out stereotypic behaviors. Wild animals also often show repeated escape behaviors when caged, and which seem to be exacerbated when an animal has "known" freedom as opposed to being born and bred in captivity. Examples might include polar bears and wolves pacing in their concrete pens, horses weaving at the doors of their stables, rabbits pawing at the backs of their cages, wild birds looking for an escape route from their cages. These poor environments lead to psychological responses that are internally driven, but physiological changes are less obvious. We can start to examine aspects of the environment that may be better for animals—"asking" them what they prefer by looking at what environments they choose to be in and how hard they will work to get there. For example, rodents work hard in choice tests to reach a particular environment or goal (solid versus grid floors, certain substrates). A word of caution, however—they may not choose what is good for them in the long term, nor does it really tell you what they want since we may not offer that choice.

Adverse physiological circumstances would include poor health, for example, due to an infection that in humans leads to feelings of poor well-being such as discomfort or malaise. Similar signs are also seen in animals such as when their behavior changes to inactivity, poor appetite, and possibly a change in disposition from docility to aggression. Pain in animals, for example, horses with colic, animals with fractured bones or slipped discs or arthritis, cats with an aortic thrombus, animals with unrelieved post-surgical pain, all lead to various changes in behavior and physiology as a result of poor well-being. Animals that carry out stereotypic behaviors to the point of causing tissue damage to themselves will obviously suffer additionally.

"... and which is determined by the cognitive capacity of the species and the individual being"

The development of the central nervous system is manifestly different between species. Consider the development of a key area such as the cerebral cortex. The evolutionary older part of the brain dealing with emotions, the limbic system, is present in all sentient species. Perhaps it is the interaction between the cerebral cortex and other areas of the brain (e.g., hippocampus) that determines the level of cognitive ability and hence the ability of an animal species to suffer. However, it is also apparent that individuals within a species will have had different life experiences, and this too will determine their ability to suffer. At one end of the spectrum, a human in a permanent vegetative state will be unable to suffer because their cerebral cortex has been irreversibly damaged; for others it may have not developed (anencephalics) or development has been restricted or retarded (e.g., mentally impaired through hypoxia at birth). At the other end, one might have a highly sensitive and imaginative person who will suffer more mental anxiety then most. Animal species represent a range of development of the central nervous system, but those that are self-aware (the ability to become the object of one's own attention) and self-conscious (the ability to be aware of one's own existence especially in relation to others and over time) may suffer more through an anticipation of the future based on a mix of past experience, natural instincts, and intuition. Thus, a captive animal that has known what it is like to live in the wild may have internalized that experience and so suffer more when kept confined, compared to an animal that has not had that experience. The internal basic instincts to carry out certain behaviors are still there, but now there is the extra dimensions of physiological integrity, prior experience, and memory.

Animal Well-Being

So how does this impact on our assessment of animal suffering (i.e., on animals' welfare or an individual animal's well-being). The following definition can be used to define welfare and so help decide what might be good and poor welfare.

"Welfare is dependent on and determined by an animal's physiological and psychological well-being in relation to its cognitive capacity and life's experience."

At one level an animal's well-being is reflected in its subconscious attempts to cope with its (aversive) environment—the homeostatic reflex—through activation of the autonomic nervous system, the hypothalamic—pituitary axis and adrenal gland. But this is not the stuff of suffering that I am dealing with; suffering is more an animal's conscious attempts to deal with its specific predicament. When an animal feels threatened in some way it usually tries to take avoidance action. This self-preservation is universal in vertebrates as far as we can tell, and has been preserved through our evolution. Many laws and guidance notes state something like, "It should be assumed that persistent pain or distress in animals leads to suffering of animals in the absence of evidence to the contrary" (OECD 2001).

See also Cognition—*Animal Consciousness*
Cognition—*Cognitive Ethology: The Comparative Study of Animal Minds*
Emotions—*Emotions and Affective Experiences*
Welfare, Well-Being, and Pain—*Behavioral Assessment of Animal Pain*
Welfare, Well-Being, and Pain—*Behavioral Correlates of Animal Pain and Distress*

Further Resources

Dawkins, M. S. 1992. *Animal Suffering: The Science of Animal Welfare*. 2nd edn. London: Chapman & Hall.
Dawkins, M. S. 1993. *Through our Eyes Only. The Search for Animal Consciousness*. San Francisco: W.H. Freeman Spektrum.
OECD 2001. Environmental Health and Safety Publications Series on Testing and Assessment. No. 19: *Guidance Document on the Recognition, Assessment, and Use of Clinical Signs as Humane Endpoints for Experimental Animals Used in Safety Evaluation Environment Directorate.*
   This publication is available electronically, at no charge. For the complete text of this and many other Environmental Health and Safety publications, consult the OECD's World Wide Web site (http://www.oecd.org/ehs/) or contact: OECD Environment Directorate, Environmental Health and Safety Division 2 rue André-Pascal, 75775 Paris Cedex 16 France; E-mail: ehscont@oecd.org

David B. Morton

■ Welfare, Well-Being, and Pain
Behavioral Assessment of Animal Pain

Can Animals Feel Pain?

The pain system acts to alert us to harmful or potentially harmful events in the environment or to internal damage to our body, using sensory receptors called *nociceptors*. Nociceptors respond to events that could inflict damage, such as intense heat, physical pressure, or dangerous chemicals. For example, putting your hand into very hot water, dropping a heavy object on your foot, or spilling acid on your skin, would all result in the activation of nociceptors. When activated, nociceptors send warning messages through nerve fibers to the brain. Interpretation of these messages by the brain allows the position and amount of damage to be identified. At the same time the unpleasant feeling that we call pain is felt. These mechanisms are protective and allow injury to be avoided whenever possible. The few unfortunate people born without the ability to feel pain are regular visitors to accident and emergency departments, and many die young.

Pain

Kenneth M. D. Rutherford

Most people experience pain due to injury or disease at some point in their lives. Although we can directly experience only the pain we ourselves feel, we all accept that someone grimacing or crying out is also suffering, especially in a situation we ourselves have found to be painful. While humans in pain can tell others about their feelings, such communication is not possible in animals. Yet humans feeling pain do not just say, "it hurts," and continue as normal; their behavior is also altered. This is also true of animals, and alterations in animal behavior—although sometimes different than changes seen in human behavior—can be used to assess the degree of pain that an animal feels.

Pain, although unpleasant, is a useful feeling. It alerts the individual to damage, or the threat of damage, to the body, causing various alterations in behavior that are designed to limit injury and create the right conditions for recovery. Pain also influences learning and memory, so that situations or events that have caused pain in the past are well remembered and avoided. Pain therefore has both short-term and long-term influences on behavior.

Although pain can be protective, we should attempt to reduce the pain felt by animals in our care, such as our pets, farm animals, or laboratory animals. The essay, Behavioral Assessment of Animal Pain, describes some of the behaviors shown by animals in pain. It will highlight the fact that these behavioral changes can be used to recognize and measure the pain an animal is experiencing. Pain assessment is used by veterinarians to help detect and treat pain and also by scientists to investigate whether or not particular practices cause pain—informing debate about their ethical acceptability.

Unlike most humans, animals lack the ability to speak, so it is not possible to get direct evidence that they feel pain. However, there is a lot of other evidence: Vertebrate animals have similar sensory systems to humans and show similar physiological responses to injury as humans, such as increases in heart rate and the release of stress hormones. They also show changes in behavior that can be similar to behaviors shown by humans in pain and will make strenuous efforts to avoid places or events that have caused them pain in the past. Based on this evidence, scientists conclude that some animals are capable of feeling pain.

Precisely which animals can feel pain and which cannot is far harder to assess. Although the ability to sense and avoid harm in the environment may be a basic protective mechanism in the animal kingdom, it is uncertain in which species that mechanism might involve conscious feelings and the potential for suffering. Most people believe that primates, such as the chimpanzee, and pets, such as cats and dogs, are just as capable of feeling pain as they themselves are. Equally, most people swatting a fly do so in the belief that it is not able to suffer. However, it is important to avoid the tendency to believe that animals that look more human or are "cuddlier" have a greater ability to feel pain than other animals. It may simply be the case that pain-related behavior in certain animals, such as insects, fish, or birds, is harder to identify from a human perspective. In these groups, detailed study of both normal and pain-related behavior is very important. There is growing evidence that nonmammalian vertebrates, such as reptiles, amphibians, fish, and birds, are capable of feeling pain. Insects, such as the housefly, being invertebrate, are simpler animals, yet have a basic sensory system and show withdrawal reactions to potentially harmful stimulation. However, they can continue functioning with major injuries, such as missing limbs, which would often incapacitate other animals, suggesting that they may not be feeling pain in these situations.

What Behavioral Changes Are Caused by Pain?

Behavior can change in different ways when an animal feels pain. Normal behaviors may be seen far more or far less than usual, or unusual behaviors, not seen in animals free from pain, may be shown. Although responses to pain can vary between species, breeds, and even individuals, some shared responses are seen. Although many of the changes are protective, severe or long-lasting pain can cause changes in behavior that may not have any function.

Following sudden pain, the immediate behavioral response is withdrawal from the source of the pain. The quickest part of this withdrawal may occur as a reflex, where the alarm signals from the nociceptors cause a response in fractions of a second, before the feeling of pain even occurs in the brain. Following this, a more organized behavioral response, controlled by the brain and reflecting conscious awareness of pain, is seen. In some species this may involve escape behaviors, whereas other species may freeze. Vocalizations will often be heard at the time of injury, although the degree of vocalization will vary a lot between species (e.g., dogs will generally vocalize much more than cats).

If the immediate responses to injury fail to stop pain from occurring, and pain persists, other changes in behavior are seen. An important feature is absence of normal behavior, and the return to normality can be a good indicator that pain has passed or that treatment has been effective.

A common response to pain is to focus attention (e.g., by licking, rubbing, or scratching) on the site of injury. Animals in pain may be seen to tremble, shiver, or grate their teeth and to whimper or groan. Some animals may become completely inactive and will feed less or stop feeding altogether. Injured animals will make efforts to protect the damaged area by guarding it from painful stimulation. For example, limping is a form of guarding behavior, where an injured limb is protected from further damage. Pain will also commonly alter an animal's posture when it is inactive, since the animal adopts the position that causes it least pain. Posture may generally appear hunched or rigid, but the precise change will depend on the location of the problem. The guarding response may involve aggression toward people if they approach the animal or touch the painful area.

If pain lasts for a significant length of time, such as during long-term disease, broader changes in behavior may be seen. Sleep patterns may be disrupted, and the animal may withdraw from social interaction. Often an animal in pain will alter its interaction with, or response to, a familiar person. For instance, a previously playful dog may become unresponsive, or a normally placid cat may show uncharacteristic aggression. Animals in pain will often pay little attention to their environment or to grooming themselves and may appear restless. Such changes may be part of a general change in temperament, and these animals are commonly described as appearing depressed, anxious or fearful. Some animals suffering from chronic pain may show stereotypic behaviors, such as rocking or circling.

What Factors Influence Pain-Related Behavior?

Behavioral responses to injury can be highly variable. Some individuals will show a larger change in their behavior than others with the same injury. These differences could be because some individuals actually feel more pain than others. Sex, age, and previous experience can all create variability in the experience of pain. There are also factors that affect the extent to which particular individuals or species show that they are suffering. The variability in response can hinder the recognition of pain, so it is important to consider these factors.

The pain system is extremely complex. There are systems in the body that can shut off the feeling of pain, even when pain messages are being carried to the brain. The presence of such systems may seem strange given the value of feeling pain. However, during times of emergency, a change in behavior could interfere with more important responses. For instance, if an animal were injured during an attack by a predator, it would not be beneficial to respond to the injury immediately. The first priority must be to escape from the predator, and feeling pain could interfere with this. The system for shutting off pain is only temporary; once the emergency has passed, the pain returns. This system may also be triggered in situations that the animal views with fear. This means that some animals may show altered pain behavior in unfamiliar environments or when isolated.

Pain behavior will be strongly affected by a species' normal social organization. Even when assessed in artificial surroundings, animals may retain ancient protective responses that would have been of value in the wild. Prey species may be less likely to openly display their pain since such a display could act as a sign of weakness to be exploited by a predator. Social animals may be more likely to show altered behavior, either as a signal to other group members or because their sociality affords them some protection from danger. Animals such as the dog might view their owner or other humans as part of the social group, and veterinarians often find that behavior will differ depending on whether or not the owner is present.

The source and severity of pain will greatly influence the way behavior is altered. A sudden injury to the skin, such as a burn, will cause a very different response than a gradually increasing internal pain such as a stomach upset. Injury to different sites around the body may cause more or less pain because different areas of the body have different numbers of nociceptors. The site and severity of an injury or disease can therefore give clues as to the magnitude and type of pain-related behavior that might be expected.

The role that thought might play in pain perception is unclear. Humans can think about pain in a way that most animals probably cannot. However, this could be either a good thing or a bad thing. Although in some cases fear can decrease pain, anxiety may actually increase the pain felt. A trip to the dentist may be made worse by our unease about what will happen; however, our previous experience also tells us that any pain is ultimately for the best and will normally pass quickly. This helps us to cope with the situation more effectively. The extent to which an animal's ability to think about pain or to perceive time passing affects its degree of suffering is unknown. So, although an animal's ability to think may affect the way that it responds to pain, animals that are judged by humans to possess less mental complexity cannot necessarily be judged to suffer less when in pain.

When Is It Necessary to Measure Animal Pain?

Companion Animals

Veterinarians will frequently deal with ill or injured animals that need an operation, so pain assessment following surgery is very important. Although pain in nature acts to aid recovery, there is no evidence to suggest that removing pain following an operation harms the animal. In fact, when pain relief is administered following surgery, both humans and animals tend to recover more quickly. Following an operation, the time taken before dogs will start feeding and become willing to interact with a human is a good indication of the duration of pain they feel. They also show changes in posture, temperament, and the response to touch around the site of the incision that can provide an indication of the pain they are still feeling.

Another main use of pain assessment in veterinary practice is to identify the degree of long-term pain felt by animals with problems such as arthritis or cancer. An accurate pain assessment allows the best treatment to be identified and administered. It can also indicate when euthanasia may be the most humane option for the animal in question.

Farm Animals

Farm animals experience pain through disease or injury like other animals. However, they may also experience pain due to husbandry procedures, such as beak trimming, branding, castration, and de-horning. For example, it is common practice to remove a small portion of the beak of chickens kept for egg production to stop birds from injuring each other when they fight. Following the operation, birds show guarding behavior, by decreasing their use of the beak to feed, drink, and preen, and also show alterations in sleeping, eating, general activity, and social behavior that may indicate long-term pain. Male pigs, cattle, and sheep are all commonly castrated on farms, and there is behavioral evidence in all three species that this causes pain. Young pigs will show an increased level of high-pitched squealing, while calves and lambs show a variety of abnormal behaviors and postures following the procedure.

A lamb showing an abnormal posture following castration.
Courtesy of Joyce Kent.

Scientists interested in animal well-being study such practices with the aim of reducing the pain that farm animals suffer. In chickens, behavioral observation following beak trimming has identified the methods that cause least pain. Similarly, in lambs, measurement of subtle behaviors and abnormal postures has allowed the best ways of limiting pain during castration to be identified.

In both cases, the best way of decreasing the pain felt is by not doing the procedure at all. However, this may not always be possible. Pain assessment can allow scientists, farmers, and those in the government to decide whether any benefits of the procedure outweigh the pain felt by the animals and to pick the most humane methods possible where there is no alternative.

Laboratory Animals

Many animals, most commonly rats and mice, are kept in laboratories for use in scientific research. This research may involve pain either as a side effect of a particular treatment or as a deliberate effort to study pain mechanisms. In laboratory science, pain assessment can be used to assess the severity of particular procedures. This information can then be used to decide whether the procedure is justified in terms of its benefits balanced against the pain caused to the animal. Pain assessment can also allow the best pain relief treatments to be used in experimental work. Animals are also used to test new pain relief drugs for humans and much of this work involves behavioral tests.

How Is Animal Pain Measured?

To allow these various pain assessments to be carried out effectively requires that pain be accurately measured on a scale. The presumption behind pain scales is that larger departures from normal behavior indicate the presence of more pain, and that the return to normality reflects the loss of pain. However, pain is not a physical thing like length, which can be measured with a ruler; so how can we measure pain?

In the past, assessments of animal pain have been based on the personal judgment of a veterinarian or owner. Although this may be a relatively accurate judgment, different people sometimes differ in their assessments depending on their experience and potentially on how they themselves are feeling at the time.

To attempt to standardize pain measurement, various pain scales have been devised to provide a way of assigning a number to an animal's pain. The simplest type of pain scale is called a "Visual Analogue Scale" (VAS). This involves an observer rating the degree of pain they feel the animal is in by marking a position along a single line, typically ranging from "no pain" to "worst pain imaginable". The measurement of pain is the distance along the line. This method is simple to use, but lacks detail, and different users may still vary in the scores they give to individual animals. To overcome this problem, scientists take a more methodical approach where an animal is carefully observed, allowing a catalog (called an *ethogram*) of all its behaviors to be produced. Ethograms can be used to measure both normal and pain-related behavior. For instance, we can count how many times an animal wags its tail in a set period of time, or measure how long it spends in a particular position, such as lying on its side.

Another way to identify pain-related behaviors is to gather the opinions of people familiar with animals in pain, such as veterinarians. Although there will be some variation in the behavioral indicators individuals use, there will be common factors that they all use. In this way the knowledge and experience of experts can be transferred to those with less experience.

Such different methods can allow behavioral indicators of pain to be identified. These can then be used in pain scales that are practical for everyday use and are more informative and reliable than VAS scales. In these "Numerical Rating Scales" (NRS), pain is measured by assigning scores on various different parameters. These different scores are added up to give a final pain score. Different NRSs can be constructed for different species and sometimes for different types or sources of pain. Sometimes the scale may also include nonbehavioral parameters, such as heart rate, breathing rate, body weight loss, body temperature, and pupil dilation.

Animals Can Feel Pain

There is a lot of evidence, including that from behavioral observation, that animals can feel pain. Pain changes motivational priorities and causes the animal to alter its behavior, often in a beneficial way. However, although function can be attributed to feeling pain, we should still make all possible efforts to remove unnecessary pain. Freedom from pain is a crucial part of animal well-being; yet unfortunately, animals often suffer pain as a direct result of human action. Measurement of behavior during injury, disease, or following operations has an important role in the assessment of animal pain. Different species may show different types of pain-related behavior, so a good knowledge of both normal and pain-related behavior in different species is required to assess pain accurately.

Pain often has to be measured in humans where verbal communication is not possible or reliable, such as in infants, the mentally handicapped, patients in intensive care units, and elderly people suffering from senility. Although undertreatment of pain has been common in these individuals, the situation is gradually improving, due in many cases to measures of human pain-related behavior. In the past, animal pain has also been undertreated, and behavioral measures have a large part to play in future improvements in the detection and alleviation of animal pain. There is still a lot to be learned about the abilities of various animals to feel pain and how they respond to it. Where there is doubt about a species' capabilities, it is best to err on the side of caution and presume that it is capable of feeling pain.

See also Welfare, Well-Being, and Pain—*Behavioral Correlates of Animal Pain and Distress*

Welfare, Well-Being, and Pain—*Veterinary Ethics and Behavior*

Welfare, Well-Being, and Pain—*Psychological Well-Being*

Further Resources

Bateson, P. 1991. *Assessment of pain in animals.* Animal Behaviour, 42, 827–839.

Flecknell, P. A. & Molony, V. 1997. *Pain and injury.* In: *Animal Welfare* (Ed. By M. C. Appleby & B. O. Hughes), pp. 63–73. Wallingford, U.K.: CAB International.

Flecknell, P. A. & Waterman-Pearson, A. 2000. *Pain Management in Animals.* London: W. B. Saunders.

Hansen, B. 1997. *Through a glass darkly: Using behavior to assess pain.* Seminars in Veterinary Medicine and Surgery (Small Animal), 12, 61–74.

Holton, L., Reid, J., Scott, E. M., Pawson, P. & Nolan, A. 2001. *Development of a behaviour-based scale to measure acute pain in dogs.* Veterinary Record, 148, 525–531.

Rutherford, K. M. D. 2002. *Assessing Pain in Animals.* Animal Welfare, 11, 31–53.

Kenneth M. D. Rutherford

■ Welfare, Well-Being, and Pain
Behavioral Correlates of Animal Pain and Distress

Humans use nonhuman animals in many ways, such as for food and medical research; as a result, humans have the responsibility of ensuring that animal pain, distress, and suffering are minimized or prevented. Animal behavior plays a very important role in determining both if an animal is experiencing pain or distress and whether the methods of eliminating the pain and distress are successful.

As humans, we know when we experience happiness, fear, anxiety, or pain. When other people have similar emotions, we understand those feelings through the words they use to communicate the experience, facial gestures like a smile or grimace, or changes in their body posture, for example holding a damaged limb. We might even get an idea of the intensity of the experience by the shape, size or intensity of the gesture or facial expression. We can never know exactly what the other person feels because their experience is subjective—in other words, they are the only individual who actually feels it. The experience cannot, therefore, be directly observed and measured; however, we can get an idea from similarities between their behavior and verbal descriptions and our own.

How can we know whether animals experience pain or distress? Are animals like humans in the sense that they can feel pain or fear for example, or are they unaware of their experiences? To find solutions, we need to ask two further questions. First, do animals have the capability of experiencing pain or distress—does their anatomy, and the way their bodies are designed and their brains are organized mean that they can sense pain or experience unpleasant and distressing emotions like fear? Second, how can humans recognize when an animal is in pain or distress—what observable evidence can be measured? After all, animals cannot tell us what they are feeling.

A cat in distress.
Courtesy of Leslie A. King.

The idea that animals are capable of experiencing pain or distress was controversial for a long time. René Descartes, a highly influential philosopher, argued that animals were like machines—if you injured them they might be damaged, and they may respond by calling out or moving, but this would not matter to the animal itself because it did not have a mind, could not think or feel, and therefore could not suffer. In the early twentieth century, many scientists also thought that animals were the kind of creatures that could simply detect things (stimuli) in their environment and respond to them, but could not have an inner "experience" of that stimulus—the animal did not know or feel, just showed an unconscious reflex response to the outside stimulus. If an animal could not feel pain, there would be no need to alleviate it when the animal experienced injury, for example, by giving anesthetics during and analgesics (pain-killing medication) after surgery. Furthermore where humans used animals, such as in farming or in biomedical research, there would be no need for consideration of animal welfare. However, was Descartes right? Are animals just like machines, or are they capable of feeling pain and distress, and, if so, how can this be detected or measured?

First, it is important to define the concept of animal pain. Pain can be defined as a strongly unpleasant or aversive experience that is associated with either real or potential damage to the body. Damage to body tissue leads to injury, which harms the animal; the very unpleasant or "noxious" experience that goes along with the injury is the sensation of pain. Modern scientific research has shown that mammals and birds have the same type of anatomy for sensing and responding to pain as humans. Their sensory connections, or neurons, throughout their bodies work in the same way. The experience of pain ensures that the human or animal detects and moves away from, and in the future avoids, things that could increase the likelihood that s/he would die or be permanently harmed. For example, when a person touches a hot stove, s/he immediately withdraws the hand. The same thing happens if an animal's tail is placed on a hot surface—the tail is immediately withdrawn; the presence of pain and quick withdrawal prevent further injury.

Neurons allow messages about pain sensations to be passed from the area on the body where the injury occurs, all the way to the brain, where they are processed and cause the animal to move away from whatever caused the injury. The stimulus that causes injury, for example, by cutting or damaging skin, is detected by *nociceptors*, specialized cells designed to respond to tissue injuries. The signal generated by the nociceptor is passed along a set of connected neuron cells to the brain. Depending on which route it takes through the body, it may cause a sudden, automatic response, or reflex. However if the signal reaches certain parts of the brain, the animal will become aware of the sensation and experience it as pain. By withdrawing from whatever causes pain, the animal is less likely to be harmed; if the animal has already been harmed, the pain sensation should cause a change in behavior, to protect the animal from further injury or to minimize the harm done. Imagine trying to hold a hot cup of coffee—eventually you would have to change your behavior and put it down when the pain from the heat became too great—the pain indicating that your hand was about to be burned.

Animals often indicate pain or discomfort by changing their behavior. First, it is important to know what the normal behavior of a species is so that changes in behavior can be detected. Some changes in behavior due to pain and distress are obvious; for example, the animal may not put weight on a damaged leg while walking, or will develop a limp. They may not perform certain behaviors in the way that they used to, for example, remaining motionless in the corner of a cage after surgery or after administration of toxic chemicals. They may change the way that they do certain activities; for example, dogs used in laboratory experiments may eat less or change their posture, taking a hunched or huddled position when in pain. Reductions in activity, alongside a staring coat and changes in body temperature, may indicate suffering related to infection or disease. Knowledge of how different species respond to pain allows these observable changes in behavior, called *clinical signs*, to be detected and analgesic drugs to be administered to reduce the pain. Often the behavior will return somewhat to normal when pain relief is given. When animals used by humans are subject to procedures that might cause pain, as in some forms of biomedical research, analgesics and anesthetics should be used routinely to ensure that animals do not suffer, and the animal should be regularly checked for clinical signs to make sure they are not in pain. Measures of behavior and health, like the amount of movement, weight, and heart rate, can be used to quantify and measure the extent of pain experienced by individual animals. These "scorecards" can indicate when animals may be suffering, and therefore, they can assist pain and distress prevention. These measurements also inform us whether certain procedures cause pain or distress so that, before more animals are used, alternative methods can be thoroughly investigated and introduced, thus meeting our ethical and legal responsibilities to prevent animal suffering.

Humans experience a wide range of sensations other than pain. Do animals also experience other unpleasant states like fear? How can these be detected? Scientists studying animal distress or suffering propose that both human and nonhuman animals have evolved emotional experiences for similar reasons to those discussed for pain: Unpleasant sensations like fear or hunger tell the animal that the situation is dangerous, or the resources it needs cannot be found there. The experience makes the animal select a different behavior that reduces the risk, for example, moving away and avoiding whatever made the animal fearful, or enabling it to move to an area where it is more likely to obtain a resource that it needs, such as food. As with pain sensation, mammals, birds, and many other vertebrates are known to be "hard-wired" in the same way as humans to sense things in the environment and process this information in parts of the brain that determine how important it is to the animal to gain or avoid it. If something is detected that has caused damage in the past, the animal will experience the sensation of fear, and this will encourage the animal to move away. If the animal finds itself in a place where danger has occurred before, even if the scary thing is not present, the animal will experience anxiety. If the animal can see food but cannot reach it, s/he will become frustrated. Distress or suffering can be thought of as resulting from the intense or prolonged experience of these unpleasant emotional states associated with, for example, pain or things that cause fear.

Because emotional distress is subjective, it cannot be measured or observed directly. However, behavior can give a good indication of what the animal is experiencing. Certain behaviors occur in the presence of clearly dangerous situations; for example, a mouse may freeze motionless in fear if it sees a predator. If this same behavior is seen in other circumstances, such when being handled by a human, it could mean that this situation also causes fear. Huddling or attempts to escape from a place where experimental procedures occur may reveal fear or anxiety; some postures, such as a dog's growling or baring of teeth, may indicate that the animal feels threatened. The strength of a feeling can be measured by assessing how important it is to the animal to escape or avoid a particular stimulus, such as a human handler. These methods assess the amount of work an animal will do, or the speed at which it escapes. If something is particularly aversive, the animal will work to its maximum

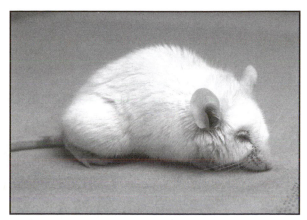

A mouse in distress.
Courtesy of Leslie A. King.

ability to keep away from it; if it is less important, the animal will expend less effort. Different methods of handling or animal management can be compared to find out from the animal's perspective which is less unpleasant.

These methods can also be used to assess what resources are important to animals, and determine what things captive animals need in their enclosures, so that they do not suffer in the first place. The scientific assessment of animal welfare has been established in the last thirty years to assess the needs of animals used for food and clothing. Some examples of this research illustrate that farmed animals have quite important and specific requirements that may not always be provided in their usual housing. For example, female pigs (sows) are housed in long thin metal "farrowing crates" just before they give birth. Although these do not allow the pig to turn around and often do not contain bedding, research has shown that just before they give birth, sows will work extremely hard for nesting material like straw, and for the opportunity to make a nest in which to have their piglets. Chickens are more distressed when caught by humans than by mechanized catching machines, even though the machines are large and noisy. Mink reared in cages for their fur will work so hard for water to swim in that they will push a door weighted with three times their body weight.

Animals who are housed without the resources or space they need may develop abnormal behaviors that are highly repetitive, which may show that the animal is distressed. These "stereotypies" are seen in some captive animals, like some zoo polar bears and tigers who pace back and forth repeatedly in the same part of their enclosures for long periods of time. They are also seen in some rats and mice housed in standard laboratory cages. These behaviors can occur when the animal needs to do certain things, like hiding in a tunnel or traveling over long distances, but cannot because the cage does not provide the resources for them to do so. These behaviors have been linked to permanent dysfunctional changes in animals' brains, so that animals can no longer respond correctly to things in their environment, like food, or the needs of their offspring, because they are stuck performing repetitive, abnormal behaviors. In some cases, the stereotyping animal may injure itself, causing physical harm through biting skin or whiskers, or neglecting its offspring, leading to mortality. However, sometimes, providing resources which the animal needs, such as tunnels and burrows for caged mice, or social companions for chimpanzees, can reduce or remove abnormal behavior.

In conclusion, we can tell a lot about whether an animal is suffering through behavior. Interpretations of behavior must be cautious since there could be many explanations for why an animal acts in a particular way; however, abnormal behaviors or those that are associated with fear and pain should be used as a warning sign that an animal may be suffering. Sometimes taking physical measurements such as heart rate, or measuring the amount of hormones in the blood associated with stress responses, can help in understanding whether a behavior shows that an animal is in distress. But do all animals, even those that are newly born, or insects and invertebrates like ants, feel pain? Should we worry when stepping over the cracks in the pavement, in case we squash a bug and cause suffering? In the past, even newborn human babies were not believed to be capable of experiencing pain; however, as scientific knowledge has grown, more evidence has shown that human babies and newborns of many species can feel pain. We are less sure about the ant; they may have the same basic neuronal "wiring" as humans and higher mammals, but it is unclear that they would be aware of sensations. Mammals, birds, and fish clearly have complex neural systems, and although we cannot be sure they are fully aware in the way that humans are, the likelihood is that they are *sentient*, or capable of feeling pain and fundamental kinds of unpleasant distress like fear and anxiety. Indeed, animals are also used in laboratories in order to study pain, fear, and anxiety. The ability of nonhuman animals to experience pain, fear, and distress means they are not like machines: Their ability to feel means humans have a responsibility to ensure that their suffering is minimized or prevented.

See also Welfare, Well-Being, and Pain—*Behavioral Assessment of Animal Pain*
Welfare, Well-Being, and Pain—*Psychological Well-Being*

Further Resources

Appleby, M. C. & Hughes, B. O. (Eds.) 2000. *Animal Welfare*. Wallingford, UK: CABI.
Bekoff, M. 2000 *Strolling with our Kin*. Jenkintown, PA: American Anti-Vivisection Society.
Dawkins, M. S. 1998. *Through our Eyes Only: The Search for Animal Consciousness*. New York: Oxford University Press.
The Humane Society of the United States, 2000. *Science and Conscience*. East Haddam, CT: NAHEE.

Lesley A. King, Kathleen Conlee & Martin L. Stephens

■ Welfare, Well-Being, and Pain
Carnivores in Captivity

Monotonous pacing is a remarkable and disturbing sight in zoos. It is a *stereotypy*—a repeated, unvarying behavior with no apparent goal or function, and pacing is particularly favored by carnivores like the cat, bear, and weasel groups (other mammal families instead specializing in other forms such as repeated head nodding or rhythmic object biting). The Dutch even term such pacing "ijsberen," literally "to polar bear," sometimes using this to describe agitated, restless people. Welfare research shows that such stereotypies generally signify that something is missing from the environment—that the animal is denied some opportunity to forage, build a home, socialize, or escape from threats, so that it becomes stuck in a behavioral rut, repeatedly attempting to carry out the thwarted activity. Scientifically identifying the precise roots of a stereotypy can be difficult in zoos, however. Unlike animals in research laboratories or on farms, usually only a few individuals per species occur at any one site, making statistically sound conclusions difficult. Furthermore, it is often impractical to alter zoo animals' environments in substantial ways (imagine running an experiment to find out if caged tigers benefit from hunting, for example!). But one way to investigate why zoo animals respond to captivity in the ways that they do, is to unravel why different species vary so much in their responses to zoo life.

The pacing of captive carnivores had long been thought to derive from frustrated hunting. Its energetic nature and tendency to be most common and rapid just before feeding time, led many researchers to speculate that it derived from the urges these animals may have to hunt and chase prey. Pacing was hypothesized to be an outlet for these natural activities so impossible in the typical zoo. But other researchers had other ideas. They observed how the behavior hugged enclosure boundaries, and suggested that instead it represented territorial patrolling, or the daily forays that carnivores would naturally make over their home ranges. So how can we tell which idea is right? And could constraints on natural behavior lead to further signs of poor welfare, over and above pacing? My colleague Ros Clubb and I were sure that species differences would yield the answer. Polar bears may be infamous for their repetitive pacing in zoos, but their cousin the brown or grizzly bear seems far more relaxed and well-adapted in captivity: In zoos, these animals pace only rarely, and in further contrast to the polar bear, they breed well, showing good mothering abilities and excellent cub survival rates. If species differ that much in how they react to zoo life, could looking at their natural behavior in the wild explain why? We reasoned that if the foraging hypothesis were correct, species that naturally hunt for 100% of their prey, or that track and pursue them over great distances, should fare least well in captivity, while in contrast, species that are naturally omnivorous (like racoons), carrion-eating (like hyaenas), or that forage by waiting, still and silent, for passing prey (like lynxes), should all adapt well to the captive life. However, if the ranging hypothesis was instead the correct idea, then it should be species that naturally have large home ranges that show most pacing in zoos, and perhaps other signs of poor welfare, too.

Testing this idea involved over 2 years' worth of intense library work. We needed data on stereotypy, on other variables affected by welfare (we chose infant mortality, since plenty of data were available on this), and on carnivores' natural time budgets in the wild. We found good quality reports of pacing in over 100 studies of over 300 stereotyping individuals, spread over 40 zoos and representing 35 species. From this, we could calculate the mean proportion of time spent stereotyping by the pacing individuals within each species. We then collated figures from the zoo community's International Zoo Yearbooks to calculate infant mortality; over the last decade or so, these volumes provided data on around 26,000 births for our target carnivores, from about 500 zoos worldwide. Finally, we needed to quantify these animals' typical activities in the wild. An exhaustive trawl through many books, and the last 4 decades' worth of relevant biology journals, generated over 500 good quality sources of data. From this, we could calculate each species' typical natural home range size, time spent hunting, time spent active, distance covered while hunting, and several other variables. At last we were ready to test the two ideas. To our surprise, natural foraging did not predict captive welfare. Naturally active hunters were no more likely than other species to be very stereotypic when caged, or to have high captive infant death rates.

Instead our data backed the second hypothesis: Species with large natural home ranges, and that naturally travel great distances each day, were the most likely to produce highly stereotypic individuals in zoos and have relatively low cub survivorship. Thus grizzlies thrive better than polar bears in zoos because in the wild, their typical home range is around 182 km^2 (70 mi^2) a mere fraction of the 79,500 km^2 (30,700 mi^2) covered by the typical polar bear!

This result demonstrates, for the first time, that particular natural lifestyles can make animals consistently more or less vulnerable to welfare problems in zoos. For carnivores, this allows zoos either to better select the species they keep (if naturally small-ranging species are inherently likely to fare better, maybe zoos should specialize in these?), or to alter carnivores' husbandry to incorporate features of their naturally wide-ranging lifestyles. This of course raises the key question: What exactly is it about having a naturally large home range that is so important? The simple answer is that we do not know, but we have a number of ideas. It could be that naturally wide-ranging carnivores simply need more space in captivity, but other things may well be important. For instance, wide-ranging carnivores' lives in the wild are typified by enormous day-to-day and season-to-season environmental variety: Over time, they shift between different subsections of the vast areas they call home, and so naturally do not spend month after month in the same locale. Perhaps if zoos can provide sustained, relevant environmental variability, they can help adapt these animals better to the captive life.

See also Welfare, Well-Being, and Pain—*Sanctuaries*
Welfare, Well-Being, and Pain—*Psychological
Well-Being*

Further Resources

Clubb, R. & Mason, G. 2003. *Captivity effects on wide-ranging carnivores.* Nature, 425, 473–474.

Clubb, R. & Vickery, S. (in press). *Why do carnivores pace? Insights from comparative and experimental studies.* In: *Stereotypies in Captive Animals* (2nd edn.) (Ed. by G. Mason & J. Rushen). Wallingford, UK: CAB International.

Mason, G. J. 1994. *Age and context affect the stereotypies of caged mink.* Behaviour, 127, 191–229.

Wielebnowski, N. C. 1996. *Reassessing the relationship between juvenile mortality and genetic monomorphism in captive cheetahs.* Zoo Biology, 15, 353–368

Wielebnowski, N. C., Fletchall, N., Carlstead, K., Busso, J. M., & Brown, J. L. 2002. *Noninvasive assessment of adrenal activity associated with husbandry and behavioral factors in the North American clouded leopard population.* Zoo Biology, 21, 77–98.

Georgia Mason

■ Welfare, Well-Being, and Pain
Enrichment for Chickens

Just like people, chickens seek stimulation. However, they are often housed in barren, monotonous environments that provide little to occupy their interest. This practice was claimed to eradicate fear and stress, but, in reality, it is misguided and dangerous. Such environmental impoverishment compromises normal behavioral development. For instance, it reduces the opportunity for the birds to perform natural behaviors, like exploration and foraging, and it often leads to increased fearfulness, apathy, feather pecking, and cannibalism. These consequences seriously damage the birds' well-being (welfare is the European term) and productivity. First, for example, frightened chickens often run into obstacles or pile on top of each other; these inappropriate reactions may injure or even kill the birds or their companions. Fearful birds (those predisposed to be easily frightened) also show poor egg production and growth, and the quality of their eggs or meat is reduced. As well as compromising welfare, fear imposes a substantial economic burden. Second, chickens often peck and pull at other birds' feathers. In severe cases, feathers are pulled out and the victims end up virtually denuded, thereby increasing their susceptibility to injury. Feather pecking can also lead to cannibalism and a painful death; the birds literally eat each other alive. The European community will ban battery cages starting in 2012, but this behavioral vice is a major obstacle to the adoption of alternative housing systems, like free range, because it is difficult to control when birds are kept in large groups.

Consumers now expect their food to be produced with greater respect for animal welfare, but the farmers also need to make a living. Therefore, to ensure sustainable agriculture, we must find practical, affordable ways of minimizing fear and feather pecking. These behavioral problems are governed by both genetic and environmental factors. Therefore, breeding chickens that show less fear and feather pecking might yield rapid improvements, but it would be necessary to monitor the birds carefully to ensure that there were no undesirable associated effects. For present purposes though, I will concentrate on environmental enrichment. Providing perches or nest boxes help satisfy requirements for roosting or for seclusion during egg lay, but traditional enrichment involves increasing general stimulation by exposing the animals to novel stimuli and giving them objects they can manipulate or play with. This procedure can reduce fear and improve chickens' health and productivity. However, care is needed. Some studies yielded inconsistent results and many so-called enrichment stimuli were soon ignored; some of them actually increased aggression. These inconsistencies probably reflect the fact that the stimuli were chosen according to human preconceptions rather than a critical consideration of what a chicken might find enriching. We addressed four main questions in our efforts to identify practical, effective enrichment strategies.

1. Is Enrichment Attractive?

First, we determined whether or not chickens are attracted to enriched environments. Rather than offering them a straightforward choice, we asked if enrichment could alter chicks' previously established environmental preferences. We determined the least preferred side of their home cage during the first 5 days of life and then introduced a ball, some rubber tubing and a bunch of string into it. Many chicks showed significant shifts in preference to the enriched half. These findings are consistent with reports that hens looked through spy-holes to view novel stimuli, and that broiler chicks moved readily from their home pens into a nearby area when it contained novel objects. Collectively, the findings suggest that chickens do find enriched environments attractive.

2. Can Playing the Radio Help?

Music is thought to help chickens to thrive. This concept is difficult to test experimentally because of the diffuse nature of sound. Therefore, we surveyed more than 100 British poultry farmers to determine the incidence and perceived benefits of playing the radio to chickens. Forty-six percent of them played the radio routinely to their flocks. Encouragingly, 96% of them believed that the chickens were calmer, 52% said that music reduced aggression, 20% reported improved health, and 16% claimed increased egg production. Thus, both welfare and economic benefits were apparent. There are various explanations for these beneficial effects, and none are mutually exclusive. First, farmers who play the radio may be more concerned about their flocks' welfare in general and consequently adopt better practices. Second, music might partially satisfy the birds' need for stimulation. Third, birds reared in quiet environments show pronounced panic and stress responses when they hear unfamiliar or infrequent sounds, like sneezing, shouting, or a slammed door. Panic reactions can cause injury, and stress-induced activation of defense mechanisms requires the utilization of bodily resources that could otherwise be used for egg production or for growth. Playing the radio may minimize these undesirable effects by helping the birds to learn that new sounds are not necessarily dangerous.

The farmers reported that chickens preferred easy listening and chart music, though this probably tells us more about the farmers' likes and dislikes. However, we could determine how much and what type of music the chickens prefer by giving them the opportunity to switch it on or off in a simple operant paradigm. Chickens readily learn to peck illuminated keys for various rewards so the on/off switches of the radio could be linked to dedicated keys.

Whatever the underpinning mechanism, switching on the radio must be one of the easiest ways of enriching the environment for the chickens and for the farmers.

3. Are Familiar Odors Reassuring?

The possible existence of a sense of smell in chickens was ignored for many years. However, we have now established that chickens respond to a variety of olfactory cues; for example, they avoid odors of predators or blood, they are reluctant to accept familiar food treated with a novel odorant, and they use odors as "learning" cues in conditioning experiments. This sensory capability might have important implications for their welfare. During their lifetime, chickens encounter several unfamiliar stimuli, like new places, objects, and food; these can all elicit damaging fear responses. Since familiar visual or auditory cues are attractive and can reduce fear in unfamiliar situations, we reasoned that familiar smells might also be reassuring. Encouragingly, chicks reared with an odorant were strongly attracted to it in an otherwise novel environment; this effect was common to at least four breeds, three methods of presentation, and four odors (vanillin and the oils of orange, geranium, and clove). Unless an odorant was irritating, unpleasant, or toxic, I suspect there might be no limit to this phenomenon. Recently, chicks were reared in cages containing vanillin. After a few days two cagemates were placed close together in a novel arena containing food and either the familiar odorant or a color-matched solution of odorless food dye (control). Fear inhibits social dispersal, exploration, and feeding, so frightened chicks would not be expected to move apart or feed until their fear levels had dissipated sufficiently. We found that vanillin chicks moved apart sooner and further and were more likely to feed than the controls. Furthermore, chicks ate a novel food sooner if it were associated with a familiar smell. These results show that chicks form olfactory memories, and that familiar odorants can act as reassuring agents in otherwise unfamiliar situations.

By reducing fear and making new environments and resources more attractive to the birds, this sort of aromatherapy could be a valuable husbandry tool. Since olfactory memories can be formed in the chick embryo, identifying the developmental stages at which attachments to odorants are most strongly established presents an exciting research opportunity.

4. Can We Develop Enrichment Devices That Sustain Interest and Reduce Injurious Feather Pecking?

Many of the wide range of enrichment stimuli that have been used (e.g., bells, beads, toys, stones, flowers, commercial devices) were ignored by the chickens, perhaps because insufficient attention had been given to establishing the birds' preferences. Therefore, we aimed to identify practical enrichment devices that sustained the birds' interest and reduced the occurrence of harmful behaviors like feather pecking.

We began by simultaneously presenting pairs of chicks from two commercial breeds with selected pecking stimuli in their home cages, either continuously for 5 days or for varying periods on 5 consecutive days. They always pecked much sooner and more often at string than at beads, chains, feathers, or baubles; this preference became stronger over time. The chicks may have been attracted to string because it resembled some inherently supernormal stimulus, like straw, twigs, or worms. However, they manipulated the string in a different way than the other stimuli. As well as pecking and pulling at it, they drew the string through their beaks and teased the strands apart; almost as if they were preening them. Thus, pecking at string may provide the most reinforcement.

Chickens have very good color vision so color is likely to be an important attribute of any enrichment device. In previous studies, color preferences varied according to the type of stimulus and the experimental context. However, we have established that chicks and adult hens consistently paid more attention to white or yellow strings than green, blue, or red ones. These findings support suggestions that chickens find blue objects aversive, and that red often serves as a warning signal that causes alarm and avoidance. Furthermore, although visual complexity is thought to be attractive, simple white or yellow strings were pecked sooner and more often than combinations of white and yellow or of all five colors.

Chickens are often strongly attracted to small spherical objects and to shiny material, so, in another test of stimulus complexity, we incorporated small silver beads in the string devices. Contrary to expectation, the chicks pecked much more at plain than beaded devices. This might mean that chicks prefer simple devices to complex ones, but it is more likely that the beads made the string less attractive by interfering with the chicks' ability to tease the strands apart.

Moving objects are thought to be more likely to stimulate play than stationary ones in many species, but we found that chicks pecked

Enrichment has been found to be vital for a number of animals. This chicken enjoys an enrichment tool of string to peck, thus reducing stress and avoiding feather pecking.
Courtesy of R. Bryan Jones.

less at string that was moved occasionally than they did at stationary devices, perhaps because unpredictable movement caused slight alarm and avoidance.

Of course, the string devices must retain the birds' interest for lengthy periods if they are to be considered effective forms of environmental enrichment. When small groups of chicks were given chains, beads, and string continuously from 1 day of age, they soon lost interest in the first two stimuli. However, although interest in string declined gradually, it was still being pecked after 17 weeks. In this context, we might consider developing an automated system that could detect waning of interest in the devices and then maybe raise them briefly or move them to a nearby location in an attempt to rekindle interest.

Our next two studies tackled the important question of whether the provision of string could reduce the incidence of damaging feather pecking. First, groups of chicks from an experimental breed known to show high levels of feather pecking were given string from 1 day to 6 weeks of age, whereas control groups received none. The results were exciting; considerably less feather pecking was shown in the continued presence of string. Second, groups of caged laying hens were given string continuously either from 1 day to 30 weeks of age, from their transfer from rearing to laying cages at 16 weeks, for just 1 day every 4 weeks, or not at all. Pecking-related feather damage was significantly reduced in those hens that had been given access to string. Unexpectedly, this effect was apparent regardless of when the devices had been provided. The string was very attractive to the birds; they manipulated it so much that it looked more like a ball of wool after a few weeks.

Conclusions

1. Chicks and adult hens are attracted to enriched environments.

2. Farmers reported many benefits of playing the radio to their flocks. This is one of the most practicable ways of enriching the birds' environment as well as providing some pleasant stimulation for the farmer.

3. Chicks can remember odorants, and the presence of a familiar smell reduced fear in otherwise unfamiliar situations.

4. Finally, a stationary bunch of plain white string is very attractive to chicks and adult hens of at least four breeds. It gives them more opportunity to engage in behaviors such as exploration and foraging, which are fundamental to their nature. By maintaining lengthy interest and reducing the expression of damaging feather pecking, string devices satisfy critical requirements of effective environmental enrichment. They have the added advantages of low cost, ready availability, and durability.

5. Rather than relying on human preconceptions of what an animal might find enriching we must establish the chickens' preferences for putative enrichment stimuli and then determine their effectiveness in reducing the expression of harmful behaviors.

See also Welfare, Well-Being, and Pain—*Back Scratching and Enrichments in Sea Turtles*
Welfare, Well-Being, and Pain—*Enrichment for Monkeys*
Welfare, Well-Being, and Pain—*Enrichment for New World Monkeys*
Welfare, Well-Being, and Pain—*Experiments in Enriching Captive Snakes*
Welfare, Well-Being, and Pain—*Sanctuaries*

Further Resources

Chamove, A. S. 1989. *Environmental enrichment: A review.* Animal Technology, 40, 155–178.

Jones, R. B. 1996. *Fear and adaptability in poultry: Insights, implications and imperatives.* World's Poultry Science Journal, 52, 131–174.

Jones, R. B. 2001. *Environmental enrichment for poultry welfare.* In: *Integrated Management Systems for Livestock* (Ed. by C. M. Wathes), pp . 125–131. Penicuik, Scotland: British Society for Animal Sciences. (Occasional Publication 28.)

Jones, R. B. 2002. *Role of comparative psychology in the development of effective environmental enrichment strategies to improve poultry welfare.* International Journal of Comparative Psychology, 15, 77–106.

Jones, R. B., Facchin, L. & McCorquodale, C. 2002. *Social dispersal by domestic chicks in a novel environment: Reassuring properties of a familiar odourant.* Animal Behaviour, 63, 659–666.

Newberry, R. C. 1995. *Environmental enrichment: Increasing the biological relevance of captive environments.* Applied Animal Behaviour Science, 44, 229–243.

R. Bryan Jones

Welfare, Well-Being, and Pain
Enrichment for Monkeys

Enrichment is enhancing an animal's environment. When an animal is kept in a cage that is different from the environment in which it naturally lives in, its behavior changes, often undesirably, to accommodate those differences. It may become lethargic or even bite itself. Any goals animal behaviorists might have for the animals influence the characteristics they wish to eliminate. If they want an animal to appear happy in captivity, behaviorists would do different things for that animal than for an animal that they wished to return to the wild and to become a hunter. Toward their goal, they can change many aspects of the cage environment.

This essay looks at three areas which are addressed by enrichment studies: (1) the problems produced by small enclosures and how to address those problems, (2) the goals of enrichment, and (3) evaluation of enrichment. An example of the most successful enrichment technique will serve as an illustration.

The basic problem is that caged animals cannot or do not carry out their normal range of behavior. Lions in cages do not stalk and kill prey. However, almost all aspects of caging can be examined for improvements.

Problems of Containment

Physical Cage Size

Many behavior patterns are influenced by cage size, most commonly the incidence of abnormal behavior and where an animal repeats a behavior over and over. It is hard to imagine a case where simply increasing useable cage space would not constitute enrichment. However research on the effects of crowding shows that, surprisingly, it is the number of individuals in an enclosure that is important and not the space available for each individual.

We can look at a practical level at the physical cage and at how the cage can be used, making comparisons with behavior in the wild.

Walls

Architects who commonly design enclosures choose bare, flat walls because they are easy and cheap to construct and clean. However many animals cannot use flat walls. To measure the effects of vegetation on small South American callitrichid monkeys, climbing plants were grown on three types of garden trellis fixed on the walls of a cage. Monkeys spent the time on the large-mesh rigid wooden trellis which had the least plant growth. The presence of plants had no effect at all on behavior, but the presence of the trellis did enrich the environment by allowing the monkeys to make use of the previously unusable wall space.

Floor

The floor and areas just above the floor are often kept clear for easy cleaning. In the large enclosures commonly uses for monkeys, the floor makes up about 40% of the total surface area and as much as 66% of all the flat surface area.

However in the wild, arboreal monkeys rarely come to the ground. In captivity these same monkeys almost never visit the floor (1% of the time) when it is bare but do so more frequently (15% of the time) when the ground is covered with some leaf-like material such as sawdust, wood chips, or straw. This 15% is not "normal" or "similar" to that seen in the wild, but it has other benefits.

The effect of a floor covering on terrestrial monkeys is even more dramatic. In a zoo in Scotland, groups of ground living primates (lemurs, vervets, stump-tailed macaques) spent only about 15% of the day on the floor when it was bare and dirtied with animal excrement. When the floor was covered with woodchips, time spent on the floor increased to over 70%, similar to in the wild.

Interior

Yet another aspect of primate cage design which has been virtually ignored is that of what goes into the interior of the cage. In laboratories, this area is often empty or has a simple metal shelf. Field studies tell us that wild arboreal callitrichid monkeys spend most of their time (89%) in dense networks of thin, flexible nonwoody tangles, traveling primarily using rigid horizontal branches, and rarely (4%) moving on vertical supports. When they do, they travel up and down using large tree trunks. Nevertheless, the typical cages used for these monkeys have many thin flexible vertical hanging branches, only a few rigid horizontal branches, and no woody tangles at all.

Sharing Space

Another way to make better use of existing space is by sharing space. This can be done by mixing species, by making use of their characteristic use of different areas, and understanding how they react to other species in comparison to their own species. Mixing species also enables two enclosures to be combined, giving more space for each species. Ground living rodents eat fallen fruit in bird, bat, or primate exhibits in zoos in Edinburgh in Scotland, Auckland in New Zealand, and Milwaukee, Wisconsin, in the United States.

While animals can share the same space at the same time, they can also share space sequentially, which is a technique rarely reported. Moving a lion from its enclosure would allow a deer access to an area with long grass. It is difficult to understand why such a procedure is not regularly used in zoos; it has many health benefits on farms and is routinely used there.

Dividing Space

Dividing existing space is especially effective in reducing aggression, particularly in the presence of food, probably because visual input is so important for feeding behavior. Hiding food or making it more difficult to process (e.g., by freezing it) reduces the competition. The use of dividers can also require the animal to cover more distance when moving from one area to another. When mice cages are divided into a maze using strips of plastic, they become less timid and more active. Similar benefits have been found in monkeys, farmed bulls, deer, and pigs.

Foraging

One can increase psychological space both by encouraging animals to make more use of existing space and by changing existing space used. If individuals have to look for hidden food, then that space is changed in that it contains more bits of information with more choices. The most effective enrichment technique so far developed is to encourage foraging by scattering small items of a desirable food into the floor covering, or for arboreal monkeys, installing holes or boxes up off the floor for foraging.

Energy

Calorie expenditure is a by-product of increasing space or using dividers. Many of the tasks devised by psychologist Hal Markowitz require subjects to walk or run from one device to another for food, thereby expending more calories. There is evidence that calorie expenditure improves mood in humans, and it may do so in caged animals.

Goals of Enrichment

Two complementary short-term aims of enrichment studies are to increase "desirable" behavior and to reduce "undesirable" behavior. It is widely accepted that activities such as coprophagy, eating one's excrement, vomiting one's food so that it can be eaten again,

---

Foraging Primates

Primates spend between 25% and 90% of their day searching for food, even in large seminatural enclosures when fed food in excess. In cages, feeding time is dramatically reduced, to between 4–6% in callitrichid monkeys, great apes, and macaque monkeys. When the concrete floor was covered with 10 cm (4 in) of wood chips with tiny seed in it, eight different primate species now spent over one third of the day foraging.

In the initial studies on stump-tail macaques, self-biting and aggression were reduced to less than half, and in adults and in juveniles, to one-tenth when wood chips were used. The time required for cleaning was reduced to less than half. The enclosures were judged to be cleaner, and they were rated as smelling less after 4 weeks with woodchips than after 1 day with a bare floor cleaned with water, detergent, and disinfectant. Survival time of Salmonella bacteria was similar to tests using chickens, in that the longer the litter was in use, the more inhibitory it was to bacterial survival.

By using information from the wild, we can enrich enclosures, evaluate what has been done, and make behavior more natural.

hair-pulling, self-injury, or stereotyped movements are undesirable. Conversely, exploration, play, affiliation, and foraging are commonly considered desirable. Other behavior may be classed as one or the other depending on its frequency: Aggression, inactivity, and feeding are desirable if moderate in frequency, but not if very frequent. There are at least three approaches in assessing desirability of behavior:

- normality,
- observer/caretaker acceptability, and
- theoretical considerations.

Probably the most common approach emphasizes normality: Behavior that approximates that found in the wild is held to be desirable. One problem of implementation is that the norms for many behavior patterns vary, sometimes widely, with group size, habitat, season, and even from day to day with local weather conditions. If arboreal monkeys rarely come to the ground in the wild, possibly because of predator pressure, is it normal/desirable that they do so in captivity and so better use the available space? It is not normal for animals to operate mechanical puzzles in the wild, but is it desirable for them to do so and thereby keep active in captivity?

The goal of enrichment should be to allow and encourage animals to show behavior patterns which are within the normal range of their wild counterparts. The Bronx Zoo has made this habitat as realistic as possible for lowland gorillas.
Spencer Platt / Getty Images.

The second approach takes into account what is acceptable to keepers and the public in zoos. The public like active and easily visible animals; but they do not like to see certain normal behavior like mating, and dislike viewing certain abnormal behavior patterns.

The third approach is to choose enrichment goals because they are identified from a theoretical rationale. This rationale might be to occupy the animals, to exercise them, or to increase more normal behaviors. The two most popular theories influencing enrichment are as follows:

1. The goal of enrichment should be to allow and encourage animals to show behavior patterns which are within the normal range of their wild counterparts.
2. Others have suggested the goal should be to produce animals which could survive and reproduce if (when) released into the wild.

To work with either of these two guiding principles, one needs to know the behavior of the animal in the wild and the extremes of behavior. How much of the time is normally spent foraging when times are good and when times are bad? Surprisingly, in both the last two conditions, foraging time is greatly reduced.

Stress

The reduction of stress is another target of some enrichment studies. Poor cage design, cramped cages, crowding, and more are believed to stress animals. The problem actually is to achieve the optimal level of stress or arousal rather than just trying to reduce it to the lowest level possible. In the normal wild animal, the levels of arousal vary with the normal challenges of life. This suggests that any optimal level is above zero, is a variable one, and most important, that the duration of the peaks of arousal is brief.

Both human and animal research suggest that levels of stress which are lower than those likely to be found in the wild lead to individuals who do not cope with subsequent stressors or adapt well even in the absence of any stress at all. They overreact to mild stimuli. Animals appear to seek out challenge when living in captivity. Animals choose to work for food despite the free availability of the same food.

A monkey sits on an ice block in a zoo in Hanover, August 7, 2003. When temperatures soared to nearly 40°C (104°F) in most parts of Germany, people and animals enjoyed refreshments. KAI-UME KNOTH / AFP / Getty Images.

Increasing Psychological Space

Many successful enrichment techniques act in a way similar to that of increasing physical space by increasing psychological space. Psychological space means how much space there is, based upon how long it takes to do something, for example, to reach for food at arm's length seems quite close. But to have to search for food at that same distance seems much further away and is psychologically at a greater distance.

Evaluation of Success of Enrichment

Assessment answers the questions: Have we done what we set out to do? Are we likely to get similar results if we do again what we have just done? And does it do it well enough to be worth the effort and cost?

Cost

The cost is most often viewed in terms of time or money, but when one form of enrichment is chosen, then other potential forms are excluded. The cost of baiting time of the artificial termite mounts in a chimpanzee enclosure in one Scottish zoo was so great, and the use of the mound so low by the animals, that the device was no longer baited by staff. The most interested individual only spent 4.2 minutes per day at the mound.

Benefit

Both the short- and long-term effects of an intervention need to be measured, and criteria for the duration of the effect need to be specified. For example, a small controlled fire in a chimpanzee enclosure was a focus of much excitement during the first presentation; but interest was not maintained after the second session. It is usually long-term changes in behavior that are of interest if to be considered effective and worthwhile.

Importance

Finally, to measure benefit one needs to decide the importance of behavior to the animal. Two extremes will illustrate the difficulty in such a decision. Certain behavior may be considered important because it normally occurs frequently, such as foraging, walking, or huddling with others. The opportunity to leap is felt to be important in squirrel monkeys since over 40% of their travel is accomplished by leaping. Other behavior may be considered important precisely because it normally occurs infrequently. Dust bathing in chickens is an example.

See also Welfare, Well-Being, and Pain—*Back Scratching and*
 Enrichment in Sea Turtles
 Welfare, Well-Being, and Pain—*Enrichment for*
 Chickens
 Welfare, Well-Being, and Pain—*Enrichment for*
 New World Monkeys
 Welfare, Well-Being, and Pain—*Experiments in*
 Enriching Captive Snakes
 Welfare, Well-Being, and Pain—*Sanctuaries*

Further Resources

Benirschke, K. 1986. *Primates: The Road to Self-Sustaining Populations.* New York: Springer-Verlag.
 Birke, L. 1988. *Better homes for laboratory animals.* New Scientist, 120, 56–60.
Chamove, A. S., Anderson, J. R., Morgan-Jones, S. C. & Jones, S. P. 1982. *Deep woodchip litter: Hygiene, feeding and behavioural enhancement in eight primate species.* International Journal for the Study of Animal Problems, 3, 308–318.
Chamove, A. S. & Moodie, E. J. 1990. *Are alarming events good for captive monkeys?* Applied Animal Behaviour Science, 27, 169–176.
Erwin, J., Maple, T. L. & Mitchell, G. & Maple, T. L. 1979. *Captivity and Behavior of Primates in Breeding Colonies, Laboratories and Zoos.* New York: Van Nostrand Reinhold.
Fa, J. E. & Southwick, C. H. 1988. *The Ecology and Behavior of Food-Enhanced Primate Groups.* New York: A. R. Liss Inc.
Poole, T. 1999. *The UFAW Handbook on the Care and Management of Laboratory Animals.* 7th Edn. Vol. 1: *Terrestrial Vertebrates.* London: Blackwell Science.
Segal, E. 1989. *The Psychological Well-Being of Primates.* Philadelphia: Noyes Publishing Co.
Smuts, B. B., Cheney, D. L., Seyfarth, R., Wrangham, R. W. & Struhsaker, T. T. 1986. *Primate Societies.* Chicago: University of Chicago Press.

Arnold S. Chamove

■ Welfare, Well-Being, and Pain
Enrichment for New World Monkeys

Nonhuman primates are used in different areas of laboratory research. Among Old World monkeys (Catarrhine) the macaques are the primate species most used in research and, in particular, the long-tailed macaque (*Macaca fascicularis*). When it comes to New World Monkeys (Platyrrhine), the common marmoset (*Callithrix jacchus*) has replaced the squirrel monkey (*Saimiri sciureus*) in the last decade. Every possible effort has to be pursued in order to provide an adequate level of welfare for these animals. In this respect, the study of behavior can help to improve the quality of life of the nonhuman primates housed in research laboratories. It must be understood that captivity represents for the animals a great reduction of the level of complexity and stimulation they experience in their natural habitat. In nature, such complexity and stimulation come both from the social group they live in, and from the physical environment around them. Therefore those who work with primates in laboratory conditions must aim to provide the captive animals with environmental enrichments in order to increase levels of stimulation.

Two are the kinds of enrichments to be considered: social enrichments and physical enrichments. It is a good practice to devise enrichments that can mimic a natural situation and, at the same time, motivate the animals to perform as much as possible a behavioral repertoire shown in their natural environment. In order to do so, the knowledge of the ecological and ethological needs of a particular species is essential.

When it comes to social needs, it must be remembered that the majority of primate species live in very complex social systems and, therefore, the social dimension is absolutely vital for their well-being. This is certainly true also for the New World monkeys used in laboratory research. For example, studies both in the wild and in captivity have provided some information on the social organization of the common marmoset. This species live in the Atlantic forest of the north east of Brazil in family groups composed of a minimum of 3 to a maximum of 15–16 individuals. In general, in each family, there is only one adult breeding pair, who are the parents of a series of offspring present in the family. The breeding female gives birth to twins, and the gestation time is about 5 months. An important characteristic of the social behavior of this species is the communal rearing of infants. This means that in the common marmoset not only the mother cares for the offspring, but the father and the older siblings carry and feed the younger members of the family, too, offering them solid food, such as insects. These aspects have to be taken into account when housing a colony of marmosets in a laboratory. For example, in order to provide an adequate captive social environment, it is good practice to house the breeding pairs with their offspring until the latter reach approximately 18–24 months. It is also advisable to allow older siblings to care for the younger ones. As a matter of fact, experimental data have shown that adult marmosets will be better parents if, when younger, they had the opportunity to care for the younger twins. It follows that the cage should be large enough to house from four to five individuals.

Observations carried out on wild populations have shown that the marmosets are very territorial monkeys: When different families meet in the forest, the marmosets engage in a series of facial displays, body postures, and vocalizations aimed at scaring and chasing away the members of the other family. Real physical aggressions rarely occur. For this reason, in captivity, where the animals cannot escape, members of different families must never come in direct physical contact. Also the constant view of members of different families can be a source of stress, and so cages housing different groups can be hidden by the use of hanging curtains. However, acoustic communication between different families can be beneficial. In the wild, marmosets show a repertoire of about 30 different sounds. Among these sounds there are some whose function is for communication between different groups. For example one call, the "phee-call," is emitted in the forest by members of different groups to communicate each other's families' location. This is an essential aspect of the social life of this species, and it can be reproduced in captive colonies. Different families housed in different rooms continuously exchange these calls.

The physical environment must also provide the monkeys with adequate complexity and variety, mimicing the natural situation as much as possible. For example, in nature, marmosets communicate their social status and the ownership of a territory by marking the trunks and branches of trees with urine and gland secretion. This gives them a very important sense of identity. It is essential to provide the marmosets in captivity the opportunity to show such behaviors; therefore, wooden perches, to mimic branches of different sizes, must be provided within cages. The presence of a wooden nest is necessary. The nest gives the members of a family the possibility to sleep together in close contact and to feel safe from possible predators (the monkeys retain the fear of predators in captivity, even if they have never seen a real one). Furthermore, the wood of the nest becomes imbued with the smell of the members of that particular family, confirming every night the sense of belonging of each individual.

Feeding enrichments are also important. In nature, monkeys spend a relevant part of their time looking for and processing food. This is an important activity because it has also the function of exercising the cognitive abilities. It is possible in captivity, without much effort of either time or money, to provide the monkeys with such behavioral opportunities. Food can be scattered on the bottom of the cage, or made more difficult to obtain by using a puzzle feeder. Experimental data have shown that, when given the choice, monkeys in captivity prefer to work for their food, rather than obtaining it and consuming it easily.

See also Welfare, Well-Being, and Pain—*Back Scratching and Enrichment in Sea Turtles*
Welfare, Well-Being, and Pain—*Enrichment for Chickens*
Welfare, Well-Being, and Pain—*Enrichment for Monkeys*
Welfare, Well-Being, and Pain—*Experiments in Enriching Captive Snakes*
Welfare, Well-Being, and Pain—*Sanctuaries*

Further Resources

Davis, H. & Balfour, D. (Ed.) 1992. *The Inevitable Bond*. Cambridge: Cambridge University Press.
Dawkins, M. S. 1980. *Animal Suffering: The Science of Animal Welfare*. London: Chapman and Hall.
McDonald, D. W. (Ed.) 2001. *The New Encyclopedia of Mammals*. Oxford: Oxford University Press.
Wise, S. M. 2002. *Rattling the Cage. Science and the Case for Animal Rights*. New York: The Perseus Press.

Augusto Vitale

■ Welfare, Well-Being, and Pain
Experiments in Enriching Captive Snakes

Environmental enrichment of captive animals is designed to enhance the quality of their care by providing stimuli that encourage the animal to act in more natural ways. Animals in captivity often exhibit unnatural (stereotypical) behaviors as a result of their captive environment, which in many instances does not resemble their natural habitat. Environmental enrichment can use food, modify shelter, or employ training techniques to stimulate movment, exploratory behavior, or enhanced social interactions between individuals. Enrichment, however, has largely been limited to mammals (especially primates) and birds, while reptiles and amphibians have mostly been ignored.

Enrichment is an important factor in the care of captive animals because an unenriched habitat may be very stressful on the animals and may cause significant behavioral problems. Stress results from chronic or acute discomfort, noise or other disturbances, aggression from other animals, confinement, or an inability to predict or control significant events in that animal's life. Mammals and birds often show that they are stressed by pacing, pulling out hair or feathers, obsessive grooming, or improper social or aggressive behavior. Stress in reptiles, however, may be extremely hard to measure since reptiles often spend much of their time simply remaining still. Although it is hard to determine whether reptiles such as snakes are stressed, it seems reasonable that handling it improperly or keeping it in an inappropriate cage could lead to a stressed snake. The common perceptions that snakes are highly adaptable to captivity and that they do not need much in the way of a captive environment (some containers severely restrict movement of the snake) have influenced the way captive snakes are raised and housed.

Many captive snakes are housed in small containers that contain some form of bedding (newspapers or aspen shavings) and a water bowl. These containers are often very small and shallow, leaving the snake with little options for exploring or movement. Enrichment of captive snakes does not need to be sophisticated or take much effort. Anything that allows the animal to experience some control over its environment is believed to be enriching. For example, light or temperature can be manipulated so the snake can choose to go to an area with more light or hotter temperatures. Since snakes have highly developed chemosensory systems, smell can play a large part in environmental enrichment. The cages, themselves, can be modified to provide the animal with a dark refuge to hide in, and various levels to climb onto.

Although some people may feel that snakes are too primitive to have psychological needs, they do seen to have an innate drive to search their environment for food, shelter, and mates. If captivity deprives snakes of even the smallest chance for exploratory activities, the resulting stress may cause the snakes to become too active or inactive, refuse to eat, become aggressive, or physically deteriorate. In one experiment, when the cages of captive snakes were cleaned, or the contents rearranged, the snakes increased the number of tongue flicks, which may suggest stress. Handling of snakes may simulate dominance behavior over the snake, which may also stress the animal.

Enrichment of snakes is an extremely new field, and it is important to determine which, if any, enrichment opportunities work for snakes. There is little information about whether snakes have preferences in their food, shelter, and use of space, and whether they will make choices when presented with several options. A series of experiments were conducted on corn snakes (*Elaphe guttata*) and king snakes (*Lampropeltus getulus*) to determine whether snakes have preferences, and whether they behave differently in an enriched environment.

Corn snakes and king snakes, both nonvenomous species, belong to the common snake family, Colubridae, which also includes milk snakes, bull snakes, garter snakes, and racers. Adult corn snakes and king snakes can live for several years and reach almost 2m (6 ft) in length. Juvenile corn snakes and king snakes were also tested to see if they had different preferences than adults. In their regular cages, all snakes have a water bowl, shelter, and shredded pine shavings for substrate. One experiment was designed to see if, given the opportunity, snakes showed a preference for where they spend time in their cages. All snakes, regardless of species or age, preferred hiding in their boxes or under the substrate than being exposed on the surface of the substrate. When different types of shelters were placed into the juvenile snakes' enclosures, they also spent time in the new shelters, compared with the old shelter, which suggests that they preferred being hidden to being exposed, and they used several types of shelter throughout the day.

The snakes were also tested on their preferences for substrate, since in the wild they have several types of substrate they can choose among (eg., leaves, grass, logs, etc.). When presented with a choice of substrate (shredded paper, aspen, or paper towels), the snakes again showed preferences—this time, for the shredded paper. It could be that the shredded paper best mimicked the leaves these snakes would find in the wild. When the walls of one half of their enclosure were covered with cloth, the juvenile and adult snakes spent most of their time on that side of the cage, which seems to suggest that snakes, again, prefer to be hidden, this time from anyone who would look through the glass walls of the enclosure. Finally, snakes were given the opportunity to explore a box that had several chambers, each with a different scent. Here, the two species differed in where they spent their time. Corn snakes spent most of their time in the chamber that contained food and in the chamber that contained the scent from another corn snake. King snakes spent most of their time in the chamber that contained shavings from their own aquarium. It could be that corn snakes, being somewhat more sociable than king snakes, preferred the company of their own kind (after eating the food in the food chamber). King snakes, being a more aggressive species, seemed to prefer remaining in an area containing their own scent.

Top view of the snake observation box, with various chambers that the snake can crawl into. Each chamber contains a different scent.
Courtesy of Heidi Marcum.

Although enrichment of snakes is still in its infancy, the evidence seems to show that snakes do choose their locations within their enclosures, that they prefer not to be exposed, and that they also use different areas within their enclosures throughout the day. These findings suggest that snakes can benefit from environmental enrichment, and that conditions that simulate the real environment within the enclosure leads snakes to behave more as they would in the wild.

Further Resources

Greene, H. W. 1997. *Snakes: The Evolution of Mystery in Nature.* Berkeley, CA: University of California Press.

Love, B. & Love, K. 2000. *The Corn Snake Manual.* Mission Viejo, CA: Advanced Vivarium Systems.

Siegal, R. A. & Collins, J. T. (Eds.). 1993. *Snakes: Ecology and Behavior.* New York: McGraw-Hill.

Heidi Marcum

■ Welfare, Well-Being, and Pain
Feather Pecking in Birds

Feather pecking is a major welfare problem in birds. Feather pecking consists of pecking one's own feathers or pecking at the feathers of other birds. Many pet owners, farmers, and zoo keepers are confronted with feathers on the floor and the appearance of "naked birds." Such a discovery usually triggers many questions: Is my bird in physical or psychological pain? What am I doing wrong? What caused this? It took us 7 years of research to introduce the first breakthrough in this field: Feather pecking finds its origin in the emotional brain of birds. Birds (here laying hens) that show much feather-pecking behavior is characterized by low levels of the neurotransmitter serotonin in their emotional brains.

It has been estimated that there is a 10% incidence of feather-pecking behavior in captive birds. In principle, any captive bird has the potential to (self) feather peck, but the grey parrot and the cockatoos are most affected and best known by the public. What is less known is that there also is a huge problem in chickens, turkeys, and pheasants. Apart from its economic costs, the pecked (naked) birds have an increased risk of infection and too low body temperature. Furthermore, feather pecking of other birds can cause injury or lead to cannibalism and the painful death of birds. The current "remedial" measure, that is, beak trimming in laying hens, is not free of welfare or well-being problems because beak trimming is a painful procedure for the bird. The absence of a welfare-friendly solution forced us to think about the internal physiological mechanisms which are responsible for the development of feather-pecking behavior.

There are many causes of feather pecking, most of which are related to the management of birds (e.g., housing, diet, rearing conditions), but some experts believe that the causes of feather pecking are psychological in origin. Feather pecking is generally accepted to be a stereotypic behavior. *Stereotypies* are characterized by repetitive, patterned series of movements expressed out of context that seem to serve no obvious purpose. Most are derived from normal behaviors related to grooming, feeding, or locomotion. Psychological feather pecking has many similarities to the human syndrome called Obsessive–Compulsive Disorder (OCD). OCD is an illness that causes people to have unwanted thoughts (obsessions) and to repeat certain behaviors (compulsions) over and over again (e.g., frequent hand washing). Most people with OCD know that their obsessions and compulsions make no sense, but they can't ignore or stop them. Obsessive thoughts make people with OCD feel anxious. Performing these stereotypic behaviors usually only makes the nervous feelings go away for a short time.

In humans, the emotional brain (namely the limbic system) is suspected as the area involved in disturbed behavior such as OCD. In birds, the brains are phylogenetically ancient, lacking a large cerebral cortex. The limbic system is extensive, and, as in humans, is thought to be the area involved. Furthermore, in humans, there is a growing body of evidence that there is a chemical imbalance in the neurotransmitter serotonin in the emotional brains of patients with OCD. Serotonin and other neurotransmitters travel from nerve cell to nerve cell across fluid-filled gaps called synapses. Neurotransmitters enter another cell through a special area of the cell membrane called a receptor. The neurotransmitter serotonin is thought to be involved in regulating everything from anxiety to aggression to sleep. Selective Serotonin Reuptake Inhibitors (SSRIs) stop nerve cells that have just released serotonin from absorbing it back into the cell. This makes serotonin readily available at synapses for other neurons. The most effective medications for OCD are the SSRIs, which include the well-known drug Prozac®. This indirectly suggests that serotonin levels in the emotional brains of patients with OCD are too low. We were interested whether a similar mechanism is involved in the development of feather pecking behavior in birds.

Experiments

Two lines of the adult white leghorn laying hens (1, 2) that strongly differ in the amount of feather pecking were used: a high feather pecking line (HFP) and a low feather pecking line (LFP). Hens (chicks) of these two lines have been used for the behavioral and neurobiological experiments described here.

Neurotransmitter Involvement

The involvement of serotonin in feather-pecking behavior was investigated. First, serotonin-turnover was measured in HFP and LFP chickens. Second, a serotonin-release inhibitor was used to decrease the levels of serotonin. Third, tryptophan (a serotonin-precursor) was added to the food to increase central levels of serotonin. After manual restraint, that is, restraining a bird on its side by hand for 5 minutes, the birds were decapitated. Emotional brain (including hippocampus and amygdala) samples were injected onto a reverse-phase/ion-pair high performance liquid chromatography (HPLC) setup with electrochemical detection for the measurement of serotonin and 5-HIAA (serotonin metabolite).

Results

Serotonin turnover (5-HIAA/5-HT) in the emotional brain of HFP chicks was much lower than in the emotional brain of LFP chicks. The injection of a serotonin-release inhibitor resulted in a further decrease in serotonin-turnover in the HFP chicks which was accompanied by a significant increase in feather pecking behavior. Such effects were not observed in the LFP line (see the first table). Thus, low serotonin levels in the emotional brain seem to be crucial for feather pecking. Oppositely, if serotonin levels were further increased by tryptophan, as expected, a decrease in feather-pecking behavior was observed in the HFP birds (see table).

From our studies it has become clear that there is an inverse causal relationship between serotonin-turnover in the emotional brain and feather-pecking behavior. In other words, the lower the serotonin level, the higher the amount of feather pecking. This is in agreement with the use of SSRIs (which increase serotonin availability) in the successful treatment of OCDs. In humans, there is a OCD-like disease called "trichotillomania" (another word for compulsive hair-pulling syndrome), which has many similarities with feather pecking in

Feather-Pecking Behavior in Chickens

Effects of different treatments, that alter serotonergic neurotransmission, on feather pecking behavior in chickens

Treatment	Feather-pecking Behavior	
	HFP Selection-line	LFP Selection-line
no treatment	↑	↓
serotonin-turnover ↓ (release-inhibitor)	↑↑	±
serotonin-turnover ↑ (tryptophan)	↓↓	±

birds. According to the criteria for the most common mental disorders, as published in *Diagnostic and Statistical Manual of Mental Disorders*, 4th edition (DSM-IV), trichotillomania is a OCD-like disease and classified as an "Impulse-Control Disorder." An important criteria is the inability to resist an impulse or psychological drive to act in a way harmful to oneself or others. In human and animal literature, there is overwhelming evidence that lowered indices of serotonin function are associated with behaviors characterized by impulsivity. In humans, approaches for studying serotonin function have included neuroendocrine challenge studies, and studies of serotonin receptors and transporters on platelets. More recent methods have focused on second messenger signaling, genetic polymorphisms, and application of brain imaging techniques. These studies have been generally consistent in finding low serotonin function in populations with increased expression of impulse-control disorders.

The present results nicely show that, in birds, similar neurotransmitter problems with serotonin mechanisms can be observed as found in humans. Altogether, we conclude that psychological feather pecking in birds finds its origin in the emotional brain, more precisely: Low levels of serotonin produce a mental state in the bird that makes the bird unable to resist an impulse or psychological drive to feather peck.

See also Welfare, Well-Being, and Pain—*Animal Welfare and
          Reward*
          Welfare, Well-Being, and Pain—*Psychological
          Well-Being*

Further Resources

Bordnick P. S., Thyer B. A., Ritchie B. W. 1994. *Feather picking disorder and trichotillomania: An avian model of human psychopathology.* Journal of Behavior Therapy and Experimental Psychiatry, 25(3), 189–196.

Korte, S. M., Ruesink, W., Blokhuis, H. J. 1999 *Heart rate variability during manual restraint in chicks from high- and low-feather pecking lines of laying hens.* Physiology & Behavior, 65, 649–652.

van Hierden, Y. M., Korte, S. M., Ruesink, E. W., van Reenen, C. G., Engel B., Korte-Bouws, G. A. H., Koolhaas J. M., Blokhuis H. J. 2002. *Adrenocortical reactivity and central serotonin and dopamine turnover in young chicks from a high and low feather-pecking line of laying hens,* Physiology & Behavior, 5, 653–659.

S. Mechiel Korte, Yvonne M. van Hierden & Ingrid C. de Jong

■| Welfare, Well-Being, and Pain
| *Obsessive–Compulsive Behaviors*

In my work as a clinical animal behavior consultant, I see a wide variety of behavior problems affecting dogs and cats. I have found helping pet owners solve behavior problems of their pets to be highly rewarding, primarily since in many cases, solving the behavior problem prolongs the pet's life and strengthens the bond between the pet and the owner. In this essay, I will focus on a particular type of behavior problem know as Obsessive–Compulsive Disorder (OCD). These disorders have been shown to occur in a numerous species, including dogs, cats, and humans. In fact, much of our current knowledge regarding the treatment of OCD for both humans and animals has come from studies conducted on animals with OCD.

Let's begin by defining a few terms. The term *obsession* actually refers to the state of being preoccupied with a particular idea or feeling. Given that this refers to an internal state, we can not really know whether a nonhuman animal is experiencing an obsession. A *compulsion*, on the other hand, refers to a repetitive movement or action. The term *stereotypy* refers to an action that is done almost the same way every time. In humans affected by OCD, these three conditions often coexist. That is, an individual experiences an internal obsession that eventually manifests as an external compulsion, and the compulsive behavior often has stereotypical qualities. In animals, we use the terms *OCD* and *stereotypical behavior* somewhat interchangeably to refer to the apparently uncontrollable manifestation of repetitive, functionless behavior.

It is important to put the behavior problem of OCD in animals in the proper perspective. The majority of behavior problems of dogs and cats result from the animal expressing behavior that is inappropriate but still normal. The most common behavior problem of domestic cats is urination out of the litter box. It is normal for cats to seek out a clean, dry, and scratchable place to urinate, and luckily they usually decide this is the litter box. However, a cat may decide that the owner's pillow is actually a better place to urinate than the litter box, perhaps because it is cleaner and more absorbent. This creates a behavior problem, even though the cat's behavior is normal. The most common behavior problem in dogs is aggression directed toward a household member. It is also normal for a dog to try to establish a clear dominance structure within their pack. Since dogs treat all members of their household as one large pack, then aggression toward a household member can emerge whenever the dominance structure becomes unclear. This is another example of a normal behavior becoming a behavior problem from the point of view of the owner. By contrast, OCD is an example of an abnormal behavior, that is, a behavior unlikely to serve any function in the animal's natural environment.

According to Judith Rapoport (1991), the obsessive component of OCD in humans most commonly involves a preoccupation with germs, disease, or death, leading to a compulsion frequently involving some form of excessive grooming behavior, such as hand washing, bathing, showering, or repetitive eyebrow pulling. In nonhuman animals affected by OCD, the compulsions are also often related to grooming behavior. Dogs with OCD may excessively lick specific parts of their body. One particularly damaging expression of this is *acral lick dermatitis*, also known as *lick granuloma*. Dogs suffering from acral lick dermatitis repetitively lick their wrist or ankle, producing hair loss, skin thickening, and sometimes full thickness ulcers. In Doberman pinschers, a variant of this theme is seen with *flank sucking*, a condition in which the dog compulsively sucks on the skin fold between the hind leg and the belly. Some dogs express compulsions that are not so clearly linked to grooming behavior, such as the compulsion to snap at imaginary flies. Another compulsion seen is that of dogs who continually circle or spin, a particularly common manifestation of OCD in bull terriers.

In cats, compulsive behavior is also often linked to abnormalities of grooming. The most common form of OCD in the domestic cat is repetitive self licking, done to such a degree that a great deal of hair is lost. Another form of compulsive behavior in cats is that of wool sucking, the incessant sucking at fibers attached to pieces of clothing, seen most commonly in Siamese cats. Cats also may express compulsions not directly related to grooming, such as repetitive howling or repetitive body movement.

The exact cause of OCD in animals is not completely clear. One suggestion is that OCD may be caused by an imbalance of serotonin, which is a particular neurotransmitter in the brain. This idea is supported by the findings that drugs such as Prozac®, that boost brain levels of serotonin, are effective at reducing or even eliminating the expression of certain compulsions. It has also been suggested that there is a genetic component influencing the tendency to develop OCD. This also seems likely, due to the fact that certain forms of OCD are especially common in particular breeds of dogs and cats (spinning in bull terriers, flank sucking in Doberman pinschers, wool chewing in Siamese cats). However, it has also been suggested that a major cause of OCD in animals is stress and conflict. This suggestion also clearly has merit.

Many of us have been to zoos and have seen polar bears swimming back and forth. As a child, I assumed this was normal behavior for a polar bear, when in reality the bears were expressing compulsive behavior. In fact, there was one period during which every single captive polar bear in the entire world was exhibiting signs of OCD. Bears were rocking back and forth in corners, swimming repetitive figure eights, and exhibiting other compulsions. The simplest explanation for the widespread occurrence of OCD in this population was that OCD was being triggered by the abnormal environment. Surprisingly, dramatic improvements resulted once management changes were put in place requiring the bears to work harder. This approach was building on the work of Hal Markowitz, who developed the idea that environmental enrichment for captive zoo carnivores could be accomplished by making them use species-typical behaviors to earn their food. The polar bears were required to search around the enclosure for their food, dig their food out of pipes, and chip away at an ice block to liberate frozen food. The main principle at work here was that the animals were now expressing behavior patterns that were natural to their species, and this largely eliminated the expression of compulsive behaviors.

The lessons learned from the work with polar bears help us understand the rationale behind the recommended treatment options for animals exhibiting OCD. The first approach should be to take steps to reduce the animal's stress, reduce sources of conflict, and provide outlets for natural, species-typical behavior. Another major treatment approach is that of drug therapy. At least some of the animals expressing compulsive behavior may very well have imbalances in brain chemicals such as serotonin, either at the root of their problem or maybe even as a result it. Drugs such as Prozac® (fluoxetine) and Clomipramine® (anafranil) increase brain levels of serotonin, and correctly dosed long-term therapy with one of these drugs is probably the best option for a companion animal with OCD who has not responded to attempts at environmental enrichment. Although one study I conducted together with Ben Hart showed lick granuloma in dogs could be controlled by highly focused behavior modification, as a rule, true cases of OCD are very unlikely to resolve by applying rewards and/or punishments.

It is important to stress that not every spinning, self-licking dog or cat is suffering from OCD. Certainly, any animal suspected of OCD must first be given a thorough medical and neurological examination. Many of the commonly expressed behavioral changes suggestive of OCD may also be caused by different underlying medical conditions. For example, a cat may be excessively licking itself because it is really suffering from an allergic reaction to fleas or mites. A dog may circle because of a disease affecting the part of the brain responsible for balance. Finally, it is also important to make sure medically normal dogs or cats suspected of OCD are not simply trying to earn attention. Dogs are especially prone to engage in the most outlandish behavior patterns, as long as they are maintained by an attention reward. Ben Hart has described one dog who repetitively barked at shadows, and another who repetitively seemed to initiate flank sucking. Both cases seemed like classical expressions of OCD, but when the owners secretly spied on the dogs, they saw that the behavior never occurred when the dogs thought they were alone. In both cases, the abnormal behavior was being maintained by an attention reward.

See also Welfare, Well-Being, and Pain—*Psychological Well-Being*

Further Resources

Ackerman, L. 1996. *Dog Behavior and Training—Veterinary Advice for Owners*. Neptune, NJ: TFH Publications.

Eckstein, R. A. & Hart, B. L. 1996. *Treatment of canine acral lick dermatitis by behavior modification using electronic stimulation*. Journal of the American Animal Hospital Association, 32, 225–230.

Hart, B. L. & Hart, L. A. 1985. *Canine And Feline Behavioral Therapy*. Philadelphia: Lea and Febiger.

Leusher, U. A., McKeown, D. B. & Halip, J. 1991. *Stereotypic or obsessive–compulsive disorders in dogs and cats*. In: *The Veterinary Clinics of North America—Small Animal Practice*, Advances in Companion Animal Behavior, 21, 401–413.

Markowitz, H. & LaForse, S. 1987. *Artificial prey as behavioral enrichment devices for felines*. Applied Animal Behaviour Science, 18, 31–43.

Rapoport, J. L. 1991. *Recent Advances in obsessive–compulsive disorder*. Neuropsychopharmacology, 5, 1–9.

Robert A. Eckstein

■ Welfare, Well-Being, and Pain
Primate Rescue Groups

Most of the primates arriving at a rescue center have been separated from their mothers and their social group during their first year of life; and they have been isolated from others of their kind and must interact with humans. At a rescue center, they have to become part of social groups and develop behaviors that normally they would not have in the wild in order to survive.

Socialization of the Young

Young primates tend to become unsocialized and may not develop the communication skills they need if they cannot interact with other young primates. Babies and young capuchins occupy most of their time playing because it is through games that primates prepare themselves for adult life, learning to develop behavior typical of their species. Games are direct training for adult activities such as aggressiveness control, sexual behavior, and so on. In general, by playing, the young monkey learns to recognize all that is or is not socially acceptable in his group. When living isolated from their social group, young primates miss this learning experience.

The young-infant stage will determine the behavior that the primate will develop during its entire life. In his normal social group, the monkey will slowly learn everything there is to know in order to be a socially accepted and successful individual—both by being with his mother and by watching the behavior of other members of the social group. This young-infant stage is essential also for a successful reproductive behavior during adulthood.

Abnormal Behavior in Captivity

The study of social conflicts is a good opportunity to clarify the function of the behaviors developed by the individuals living in highly organized social systems. It is important to consider that it is easy to confuse pathological behavior with normal behavior in captivity; for example, aggression can be a normal expression of social behavior when it is used in cooperation with other members of the group against an intruder or an enemy. Inappropriate aggression can be seen in a young monkey that has been deprived of social contact, for instance, when turning against an adult male in his social group. Capuchin monkeys often develop a conduct known as displaced aggression, directing their aggression to other individuals of the group or toward inanimate objects.

The range of behavior in captivity is very broad and varies from normal, but inappropriate or exaggerated behavior, to those behaviors not found in the wild. Certain biological and social parameters are disturbed in most captive primates, including sexual maturity, age of first birth, birth interval, life span, social structure, group size, home range, day range and behavior. Generally, they suffer in a restrictive artificial captive environment; one acknowledged indicator of poor welfare and animal stress is the presence of *stereotypic behavior*—obsessive and repetitive actions with no obvious function. Also the primates can present self-mutilation, feeding disorders, reproductive disorders, and physiological unbalances.

Some other abnormal behaviors developed in captivity are regression, passivity, aimless locomotion, lost of interest in play, lost of sexual interest, ingestion of body hair, masturbation, self-biting, eating disorders, hyperaggressivness, bizarre posture, thumb sucking, rocking, self-clinging, self-clasping, and more. Some of these abnormal behaviors are more persistent than others when responding to rehabilitation.

When a primate's level of frustration is high and can't be expressed properly, it will result in a self-aggressive or an autistic behavior. This was clearly seen in the conduct developed by the chimp Toto who lived at the Rescue Centre for 7 months after being confiscated from a circus that had kept him chained and restricted to a cage for 25 years. After the rescue, he didn't interact with his environment, and he didn't react to stimuli such as noises. He would self-mutilate, masturbate, and would incessantly jump, standing on his four limbs as if he were still chained.

In general terms, abnormal behaviors developed in captivity are due to separation and isolation during infancy, and a hostile environment that doesn't provide the stimuli to develop species-typical behavior. It is hard to accomplish a balance between biological needs and the environment. The primates must adapt their behavior in order to survive in these sometimes hostile surroundings that they can not control; for instance, they have to interact with natural predators, they have to live together with species with whom they wouldn't associate in the wild, they must develop new vocalizations, and much more.

Hierarchy

Capuchins

A social group gives certain benefits to the individuals that are part of it, but it also exacts some costs, such as a higher competition for food and females; this can be clearly noticed in groups of capuchin monkeys. Female capuchins have a clearly established feminine hierarchy, and males have their own and independent hierarchy that is dominant over the females'. The relationship between male and female is often dominant–submissive with the females exhibiting fear and avoidance of the males, and males often interfering in female conflicts.

Woolly Monkeys

Woolly monkeys develop a very complex and highly organized social structure. To live as one colony is essential to woolly monkeys' well-being. The males of the woolly monkey have higher rates of social behavior than females and form male coalitions. There is, however, no sexual dominance in the woolly monkey group. Before mating can take place, the male must first establish his position of social dominance in relation to the females. The sexual behavior of the male depends on his position in the social order.

The role of the adults is to support each other in protecting the colony; there is a certain hierarchy among them which is fairly changeable. There is often a dominant male who leads the group, and he gains and maintains his position through the trust and confidence of the other monkeys. The male marks his territory by rubbing saliva over the surface of what he claims with strong upwards thrusts of his chest. By doing that, the male is showing the other members of the group that the object rubbed belongs to him.

A woolly monkey can become leader only with the encouragement of the dominant females, who will actively support him in competition with the other males. The death of the female leader produces a change in the social structure of the group. It can even cause the breakdown of the leadership of the male, and his place can be occupied by another. It can also cause a period of aggression and fights over who can have this leadership. If no female assumes the role of leader, no male leader will be chosen, so the group will remain unstable.

Squirrel Monkeys

Among male squirrel monkeys, the more common display seen is the "penile display" which can be a dominating gesture as well as a sexual one. Rival males place themselves side by side and make growl-like vocalizations. The male of higher rank stands, putting one of his feet to the side while displaying an erect penis, and threatens the other while urinating slightly. In captivity, competition and display occur between males squirrels during the breeding season. The squirrel monkeys are very aggressive toward each other, and sometimes the group can self-destruct if they are not separated, and they can even attack and kill a newborn. (The females are promiscuous and will initiate mating with various males of their choice.) Females tend to attack other females. Squirrel monkeys are very hard to keep in captivity if one doesn't consider the high level of aggression they can develop between them, especially during mating season.

Adult–Infant Relationships

The newborn woolly monkey is accepted by the whole group, which doesn't happen with capuchins and squirrel monkeys, where the baby must be protected from all other individuals who are not relatives.

The woolly monkey adult male is affectionate and protective toward infants in the group; fathering behavior does not necessarily refer to the relationship between the biological father and his offspring. When an adult male greets a new baby, he approaches very slowly, with head bowed and emits the soft and reassuring sounds that are made only for babies. Babies may respond with head shaking.

At our Rescue Centre, spider monkeys and woolly monkeys lived together in the same territory for 6 years, interacting without major problems even though woolly monkeys were always dominant. During that period of time, two woolly monkey babies were born and were immediately accepted by the female spider monkeys who played an aunt role. They loved riding the babies on their backs, constantly protected them from the young woolly monkeys, and were always near when the mothers left them for a moment.

One of the behaviors that helps the species survive in captivity, is the one exhibited by the adult female capuchins after the arrival of a new orphan. The baby is quickly adopted by one of the adult females who can even rapidly start to produce her own milk. She immediately takes care of the young monkey who receives food, care, warmth, safety, and nurturing such as grooming, comfort sounds, eye contact, protection, and discipline as if it were with its own mother. It is very common among the capuchins to see babies adopted. All this makes the survival rates of the capuchin monkeys high.

An opposite thing happens with woolly monkeys who are not easily accepted by the adult females of the group. Babies feel afraid to approach other individuals they are not familiar with. All the woolly monkeys who arrived at the Centre as infants have needed to be hand reared for a while before being gradually introduced to a group, and are generally accepted first by the leader male and then by the rest of the group. A similar behavior is developed by squirrel monkeys, who don't easily accept a young orphan who can even be attacked by the group. Generally, when a new young infant orphan arrives, it has to be raised separately creating a new group.

Capuchins are more successful than woolly monkeys both ecologically and evolutionarily speaking. This can be proved by their current widespread geographic distribution. Their capacity for adaptation makes them survive in adverse captive conditions in which other primates wouldn't. Often one sees capuchins of a mature age living like pets, even when the conditions of the captivity aren't the best. It's not the same with woolly monkeys who generally develop chronic depression similar to that seen in children who have been rejected by their mother and which can cause their death before the primate reaches adulthood.

See also Welfare, Well-Being, and Pain—*Psychological Well-Being*
Welfare, Well-Being, and Pain—*Sanctuaries*

Further Resources

Rowe, N. 1996. *The Pictorial Guide to the Living Primates*. East Hampton, NY: Pogonias Press.
Simian Society of America. 2004. *The Primate Care Handbook*. Tucson, AZ: Author.
Williams, L. 1974. *The Woolly Monkey, A Social Study from The Monkey Sanctuary*. Looe, Cornwall, UK: The Monkey Sanctuary.
Williams, L. 1977. *The Myth of the Violent Monkey*. The Monkey Sanctuary, Looe, Cornwall, UK.

Elba Muñoz Lopez

■| Welfare, Well-Being, and Pain
Psychological Well-Being

How one feels mentally and emotionally is a critical determinant of one's welfare. Determining an animal's psychological well-being tells us what the quality of the animal's life is like, which is essential in order to provide proper and humane care for animals and to make their lives the most enjoyable possible. It is also important that we utilize a method of measurement that is based on clear and specific criteria rather than intuition or "gut feelings." An objective system of assessment is the key prerequisite for then setting specific goals for maximizing psychological well-being. There have been numerous noted cases in which intuition-guided measures intended to improve psychological well-being actually had the opposite effect, such as when animals were presumed to prefer being housed with social companions, but when put into effect led to fear, aggression, and fighting. Procedures that might promote psychological well-being in one species—or one individual—may have no effect or even a harmful effect in others. What creates happiness varies in important ways between species, sexes, age groups, and individual animals.

The Role of Feelings

The evolutionary view of mental states is that they are adaptive and evolved for the same reasons as physical traits: as a means to optimize survival and reproductive success. In conscious beings, behavior requires guides and motivation, which appear to be an important function of affective states (feelings). Affective states prioritize matters for animals, assigning survival value and focusing attention on those matters most relevant at the time for self-preservation. Feelings have pleasant and unpleasant experiential qualities. Generally speaking, pleasant feelings have evolved to be associated with and promote those things beneficial to survival; unpleasant feelings are associated with and deter those things which threaten survival.

In people, feelings contribute pleasant and unpleasant qualities to one's life experience on a virtually constant basis. When people evaluate their well-being, the ratio of their affective pleasantness to unpleasantness over time plays a central role, and studies show that emotional pleasantness is one of the strongest predictors of life satisfaction.

Most scientists now accept that animals can suffer from pain, hunger, thirst, fear, and other unpleasant feelings, as well as experience certain pleasures. Although we do not yet have any scientific method of directly determining which animals have which kind of feelings, studies in evolutionary biology, neuroanatomy, physiology, ethology, and comparative psychology all provide extensive evidence to support the contention that animals experience emotions and feeling states comparable to our own. Whether each feeling *feels* the same to animals as it does to humans—for example, does rabbit fear feel like human fear?—is not as important as the more basic issue: feelings *feel* pleasant or unpleasant for all species. It is the quality of pleasantness that appears to be a, and possibly *the*, main contributing factor to psychological well-being.

All animals alive today are the outcome of a long and gradual process of evolution. Natural selection has resulted in animals that are "designed" for living in—that is, they are *adapted to*—their natural environment. Being adapted to ones' environment means that the animal functions optimally and has the greatest chance of security and survival in that environment. It also means that one experiences the greatest comfort in that environment. This is not to say that the natural environment is always safe or the perfect environment for animals. Animals in the wild may suffer from starvation, disease, malnutrition, predation, or harsh weather conditions. Animals kept in captivity are normally provided with adequate care that prevents these potential welfare problems.

However, while captive environments may provide protection against certain natural threats, by differing from the evolutionarily adapted environment, the captive environment often poses new challenges. Welfare problems arise when we keep animals in environments that differ from or present different challenges than their natural environment. Polar bears housed in hot climates, sheep housed in isolation from other sheep, chimpanzees housed in groups where the less strong cannot escape from the stronger individuals—these man-made environments force the animal to adapt to changes not commonly found in the natural environment. These new or excessive challenges may exceed an animal's ability to adapt. In these circumstances, well-being, and especially psychological well-being, may be threatened.

The survival systems in which feelings presumably play a prominent role include the acquisition and intake of food, water, and other nutrients, temperature regulation, security and fleeing danger, exploration, play, bodily eliminations, social companionship, reproduction, and nurturing and protecting offspring. One of the most widely accepted models of how these systems work views animals as homeostatic systems. In this model, each animal tries to maintain its physiological and psychological states in limited range—that is, homeostasis. Homeostasis may be described as the steady harmonious state of mind and body. Body temperature, water levels, sodium and electrolyte concentrations, mental stimulation, and innumerable other factors must be held within a fairly specific range to assure optimal health and function. Fluctuations outside these limits threaten survival, and need to be corrected. Feelings signify changes in homeostasis, such that the animal is internally alerted and motivated to perform that behavior that best restores homeostasis.

Those factors necessary for the maintenance of homeostasis are termed *needs*. Needs exist in two forms: physical and psychological. Because of their involvement in homeostasis, needs are intimately connected to feelings, in that many needs—physical as well as psychological— are signaled to the animal by specific feelings. For example, the feeling of thirst alerts the animal to a bodily deficiency in and a need for water, and the unpleasant feeling motivates the animal to find and consume water. The feeling of a filling urinary bladder signals the animal to the build-up of bodily waste products, and motivates the animal to excrete the waste fluids. The infant mammal whose mother has left him appears to experience severe distress, which is accompanied by the infant crying out to bring the mother back to tend to, feed, and protect him. Each of these instances illustrate the connection of needs and feelings.

Psychological needs include such things as mental stimulation, social companionship, security, and a sense of control over events in one's life. Just as physical health requires the meeting of physical needs, psychological well-being requires that psychological needs must be adequately met. If an animal does not receive sufficient mental stimulation, it appears that a feeling of boredom arises to motivate the correction of this deficiency. In a way that seems to work like hunger signaling a deficiency of physical nutrients, boredom signals a deficiency of psychological nutrients. In both cases, unpleasant feelings signal a need and motivate the animal to correct the deficiency.

Because unmet psychological needs (like physical needs) result in unpleasant feelings, the presence and number of unmet psychological needs adversely affects the animal's psychological well-being. Furthermore, because both physical and psychological needs, when unmet, result in the mental experience of feelings, it is important to note that both types of needs have an impact on psychological well-being. In other words, if one is experiencing unpleasant feelings, it doesn't make a difference to the animal whether such feelings generate from physical or emotional origins—they all feel unpleasant and need to be attended to.

In short, feelings assign value to stimuli for the animal. In this way, feelings represent what *matters* to the animal. Those things that matter elicit feelings; those things that do not matter do not elicit feelings. So our effort to look at psychological well-being from the perspective of what matters to the animal is, essentially, an effort to assess the animal's feelings.

Defining Psychological Well-Being

Psychological well-being is very difficult to define. Like "happiness," there is an intuitive sense of what psychological well-being is, but because it involves the mental states of animals, including emotions, feelings, and thoughts, it defies easy description. This is in contrast to physical well-being, which lends itself to rather specific description. Defining psychological well-being is difficult in our own species (try to tell a friend how exactly she can improve her psychological well-being), but the problem is enormously compounded when applied to animal species other than our own. The principle challenge facing scientists is how to measure something that cannot be precisely defined.

In defining psychological well-being, we should first understand three basic assumptions, for which there is broad agreement among scientists.

1. Psychological well-being is, first and foremost, a matter of feelings. In fact, psychological well-being has no meaning in the absence of a conscious mind that experiences feelings. Animals without consciousness or feelings cannot have a psychological well-being. Pleasant and unpleasant feelings both contribute to psychological well-being and, as will be discussed, psychological well-being is ordinarily used to refer to the presence of positive feelings and experiences, not simply the absence of unpleasant feelings (suffering).

2. Positive (good) psychological well-being coincides with a preponderance of pleasant feelings, and negative (poor) psychological well-being coincides with a preponderance or unpleasant feelings.

3. Psychological well-being, like happiness, exists along a continuum, ranging from very good to very poor.

In a concise view, psychological well-being is how an individual animal perceives its life to be faring. The concept of psychological well-being is closely related, and may be equivalent, to other concepts such as happiness, quality of life, and life satisfaction. In non-scientific terms, psychological well-being is very similar to peace of mind and emotional fulfillment. Scientists and nonscientists often use many of these terms interchangeably.

A typical description of a state of good psychological well-being is when the animal is free from distress most of the time, is in good physical health, exhibits a substantial range of the species-typical repertoire, and is able to deal effectively with environmental stimuli. Other descriptions have included the concept of meeting needs, such that an individual's physiological, security, and behavioral needs are fulfilled.

Most modern definitions of psychological well-being have taken the position that there will always be challenges to animals in any environment and that good psychological well-being occurs when animals are able to *cope* with these challenges. Coping refers to the animal's ability to lessen the negative psychological impact of a stressful stimulus; the degree of success thereby dictates the level of psychological discomfort experienced. More simply, coping is the effort to minimize the effect of unpleasant feelings, or to manage psychological stress. Coping strategies include physiological, cognitive (e.g., reasoning, memory, and positive thoughts), and behavioral (e.g., escaping, problem-solving, scaring away the threat, seeking social support, and acquiring mental stimulation) responses. In this view, the animal's ability to adapt to threats, challenges, and adversity is the primary determinant of the level of psychological well-being. This means that psychological well-being derives from successfully coping with life's problems, not from experiencing no problems that demand coping behavior. This suggests that a challenging environment with some unpleasant events is more supportive of good psychological well-being than a benign and sterile, although physically safe and healthful, environment.

Utilizing the concept of coping, psychological well-being may be defined as the mental state of an animal as it attempts to cope with, or adapt to, challenges to its integrity and survival. This state includes how much effort is required for coping, the extent to which it is succeeding in or failing to cope, and its resultant feeling. Psychological well-being will vary over a continuum from very good to very poor. Psychological well-being will be poor when the animal is having difficulty coping or when it is failing to cope, resulting in persistent unpleasant feelings. If an individual fails to adequately cope with the environment or does so at great cost, it can be said to be under stress, and its psychological well-being may be considered poor. Conversely, when an animal successfully adapts to threats and challenges, thereby keeping the negative impact of unpleasant feelings at a minimum, then its psychological well-being may be considered good.

Scientific descriptions of psychological well-being can ultimately be reduced to a very simplified concept: the animal *feels good*. This means that there are few, or brief, unpleasant feelings, and a predominant state of comfort. In addition, there are pleasant feelings that make the animal feel good. The general concept of feeling good applies to present moment experiences ("I feel good right now") as well as feeling good about life in general ("Life is going well"). This is psychological well-being.

Because behavior is one of the main tools for coping, our observations of animal behavior become critically important in evaluating psychological well-being. Just as for humans, behavior provides important information as to how an animal feels.

Behavioral Measures of Psychological Well-being

Measurement of psychological well-being requires information known only to the animal, that is, the subjective feelings it experiences. Because it is not possible to access this information directly (i.e., to read minds), or even converse with animals about what they are feeling, we must resort to indirect measures of assessing their mental states.

People who work closely with animals and those who keep animals as pets generally feel able to tell if an animal is feeling good. Our judgment relies almost exclusively on the animal's behavior. The signs that tell us that there is something wrong are changes in the behavior which we have come to expect from the individual. It may be an obvious sign, such as howling or trying to bite when touched, or something more subtle, such as less interest in play or a decreased appetite. This nonscientific approach to judging an animal's psychological well-being utilizes an important finding confirmed by researchers: The best current method for determining an animal's psychological well-being is by observing its behavior. In fact, while measurement of physiological changes (e.g., neurochemical, hormonal, and cardiac parameters) are also a useful and important tool for assessing mental states in animals, behavioral changes are usually more sensitive indicators of distress than physiological changes.

As has been discussed, when events or stimuli cause or threaten departures from homeostasis, responses are elicited within the animal (e.g., adaptation and coping) designed to return the organism to its normal state of balance or equilibrium. Behavioral responses are one of the most important methods animals use to respond to and adapt to changes, challenges, and threats encountered in life; behavior is often an animal's first line of defense against a potentially harmful stimulus. Behavioral responses are intended to lessen any threat, maintain homeostasis, and minimize unpleasant feelings. These behaviors, which occur upon perception of a challenge or threat, include orienting responses and defensive reactions, such as postural changes, immobility, fleeing, and aggression. Because behavior is a primary mechanism for coping with adversity, behavioral changes, especially manifestations of abnormal behavior, can provide information about how successfully an animal is coping with its environment. Thus, if behavior patterns characteristic of distress or discomfort are observed, this may indicate that psychological well-being is compromised.

In evaluating psychological well-being, it is rare for any single behavioral measure to provide all necessary information. It is therefore considered best to use a variety of behavioral measures. The most commonly utilized indices are: normal, species-typical behavior (e.g., social interactions, grooming, foraging, exploring, play); abnormal behaviors (e.g., pacing, rocking, bizarre postures, self-multilation); response to environmental manipulations (enrichment techniques); avoidance and aversion behavior; preference studies in which the animal can choose the environment, situation, or stimulus that it prefers; and operant conditioning in which the animal must work for a reward. All of these indices provide information about the animal's psychological well-being. For example, avoidance behavior, especially when an animal exerts great effort to avoid something, provides information about the animal's feelings and its psychological well-being. In operant conditioning, the amount of work it is willing to do can be used to assess the importance of that reward to the animal. In addition to the mere presence of specific behaviors performed by the animal, the frequency, duration, and intensity of those behaviors are believed to be important factors in psychological well-being. An animal's behavior is not definitive proof of is mental state, but behavior is an indication of psychological well-being and, combined with other observations, may assist in an assessment of the animal's status.

Normal, Species-Typical Behavior

One of the earliest views on using behavior to measure psychological well-being involves the performance of normal, species-typical behaviors. It is widely accepted that good psychological well-being requires that animals should have the opportunity to display a relatively diverse range of their species-typical behavior patterns. Normal behaviors include such actions as foraging, hunting, eating, playing, grooming, digging, rooting, socially interacting, and many more. Other behavioral indices include level of activity, posture, facial expressions, vocalization, locomotion, food and water intake, and sleep patterns.

It is important to understand that "normal" refers to both the quality and quantity of a behavior. A normal behavior may be abnormal when performed too frequently, or for too long a duration. In this way, behavioral measures of psychological well-being work like those for physical well-being. A cough or scratching an itch is normal behavior; however, either one performed at a high frequency indicates a diminished physical well-being. Likewise for emotional feelings: Fear behavior or self-grooming are normal behaviors, but an animal displaying either at excessive frequency or duration may be experiencing unpleasant feelings detrimental to psychological well-being. Other normal behaviors that, when performed too frequently or for extended periods, would suggest an impaired psychological well-being include vigilance, attempts to flee or hide, immobilization (freezing), urination or defecation, eating, and aggressive behavior. Thus, when using normal behavior to measure psychological well-being it can be said that either too little or too much of a species-typical behavior may be indicative of a compromised well-being. Unfortunately, there currently is a paucity of research to establish the optimal quantities of each behavior pattern.

The emphasis on evaluating psychological well-being in humans as well as in animals has been on the negative influences, that is, the unpleasant feeling states detrimental to well-being. However, psychological well-being is also comprised of the pleasant feelings, which means that a proper assessment of an animal's psychological well-being involves the recognition of behavior that indicates the animal is feeling good. Many researchers believe that feeling good is best indicated by normal behavior. It appears that animals that feel good are those whose voluntary activities most closely resemble the full behavioral repertoire of their free-ranging conspecifics. Positive well-being can be identified by the presence of certain behaviors such as stretching on rising, grooming, and taking an interest in novel stimuli in their environment. Animals that had previously voluntarily participated in such activity and that abruptly stop that activity may be giving a clue of an impaired psychological well-being. Play behavior appears to be a sensitive indicator for good psychological well-being, for it occurs when needs are sufficiently met (otherwise, behavior would be directed toward meeting the needs) and when unpleasant feelings, such as fear, anxiety, and pain, are minimal or absent (otherwise, behavior would be directed toward relief of these feelings). However, animals in confinement may not be able to play, thus the absence of play cannot be regarded as indicative of welfare

Prairie dogs at play. Play behavior is a sensitive indicator for good psychological well-being because it typically occurs only when an individual's basic needs are met.
© George D. Lepp / Corbis.

status. Therefore, while the presence of play behavior may be considered indicative of positive psychological well-being, its absence cannot be used in assessing well-being.

Another behavioral measure utilizes a different aspect of normal behavior. In this case, the behavioral changes which occur in response to environmental manipulations is observed. This type of evaluation is usually utilized in environments that are suspected or known to be deficient in some way, in which the animal cannot meet its psychological needs. For example, an environment which lacks social companions for social animals, or is a nonchanging, monotonous environment, devoid of mental stimulation. The term used to describe the environmental manipulations to enhance or correct presumed deficiencies is "environmental enrichment." The effects of enrichment are measurable by behavioral changes. An increase in the frequency of species-typical behaviors is taken as one indication of an improvement in psychological well-being.

Using normal, species-typical behavior to assess mental states appears to have great value, since animals permitted to behave in their natural ways would most approximate the natural environment to which they are best adapted, and hence the most comfortable. But there are numerous limitations, and many questions remain unanswered. These include:

1. What is "normal behavior"? Few firm lines exist to separate normal from abnormal, in either quality or quantity of a behavior. For example, if self-grooming is normal behavior but excessive self-grooming is abnormal, what defines "excessive"? Furthermore, because of variations of behavior, what is normal for one animal (species or individual) may be abnormal for another.

2. Because domesticated animals are genetically quite different from the wild animals from which they arose, what is the "natural environment" that domestic animal behavior can be studied as a gold standard of behavior?

3. How much of a species-typical behavior must an animal perform to ensure positive psychological well-being? Is it sufficient for the animal merely to perform a behavior or do the quantities, frequencies, and durations of the behaviors matter? How much of the range of species-typical behaviors must an animal exhibit before it is considered to have good psychological well-being? Do all species-typical behaviors, including such

things as fear behavior and aggression, need to be performed? Do animals suffer when they cannot act as free-ranging animals do? Considering that bearing and raising offspring is a very natural behavior, but pet animals are almost always prevented from such behavior (through spaying and neutering), does the inability to perform this natural behavior adversely affect psychological well-being?

4. How do we use normal behavior to evaluate psychological well-being in animals that show no or few recognizable behavior changes in situations which presumably vary dramatically in distress potential? For example, there are no agreed-upon measures of the psychological well-being of laboratory rodents. Rats and mice do not wag their tails when they are happy; they do not have facial expressions or postures indicative of positive psychological states. As a result, it is difficult to use behavior to assess the psychological well-being of a rat, mouse, or other rodent.

Abnormal Behavior

When an animal is experiencing unpleasant feelings and its psychological well-being is diminished, its behavior is often altered, becoming abnormal. Because abnormal behavior patterns usually occur in response to chronic conflict, frustration, anxiety, or other stressors, their presence has been used as one indicator of compromised psychological well-being. In using a abnormal behavior as an indicator of psychological well-being, we can ask: What does an animal do when it is feeling bad (emotionally or physically)? One response is that the animal may perform species-typical behaviors in increased frequency, duration, or intensity, in a presumed effort to lessen the unpleasant feelings. Normal behaviors can be considered abnormal if they occur in excess quantity or in inappropriate contexts. For example, unrelenting escape behavior—normal when performed as an occasional behavior—suggests things are not going well emotionally. However, unfortunately, research to date has not demonstrated an unequivocal relationship between the presence or absence of abnormal behavior and psychological well-being.

For the same reason that it is difficult to define normal behavior, it is equally difficult do define abnormal behavior. To recognize abnormal behavior, it must be determined which behaviors are normal. This usually requires an extensive observation of the species in question and the development of an ethogram, or catalog, of behavior patterns typical for that species. Individual variation in behavior must also be taken into account.

Abnormal behavior has been defined as a persistent, undesirable action shown by minority of the population that is not due to any obvious neurologic disorder. In addition, the behavioral changes result either in only partial adaptation to the environment or are in some way maladaptive. Although some behaviors, such as extensive barking in domestic dogs, may be difficult to categorize as normal or abnormal, it generally is agreed that self-mutilation and stereotypic behavior are examples of abnormal behaviors.

Abnormal behavior has many causes, not all of which are relevant to psychological well-being. For example, if a dog were to be nonresponsive to sounds that typically induce a response in dogs, it may be caused by deafness rather than adverse mental states. In using abnormal behavior as a tool to judge psychological well-being, we are interested in those behaviors which are associated with feeling states, specifically, unpleasant feeling states. In these cases the abnormal behaviors are presumed to be the animal's attempt to cope with the unpleasant experience, or a result of an inability to cope with the adversity. Overall, various causal and motivational factors appear to be involved in the development of abnormal behavior, and there is no consensus among scientists as to their exact meaning for psychological well-being. However, there is a consensus that abnormal behaviors probably do indicate that something is wrong or threatening from the animal's perspective, and that many abnormal behaviors are likely signs of diminished well-being.

In a broad overview, generally accepted causes of abnormal behavior include fear, anxiety, frustration, social position, social separation, depression, insufficient mental stimulation, and physical health disorders. For example, monkeys kept in single cages—where they are denied social companionship—sometimes exhibit bizarre, stereotypic, and self-directed patterns of behavior. Animals may be distressed because they are unable to perform a behavior that they are highly motivated to perform or because they are forced to choose between two or more undesirable options.

Stereotypies. Stereotypic behaviors, or stereotypies, are a type of abnormal behavior believed to be used as a mechanism for coping with distress. A stereotypy can be defined as an invariant and repetitive sequence of movements occurring at a high frequency and having no apparent purpose or goal. Such behavior is not observed in the animal's natural state. Stereotypies commonly occur in many species of farm animals, zoo animals, laboratory animals, birds, and confined domestic animals such as dogs and horses. Specific types of stereotypies include pacing, circling, whirling, weaving, rocking, bar-biting, self-clasping, self-grooming, feather-picking.

Stereotypic behaviors, such as bar-biting, are a type of abnormal behavior believed to be used as a mechanism for coping with distress.
© Time Life Pictures / Getty Images.

The function of stereotypies in animals is a widely debated issue. Stereotypies often develop in situations known to be aversive and stressful, such as low stimulus levels, physical restraint, inability to escape from fearful situations, frustration, or conflict. Many stereotypies can be abolished by improvements in the environment, such as increased stimulation or places to hide. For these reasons stereotypies have long been considered an indication of poor psychological well-being caused by environmental conditions which inadequately meet the animal's needs. Stereotypies may help the animal cope with its environment by increasing the amount of mental stimulation. Animals may engage in stereotypic behaviors because repetitive movements, not unlike being groomed, cradled, or rocked, are comforting.

Conflict behavior. Conflict behaviors are a type of behavior which occur in response to chronic conflict, frustration, or anxiety and represent attempts to cope with these unpleasant experiences. Conflict behaviors occur when an animal is motivated to perform two behavioral choices which cannot be expressed simultaneously. For example, a hungry cat approaching a food dish near an aggressive dog may be in conflict between approach and avoidance. The cat may step forward, then step back, then reapproach. Under such conflict conditions the animal may also engage in an unrelated behavior, such as starting to groom itself. Such behavior is termed displacement behavior. Lastly, the conflicted and frustrated animal may lash out and attack a nearby object or companion, a behavior known as redirected aggression. Conflict behaviors are part of normal life in response to life's challenges, and therefore the presence of such behaviors do not necessarily reflect changes in psychological well-being. However, when these behaviors become frequent it is generally assumed to represent impaired psychological well-being.

Abnormal behavior: limitations and questions. Abnormal behavior is an important behavioral measure of psychological well-being. However, certain limitations restrict its use as a sole indicator. First, because diminished psychological well-being may have no recognizable behavioral measures, the absence of behavioral abnormalities is not an assurance of good well-being. Conversely, the presence of abnormal behavior does not necessarily imply that an animal's psychological well-being is impaired. Although certain behaviors, such as stereotypies, often suggest that the animal is experiencing adversity, the behavior may be a successful coping response to the situation. This presents a currently unresolved problem when using abnormal behavior as a measure of psychological well-being: When a behavior is intended to cope with adversity, how can one judge the success of coping behavior? Which kinds of coping responses show that the animal is suffering and which show that the animal is simply responding appropriately to environmental challenges? A difficulty arises in determining how much abnormal behavior an animal needs to show to indicate that it is suffering. Some researchers have proposed a quantification scheme for abnormal behavior, for example, one proposal suggests that stereotypies occupying more than 10% of the animal's waking time are indicative of poor well-being.

Preference Studies

An experimental approach to using behavioral measures to assess psychological well-being involves preference studies. By observing the choices an animal makes under specifically designed circumstances, we can learn about how the animal perceives its own environment rather than how humans perceive it. The simplest method involves offering the animal a choice between two commodities and observing which one it chooses. This form of study has been used extensively in farm animals. Pigs, chickens, calves, and other animals have been given the opportunity to choose between various cage sizes, types of litter, floor surfaces, lighting conditions, social conditions, and other types of husbandry procedures.

The effort an animal exerts (the "price it will pay") to avoid something is presumed to tell how detrimental that factor is for the animal's psychological well-being. Since animals appear to be strongly motivated by unpleasant feelings to lessen those feelings, a strong effort to avoid something implies that the intensity of unpleasantness (and thus the negative impact on psychological well-being) is high. If we measure the amount of work the animal exerts to avoid something, we can indirectly measure how unpleasant that factor is for the animal. Hunger is one of the most common unpleasant experiences used for testing. For example, a horse can be taught to push a panel with its nose for a food reward, and then can be tested as to how many times it is willing to press the bar to get food. Or the door that the animal must open to reach food can be made heavier to see how much effort the animal will be willing to exert to obtain food. If food is placed in a very cold location such that the animal must endure extremely cold temperatures to reach food, or if an animal is forced to choose between acquiring social companionship versus food and chooses social companionship, this tells us something about how important these things are to the animal. It is believed that those things that the animal is willing to work very hard for (such as water to relieve thirst), or is willing to give up some other valuable commodity such as food, then being denied that commodity is likely to cause or permit suffering. If a dog strongly desires human companionship to relieve unpleasant feelings of isolation, but is housed in a way that prevents the dog from receiving such companionship, then the unpleasant feelings persist.

By utilizing the information obtained from these various preference studies, behavioral measures provide valuable information as to the animal's psychological well-being. However, as with all methods of evaluation of well-being, preference testing has limitations.

First, animals do not always choose what is best for their (long-term) well-being. Second, when the animal is given a choice, all choices may be undesirable. Hence, a preferred choice may still result in impaired psychological well-being. Alternatively, if an animal is forced to accept the commodity it did not prefer, its psychological well-being may not be adversely affected. For example, if a dog were to choose steak over hamburger, its psychological well-being is not diminished when given hamburger.

Summary

Behavioral indicators provide useful insights into animal psychological well-being. However, it is widely accepted that there is no one best measure of psychological well-being and that assessment should utilize multiple measures. Physiological measures, physical health indices, and brain scanning studies can provide important measures along with behavioral indices.

Current evidence allows us to propose general guidelines for behavioral criteria in assessing psychological well-being in animals. The following criteria imply a good (positive) psychological well-being, meaning that the animal is feeling good, with minimal unpleasant feelings and the ability to cope with adversity:

- The animal is alert, busy, and exhibits a substantial range of context-appropriate species-typical behavior.
- The animal engages in play behavior.
- The animal displays an absence or no more than minimal levels of abnormal behaviors.
- The animal is confident; it moves around freely, is outward going, and does not display fear towards trivial nonthreatening stimuli.
- The animal is able to rest in a relaxed manner, without constant signs of vigilance.
- The animal exhibits a very low incidence of behaviors associated with the lessening of unpleasant feelings, such as limping (pain), labored breathing (inadequate oxygen intake), or escape behavior or hiding (fear).

See also Welfare, Well-Being, and Pain—*Back Scratching and Enrichment in Sea Turtles*
Welfare, Well-Being, and Pain—*Behavioral Assessment of Animal Pain*
Welfare, Well-Being, and Pain—*Behavioral Correlates of Animal Pain and Distress*
Welfare, Well-Being, and Pain—*Carnivores in Captivity*
Welfare, Well-Being, and Pain—*Enrichment for Chickens*
Welfare, Well-Being, and Pain—*Enrichment for Monkeys*
Welfare, Well-Being, and Pain—*Enrichment for New World Monkeys*
Welfare, Well-Being, and Pain—*Experiments in Enriching Captive Snakes*
Welfare, Well-Being, and Pain—*Sanctuaries*
Welfare, Well-Being, and Pain—*Stress in Dolphins*

Further Resources

Dawkins, M. S. 1990. *From an animal's point of view: Motivation, fitness, and animal welfare.* Behavioral and Brain Sciences, 13, 1–61.

Dawkins, M. S. 1993. *Through Our Eyes Only? The Search for Animal Consciousness.* Oxford: W. H. Freeman and Company Limited.

Hetts, S. 1991. *Psychologic well-being: Conceptual issues, behavioral measures, and implications for dogs.* Veterinary Clinics of North America, 21, 369–387.

Shepherdson, D. J., Mellen, J. D., & Hutchins, M. (Eds). 1998. *Second Nature: Environmental Enrichment for Captive Animals.* Washington, DC: Smithsonian Institute Press.

Franklin D. McMillan

■ Welfare, Well-Being, and Pain
Rehabilitation of Marine Mammals

Marine mammals are found throughout the world. Many times these animals are found stranded along the coasts due to illness or injury. Rescue and rehabilitation facilities around the world work to rehabilitate these animals for eventual release back into the wild. These facilities implement many training and behavior techniques used with marine mammals in a captive environment to facilitate the rescued animal's recovery. Although rehabilitators often need to handle the animals to treat them, rehabilitated animals should not become conditioned to receiving food from humans so that, upon their release, they are capable of foraging on their own and not looking for handouts from a human. Any conditioning that does occur would need to be reversed prior to release.

Most marine mammals in a captive setting, such as an aquarium, fenced in lagoon, or any location where an animal is permanently under human care, are trained using a process of operant conditioning by positive reinforcement. Basically, by positively reinforcing a behavior, the trainer has increased the probability that a certain behavior will be preformed again. Primary reinforcement is food. Secondary reinforcement is anything that is positive to the animal: a rub down, favorite toy, even being asked to perform a favorite behavior. There are two important reasons for training behaviors: first, the safety of the trainers and animals is increased with the animals being under a trainer's control, and second, the learning and executing of behaviors stimulates the animals mentally.

Almost all behaviors are trained by using a series of small steps or *approximations*. When the required behavior is executed correctly, some form of signal (usually a whistle or spoken word), called a *bridge*, is used to tell the animal the behavior was performed correctly. It is used to bridge the gap between the time the behavior was performed and reinforcement.

An example is training a dolphin to jump into the air to touch a ball. The ball is first placed on the surface of the water. When the dolphin touches the ball with its rostrum (the beak or snout), the trainer blows his whistle and the dolphin returns to the trainer for food. After several repetitions the dolphin learns that the correct behavior is touching the ball with its rostrum. As the

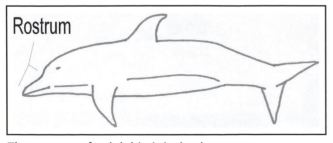

The rostrum of a dolphin is its beak or snout.
Courtesy of Brandi Sima.

training progresses, a hand signal is introduced, being paired with the behavior, just like a dog learning the signal to "sit." The difficulty of the behavior is increased by slowly raising the ball off the surface of the water. The approximations require the dolphin to jump higher and higher out of the water, until the full height is reached.

With marine mammals in a rehabilitation setting, however, rehabilitators must be careful *not* to train the animal under their care. A compromised animal undergoing rehabilitation is usually cared for in a repetitive manner several times a day. In some cases, the basics of training techniques can be used to assist the rehabilitator in the rehabilitation process, but the individual must also be careful that the animal is not trained at the conclusion of its treatment.

When an animal first arrives into a rehabilitation setting, the chance of the animal eating on its own is poor. Usually the animal must be force fed (via stomach tube or by placing a fish down past the gag reflex of the throat forcing the animal to swallow) until it starts to feel better and eat on its own. As the animal starts to recover, force feeding changes to placing the fish just inside the animal's mouth, at which point the animal takes the fish and swallows it. Over time approximations are used to increase the distance between the animal and the food fish. In the case of cetaceans (whales and dolphins) this increased distance makes the animal start to swim after the fish. As the distance between feeder and animal grows, the feeder can eventually approximate himself totally out of the water, and the animal will be foraging for fish tossed into a pool with no human present. With pinnipeds (seals, sea lions and walrus) the theory is slightly different. Most pinnipeds start their rehabilitation in a dry setting because they would probably drown if placed in a pool. As the animal recovers and strength returns, the approximations can go from force feeding, to placing the food fish in the mouth, to eating in a pool to submerging to look for fish under the water. In either case, by tossing the food into a pool, the animal does not learn to "station" in a specific location and expect food. With trained mammals, a "station" requires the animal to stay in a specific location until the trainer asks the animal to move. Sea lions sitting on platforms or dolphins waiting in front of trainers during a show are examples of the "station" behavior. In addition, the animal in reha-

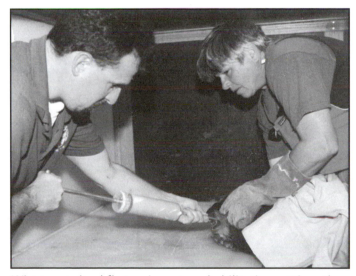

When an animal first arrives at a rehabilitation setting, the animal must usually be force-fed until it is able to eat on its own.
Courtesy of Brandi Sima.

bilitation is ideally no longer associating the presence of a human with food appearing. Eventually, live fish can be added to ensure that the animal can hunt and catch prey.

If an animal undergoing rehabilitation does accidentally become trained to perform a specific behavior, the rehabilitator must work to decondition the animal prior to its release. An animal that associates humans with food will seek out the first human it can find out in the wild. This can lead to a dangerous situation, not only for the animal but for the human as well. The National Oceanic and Atmospheric Administration (NOAA) states that illegal feeding of wild dolphins by the general public has caused dolphins that have been regularly fed to often shown aggression if food is not presented to them (Smullen 1998).

Rehabilitators face a challenging task when it comes to rehabilitating an animal for eventual release back into the wild. Besides the obvious injury or illness, rehabilitators must make sure that the animals under their care stay wild and do not become conditioned to human interaction. By knowing the basics of training and animal behavior, the rehabilitator can use the techniques needed to assist in the rehabilitation process, but insure that the animal is capable of living on its own after release.

See also Welfare, Well-Being, and Pain—*Back Scratching*
and Enrichment in Sea Turtles
Welfare, Well-Being, and Pain—*Enrichment*
for Chickens
Welfare, Well-Being, and Pain—*Enrichment for Monkeys*
Welfare, Well-Being, and Pain—*Enrichment for*
New World Monkeys
Welfare, Well-Being, and Pain—*Experiments in*
Enriching Captive Snakes
Welfare, Well-Being, and Pain—*Rehabilitation*
of Raptors
Welfare, Well-Being, and Pain—*Sanctuaries*

Further Resources

Pryor, K. 1999. *Don't Shoot The Dog.* New York: Bantam Books.
Ramirez, K. 1999. *Animal Training: Successful Animal Management through Positive Reinforcement.* Chicago: IL: Shedd Aquarium.
Scarpuzzi, M., Andrews J., McBain J., & Reidarson T. H. 2002. *The Application of Operant Conditioning During the Long Term Rehabilitation of a Gray Whale Calf.* Soundings (IMATA), 27 (3).
Schofield, D. 1995. *The Use of Operant Conditioning Techniques to Facilitate the Management of Stranded Marine Mammals Under Veterinary Care.* Soundings (IMATA) 20 (1).
Smullen, S. 1998. *Dolphin Feeding and Harassment Still Illegal*, NOAA 98-R105, website: http://www.publicaffairs.noaa.gov/pr98/jan98/noaa98-r105.html

Brandi Sima

■ Welfare, Well-Being, and Pain
Rehabilitation of Raptors

The past century's industrialization and overcrowding greatly worsened the impact of human activity on the welfare of wildlife, revealing a negative scenario: For instance, birds of prey are harmed by a wide range of events, including persecution by humans, hunting and poaching, ingesting of lead shot, colliding with vehicles, feeding on prey contaminated with pesticides, egg stealing. Fortunately, growing concern for wildlife and an understanding of the necessity to preserve natural habitats has followed. The rehabilitation of birds of prey is a consequence of this concern.

Raptor rehabilitation can be defined as the treatment and care of sick, injured, or orphaned birds until they are returned to the wild. Such a program is usually carried out at specific rehabilitation centers (RC), usually belonging to private institutions or working jointly with university veterinary clinics, or by individuals. In some countries, falconers are actively involved in raptor rehabilitation because of their knowledge and experience with falconry techniques, which can be successfully applied in such rehabilitation. The most frequent reasons for admission to rehabilitation centers are fractures caused either by collisions with vehicles, man-made structures, or by gunshot. There are many other, but less frequent, causes for admission: emaciation due to starvation or unknown reason, poisoning from agricultural pesticides and heavy metals, ocular lesions, electrocution, and parasitic infestations. These factors are not restricted to a particular bird species, and the range of species affected is obviously very wide. Unfortunately, the trauma or the debilitation is often so severe that a large proportion of birds die or must be euthanized soon after admission. Among the survivors, some birds cannot fully recover physically, preventing any possibility of successful release. As a result usually only 30%–40% of raptors are released after treatment.

There has been controversy about the biological and conservation value of rehabilitation. Skeptical people feel

Dr. Patrick Redig, founder of The Raptor Center at the University of Minnesota St. Paul campus, examines the broken wing of a bald eagle hit by a car.
© *Layne Kennedy / Corbis.*

that the low outcome does not return the large amount of labor invested and the low percentages above offer support to this opinion. There is also doubt about the real biological value of rehabilitation itself and its contribution to conservation of the populations. In contrast, people involved in rehabilitation trust in the value of their efforts, particularly from the biological and ethical point of view. Beyond bird welfare, there is the opportunity to increase our knowledge about morbidity and the improvement of therapy. There are also indirect effects coming from the enhancement of veterinary and biological skill and public education; the latter aspect in particular is important when working in parallel in habitat protection and conservation.

The goal of rehabilitation should be to durably restore a bird of prey in the wild population. The contribution to conservation requires that the bird becomes integrated into the breeding population. From this standpoint, the outcome can be considered positive only if we can assess the reproduction of the released bird. However, in natural populations, not all adults contribute to reproduction; in fact, there is constantly a nonbreeding surplus of birds used to fill in open places. On the other hand, the release of even a restricted number of birds can be a great conservation contribution to many species existing as numerically small, localized populations or at the edge of extinction.

Rehabilitated individuals have to be housed in captivity and close to humans, sometimes even for a long time. Obviously, this context is highly stressful for wild birds entering an RC and can therefore easily affect their behavior. The points that can potentially have adverse effects for a successful reintroduction both during captivity and after release will be briefly considered here.

When keeping birds of prey in captivity, every effort must be made to maintain the birds' welfare. Thus, the rehabilitation program should deal with both medical and ethological requirements. Among the many factors of concern specifically involving behavior for maintaining the birds' welfare, the imprinting phenomenon is well-known and typical of very young birds. It rarely affects adults or juveniles in rehabilitation, but is of concern to orphaned nestlings. Imprinting on humans is reduced at RCs using (cross-)fostering or, at least, rearing nestlings in group. It may have serious implications for reproductive results in birds from captive breeding projects intended for release. In contrast, habituation to captivity as a whole is dangerous even for adult birds. Although it does not require well-defined sensitive periods as imprinting, prolonged close contacts with staff must be considered as mismanagement because it can induce some form of socialization or lack of fear, particularly in juveniles. The released bird might not choose the optimal habitat and may stay in areas with human presence. Moreover, the higher the level of socialization to humans, the more frequent is abnormal aggression toward the handler or partners of the same species.

Infantilism, that is, regression to juvenile behavior patterns in adults, is fairly common in captive raptors; screaming and mantling, that is, lowering the opened wings as a mantle, are evident examples. Such a behavioral regression is likely caused by stress and weight loss. It is possible to record infantilism more often in females than in males, since juvenile behavior patterns are frequent in females' courtship repertoire. Although birds of prey are usually a nonsocial species, social deprivation can affect normal behavior. This occurs when birds, particularly if young, are kept alone or without visual contact with members of the same species for long time. The effects are proportional to the duration of seclusion, to the age of the bird, and are more evident in species with long developmental periods. Socially deprived juvenile raptors show behavior disorders and are unable to relate to members of their own species.

Prolonged captivity can reduce release success in some species because it promotes habituation to humans and reduces flight ability and responsiveness to natural prey. Experienced common kestrels (*Falco tinnunculus*), housed necessarily for more than 4 months due to several causes, performed less varied responses to the several types of prey than those in captivity for less than 3 months. Familiarity with a specific type of food results in kestrels performing the same predatory sequence when attacking any prey.

It is common to feed all species with the same type of food, often day-old-chicks, usually because of the need to save both money and staff labor. A varied diet specifically designed for any species is important, particularly in the case of prolonged captivity, and must provide certain constituents at balanced levels. We should consider that birds of prey are adapted to feed on a wide range of prey, and even the most specialized species do remain very opportunistic predators.

Although it is widely recognized that both Falconiformes (kites, eagles, falcons and allies) and Strigiformes (owls) perceive a strong releasing signal for predation by prey movements, food supply is usually made up of dead items. Young inexperienced birds of prey may have, therefore, difficulty in developing correct predatory ability. Moreover, prolonged food supply limited to one food source, particularly if "artificial," such as chicks, might have negative effects, because the raptor may learn to concentrate upon that food only. In common kestrels, even experienced birds may be affected as has been observed by a weakened response to live prey.

The use of live prey is banned or considered unethical in some countries, although they are accepted as a food source for some species. Animal rights movements concerned with animal welfare oppose the use of animals for research purposes. However, we must also consider and assess all aspects of the captive raptors' biology, and predatory ability is certainly one of the most important. While being concerned with prey welfare, it is also necessary to be concerned with the predator's welfare. There is evidence of behavior alterations among several mammalian predators in zoos, and a bird of prey may be affected from prolonged inability to perform the full predatory sequence in captivity.

Most efforts have been concerned with the captivity phase of rehabilitation, whereas the release phase has attracted less attention. Some recent studies have evaluated if a bird is ready for release from the anatomical, behavioral, or physiological viewpoints. A few studies have evaluated rehabilitated birds and used large samples, although all focused on survival rate and dispersal distances. The latter parameter is of limited utility because it cannot ascertain the bird's previous movements, particularly if survival lasts for months or years. Continuous radio tracking is very useful for monitoring released birds, but it is of limited use with fast or long-range moving birds, or in mountain areas. This problem can be overcome with GPS or satellite telemetry technology, but its high cost, combined with the difficulty of receiving the radio signals when the bird is in areas with electromagnetic interference, limits its use. Nevertheless, although these studies allow for inferring detailed information about movements, even long-range ones, we do not learn anything about behavior. This is fundamental, particularly in the early period, to assess the birds' adaptability to the wild.

Such types of investigation may be expensive and force a reduction in sample size, but they contribute to fill in the gap of our knowledge. The few studies dealing with this problem revealed, in fact, that the release success can be affected to a large extent by previous experience of life in the wild (birds entering the RC as adults or subadults have more probability for safely joining the natural population) and by the species adaptive characteristics. It is then of primary importance to consider these aspects before release in order to get a successful rehabilitation outcome.

See also Welfare, Well-Being, and Pain—*Back Scratching and Enrichment in Sea Turtles*
Welfare, Well-Being, and Pain—*Enrichment for Chickens*
Welfare, Well-Being, and Pain—*Enrichment for Monkeys*
Welfare, Well-Being, and Pain—*Enrichment for New World Monkeys*
Welfare, Well-Being, and Pain—*Experiments in Enriching Captive Snakes*
Welfare, Well-Being, and Pain—*Rehabilitation of Marine Mammals*
Welfare, Well-Being, and Pain—*Sanctuaries*

Further Resources

Csermely, D. 2000. *Behaviour of hand-reared orphaned long-eared owls and tawny owls after release in the wild*. Italian Journal of Zoology, 67, 57–62.
Csermely, D. & Corona, C. V. 1994. *Behavior and activity of rehabilitated common buzzards* (Buteo buteo) *released in northern Italy*. Journal of Raptor Research, 28, 100–107.
Humphreys, P. N. 1981. *The problem of rehabilitation*. In: *Recent Advances in the Study of Raptor Diseases* (Ed. by J. E. Cooper & A. G. Greenwood), p. 165. Keighley, Yorkshire, UK: Chiron Publications.

Kirkwood, J. J. 1991. *Introduction and rationale for rehabilitation*. In: *Raptor Rehabilitation Workshop* (Ed. by London Zoo, The Hawk Trust, & The Hawk Board), pp. 11–18. Newent, UK: The Falconry Centre.

Davide Csermely

■ Welfare, Well-Being, and Pain
Sanctuaries

There are thousands of sanctuaries for animals around the world, havens offering protection to individuals rescued from circuses and other forms of "entertainment," from medical research laboratories, from intensive farming, from the "pet" trade, and indeed from all situations in which they have been shamefully neglected or horribly abused. And for orphan animals whose mothers have been killed for food, for sport, or for any other reason. There are sanctuaries for animals of all kinds, from hens to elephants, and whereas some are small enterprises in backyards, others are big operations that require a large staff and considerable funding. Unfortunately, once a refuge has been created, it tends to grow, an indication of the number of desperate creatures needing care. Fortunately, there are also a growing number of people around the globe who are not only aware of, but care, sometimes desperately, about animal suffering.

I became involved with sanctuaries for orphan chimpanzees in Burundi in 1990. This was when The Jane Goodall Institute (JGI) initiated a conservation project there, and people began telling us about the "pet" chimpanzees in the country, many of them held in appalling conditions. One of these, whom I went to see for myself, was Whiskey. His owner came to greet me and led me through his noisy garage to a cement-floored 6 ft x 6 ft (1.8 m x 1.8 m) space that had once been a lavatory. The only light came through a hole in a corner of the corrugated iron roof. A 5 or 6-year-old male chimpanzee with a collar around his neck was chained to a pipe in the wall. Whiskey held his hand towards us, stretching as far as he could—but his chain was only 2 ft long, and we were out of reach. So he turned and stretched back with one foot. When I went in and crouched down beside him he put his arms around my neck.

Whiskey's mother had been shot for the live animal trade so that her infant could be stolen and sold as a pet or to attract visitors to a hotel or bar. He had been captured in neighboring Zaire (now the Democratic Republic of Congo), then smuggled over the border and sold in Bujumbura. At first he had been part of a human family, sitting with them at table, riding in their car, and playing with the children, until the time he was about 4 years old when

Whiskey the chimpanzee, in the squalid garage from which he was eventually rescued.

© *Michael Nichols / National Geographic Image Collection.*

they realized how strong and potentially dangerous he was. Then he was banished to his prison cell. Eventually we (JGI-Burundi) persuaded his owner to hand him over to what we called "The Half Way House." This was a small backyard facility where other ex-pets were waiting until we could raise the money for a permanent sanctuary. But in 1994, because of the ethnic violence in Burundi, we had to move all of them—20 by then—to Kenya, where a new sanctuary, Sweetwaters, had been built for them.

Then there was Little Jay, offered for sale in the main tourist market in Congo. He had a piece of thin, hard rope tied tightly around his waist, attaching him to the top of a tiny wire cage. It was very hot, and people milled around, talking and laughing, towering over the frightened and sick little chimpanzee. I went close and saw that he was emaciated and sweating, and his glazed, dull eyes indicated closeness to death. Yet when I made small sounds of chimpanzee greeting, he sat up and reached out to touch my face. Little Jay's mother had been hunted for the bush meat trade, the commercial hunting and selling of the meat of wild animals. There is so little flesh on an infant chimpanzee that the hunters usually try to sell them alive, as pets. People sometimes buy these pathetic orphans, to rescue them. But this only serves to perpetuate the trade in live animals. And when the cute infant becomes a strong and willful juvenile, he or she will often be banished, like Whiskey, to prison conditions.

Chimpanzees are classed as endangered species in almost all the countries where they live, and it is illegal to hunt and sell them without a license—but this law is seldom enforced. However, we were able to persuade the authorities to confiscate Little Jay. Graziella Cotman, with whom I had been corresponding for several years, agreed to nurse him back to health. Then all the people who were upset by the abuse of chimpanzees, but unable or unwilling to do anything about it themselves, brought others to be cared for by Graziella. Thus in 1990, Little Jay became the nucleus for what is now the largest chimpanzee refuge in Africa, JGI's Tchimpounga Sanctuary near Pointe Noire in Congo-Brazzaville. There were 115 orphans in July 2003.

That is how it always starts—an individual chimpanzee looks, from his place of fear and confusion and pain, into your eyes, and reaches out to touch you. The very first African sanctuary began when one tiny and badly wounded infant was confiscated from a hunter (who had shot her mother in neighboring Zaire) and taken to a British couple, Dave and Sheila Siddle, who run a cattle ranch in Zambia. They nursed her back to health and were given a permit to keep her. And so, of course, government officials brought them the next confiscated infant. And the next, and the next. And when people realized that, at last, there was a place where young chimpanzees would be properly cared for and loved, youngsters began arriving from other parts of the world. Of course, as the Siddles' chimpanzee family grew so did their expenses. They had to fence in a large area of their land, and build strong cages for night quarters and where the chimpanzees could be cared for if they were sick or injured. The Chimfunshi Animal Orphanage is home to 100 chimpanzees in three groups at the time of writing (August 2003).

In 1992, during a visit to Uganda, a small group of concerned people asked for my help. They were raising funds to help a growing number of orphan chimpanzees, confiscated by the government and placed in the Entebbe Zoo—where, at that time, there was neither the money nor the expertise to care for them properly. JGI paid the expenses of Christine Manning, one-time ape keeper of the London Zoo, to work in Entebbe as a volunteer. JGI-Uganda was then registered, thanks to Linda Rothen, so that we could play a more active role in caring for the chimpanzees. The accommodations at the zoo were not suitable for the chimpanzees, and finally, thanks to Wilhelm Mueller, it became possible to send one group of youngsters temporarily to a small island in Lake Edward National Park, while at the same time a new facility was constructed for the adult chimpanzees at the zoo. Meanwhile, we were searching for a suitable permanent home for the orphans. Our dream was realized in 1995 when, under the supervision of Debbie Cox, the Ngambga Island sanctuary on Lake Victoria, some 30 minutes boat ride from Entebbe, became the beautiful new home for our original group of chimpanzees and all those who arrived subsequently. This sanctuary, superbly managed by Debbie and her Ugandan staff, with help from "Monty" Montgomery, is funded by a coalition of five animal welfare organizations. A second island will soon be ready and thus prevent overcrowding on Ngamba Island.

Today there are thirteen sanctuaries in Africa that care for orphan chimpanzees, including Tchimpounga, Ngamba Island, Sweetwaters, and Chimfunshi. Three of them are in Cameroon—The Limbe Wildlife Center (28 chimpanzees), Sanaga-Yong Chimpanzee Rescue Center (18 chimpanzees), and Cameroon Wildlife Aid (37 chimpanzees). The others are the Chimpanzee Rehabilitation Center in Gambia (61 chimpanzees), Pandrillus in Nigeria (22 chimpanzees), the Chimpanzee Conservation Center in Guinea (31 chimpanzees), Baboon Island sanctuary in the Gambia, Pandrillus in Cameroon, the Tacugama sanctuary in Sierra Leone (55 chimpanzees), Sodepal sanctuary in Gabon (15 chimpanzees), and Habitat Ecologique de Liberté des Primates (HELP) in the Republic of Congo. JGI is in the process of building yet another in South Africa to care for the five chimpanzees rescued from Angola, one from Nigeria, and one donated by the Johannesburg zoo. In Zambia, Kenya, and South Africa, where there are no wild chimpanzees, the orphans are considered as refugees from neighboring countries. This gives a total of almost 600 chimpanzees in sanctuaries. Of course, the number is constantly changing as new orphans arrive and, inevitably, a few die. The best place to check on the African chimpanzee sanctuaries is the PASA (Pan African Sanctuary Association), *www.panafricanprimates.org*. There is an additional colony of chimpanzees belonging to the New York Blood Center located in Liberia. Dr. Fred Prince is working to move these ex-experimental chimpanzees to a safe sanctuary. It must be mentioned that there are also five sanctuaries that care for orphan gorillas, two in Cameroon, two in Gabon, and one in the Democratic Republic of Congo (DRC). These sanctuaries care for a total of 78 gorillas. Another newly started sanctuary in the DRC cares for four bonobos.

Often I am asked why we do not return our orphans to the wild. The answer is that we would if we could, but it is almost impossible. First, it is necessary to find an area of suitable chimpanzee habitat where there are no wild chimpanzees (who are territorial and typically kill strangers, especially males), and no people (for our orphans have no fear of humans and would wander into a village and either be hurt, or hurt someone). We are actively searching for ideal places for reintroduction in Congo-Brazzaville, for our sanctuary is horribly overcrowded. If we are successful we shall then have to ensure that our youngsters acquire the skills they need to survive in the wild. The one sanctuary that has released chimpanzees into the wild (in an area where it was—mistakenly—thought that the ex-captives would be unable to mix with wild individuals) is HELP in Congo-Brazzaville.

Conservationists often accuse us of wasting money by caring for captive individuals rather than spending our precious dollars on trying to save wild chimpanzees and their vanishing habitat. I really wish that JGI were not involved with sanctuaries, but feel I have no choice. After all, ever since I began my research into chimpanzee behavior at Gombe, I have stressed the importance of individuality. Each chimpanzee has his or her own, unique personality, and each plays an important role in his or her society. This thinking was not fashionable among scientists back in 1960, but is widely accepted today. From the beginning,

I insisted that the chimpanzees had feelings and emotions similar to ours. (After all, I had learned during my childhood that this was true for my dog.) Thus to abandon an orphaned chimpanzee would be, for me, as unethical as abandoning a small human child. However, I know also that it is desperately important to do everything we can to protect the remaining wild chimpanzees and their habitat. And so JGI struggles to achieve both these goals.

In Congo-Brazzaville we are working with the government to preserve a 40 mi^2 (103 km^2) area (and hope that soon we shall get permission to triple this) around the Tchimpounga Sanctuary. We have employed "eco-guards" from the local villages to patrol the area and to lead small groups of visitors into the reserve. This area is a unique stretch of forest–savanna mosaic along the shore of the Atlantic Ocean, and we find there are more wild chimpanzees living there than previously thought. We are also deeply committed to chimpanzee conservation efforts in Uganda—and were trying to protect wild chimpanzees in Burundi also, before we were forced to pull out.

A final and important point is that our orphans serve as ambassadors for the wild chimpanzees. Most people, even if they live near a forest, have little or no opportunity to observe chimpanzees. When visitors from the villages or from a nearby town come to see our chimpanzees, they are typically amazed to see how like humans they are. A number of people, after watching our youngsters kissing and embracing, using tools, playing, and so on, have said they will never eat chimpanzees or other apes again, and never go to a restaurant that serves ape meat. We especially encourage children to visit. And we aim to provide research opportunities for students from universities to study chimpanzee behavior.

Another criticism often leveled at those working to save orphan chimpanzees or other animals in Africa is that we have our priorities wrong—surely, in view of the poverty and suffering of the *people* of Africa, we should not be "wasting" money on animals. We realize only too well the desperate need of hundreds of thousands of Africans, and JGI is working hard to improve the lives of the people living around our sanctuaries. We are modeling these efforts on our highly successful TACARE (take care) program that has improved the lives of approximately 150,000 thousand people in the 32 villages around the Gombe National Park. This program, in addition to introducing tree nurseries, agroforestry, the most suitable environmentally sustainable farming techniques and conservation education, also provides primary health care (through the regional health authority), AIDS education, and family planning. A series of micro-credit banks enable women to start their own environmentally sustainable projects, thus earning some money for themselves, often for the first time. Gifted girls can apply for scholarships to go from primary to secondary school. Around the world it has been shown that as women's education and self-esteem improves, so family size drops. In Uganda there is also a well-developed community outreach program. And at Tchimpounga a similar program is in its early stages. We have built a dispensary, improved a school, introduced our education program Roots & Shoots, and are in the planning stages of many other projects to help people to help themselves. Of course, in and of themselves, our sanctuaries provide ongoing jobs for local people, help to boost the local economy and, when tourism is possible, bring foreign exchange into the country.

People running sanctuaries in Africa must cope not only with the expense, the lack of infrastructure, and, often, high levels of corruption, but also, in some countries, may face personal danger. Our JGI staff in Burundi, for example, were regularly awakened by gun shots, encountered military road blocks daily, and faced increasing levels of violence. It was when two of our Burundian staff were shot and killed as they walked home that we decided we needed to move, eventually to Kenya.

During the 9-year civil war in Congo-Brazzaville, our Tchimpounga project manager, Graziella Cotman, was several times threatened by looting groups of militia. Bala Amarasekaran and his wife risked their lives over and over again during the terrible civil war in Sierra Leone. They stayed in Freetown and somehow managed to care for the infant chimpanzees in their sanctuary. Betsy Brodman, working for the chimpanzee colony belonging to the New York Blood Center, stayed on with her husband in war-torn Liberia, creeping out to find food for the chimpanzees despite the danger. Only after she had watched her husband shot and killed as he knelt in front of a rebel soldier, begging for his life, could she be persuaded to leave. But not for long—she was soon back, caring for the chimpanzees again.

Unfortunately it is not only in Africa that chimpanzees desperately need the help of dedicated people. In the Americas, Europe, and Asia, chimpanzees have been mistreated, often shockingly, in zoos, circuses and other forms of "entertainment," and in medical research laboratories. Many of these were taken from Africa, snatched from the dead bodies of their mothers as infants. Others were born in captivity. We owe it to these unfortunate individuals to provide them with save havens where they can live out their lives in relative freedom once they have been rescued.

In the United Kingdom, Jim Cronin founded the Monkey World Ape Rescue Centre, which he runs with his wife, Alison. Originally this center was built to provide a home for the infant chimpanzees who were smuggled into Spain from West Africa and used as photographers' props in tourist resorts. Jim worked with a British couple who lived in Spain, the Templars, and with the police, to stop the illegal trafficking, and also with tourist agencies, persuading them to warn visitors of the cruel practice. Jim has now rescued chimpanzees (and other primates) from many parts of the world. In America, Wally Swett began taking in abused animals (mostly primates) discarded by the pet and entertainment industries. His Primarily Primates is situated in San Antonio and now provides sanctuary for several groups of chimpanzees. Patti Regan, at the Center for Orangutan and Chimpanzee Conservation, Vachula, Florida, and April Truit, at the Primate Rescue Center, Inc., Nicholasville, Kentucky, have both built small sanctuaries for ex-pet and ex-entertainment individuals.

A very hard challenge is to create sanctuaries for chimpanzees who have been used and abused in medical research laboratories. These individuals are typically full grown, and often they have been housed alone for most of their lives, so it can take years to resocialize some of them. The very first rescue of a group of ex-lab chimps, who were released onto a man-made island at Lion Country Safaris in Florida, is described by Linda Koebner in her moving book *From Cage to Freedom*. They are still there.

Years later, the Chimpanzee Health Improvement, Maintenance and Protection Act— the CHIMP Act, H.R. 3514,—sponsored by U.S. Representative James Greenwood, was passed by both House and Senate in 2000, and signifies the U.S. government's commitment to partner with the private sector to provide sanctuaries for chimpanzees retired from medical research. Chimp Haven, a nonprofit organization, has received $24 million from the National Institutes of Health (NIH) to build and manage a sanctuary on a 200-acre site of forested land, donated by the citizens of Caddo Parish in Louisiana. Up to 200 ex-laboratory chimpanzees will eventually be housed there. Chimp Haven must raise funds themselves equal to 10% of the government grant.

Unfortunately, just before the CHIMP Act was passed, some wording was added to the original draft as a result of intense lobbying from the National Institutes of Health. In its final form the Act has a clause that will enable the medical research industry to take a chimpanzee out of retirement and put him or her back into the laboratories. Many people felt strongly that this was such a repugnant idea that they could no longer support the bill, and they actively lobbyed against it. Others continued to support the bill, feeling that despite the new wording, it would still, if passed, benefit hundreds of chimpanzees. Thus the animal protection groups that had, up to that point, united in a coalition to work to retire chimpanzees from the laboratories, were divided.

I personally felt very strongly about this issue. For one thing I believed (and still do) that if the bill had not passed in 2000, it would have been years before it was introduced again. (And with the change in the United States administration and the events of 9/11, this seems almost certain.) I tried to put myself into the place of one of the chimpanzees who had been cooped up in a 5 ft by 5 ft (1.5 m by 1.5 m) cage for 10 years or more. I imagined being offered one of two choices. First choice: Wait for an untold number of years for a bill that, *if* it were passed, would totally secure my lifelong retirement. Second choice: Get out of my cage as soon as a sanctuary could be built, knowing that there was a possibility—and it is only a possibility—that I might have to go back into my small cage at some future date. I truly believe that a chimpanzee, who lives very much in the present, would opt for the second choice, especially in view of the fact that the criteria for plucking an individual chimpanzee from retirement are quite stringent. The removal provisions of the CHIMP act read as follows:

The chimpanzee may be used in research if:

I. the Secretary finds that there are special circumstances in which there is need for that individual, specific chimpanzee (based on that chimpanzee's prior medical history, prior research protocols, and current status), and there is no chimpanzee with a similar history and current status that is reasonably available among chimpanzees that are not in the sanctuary system;

II. the Secretary finds that there are technological or medical advancements that were not available at the time the chimpanzee entered the sanctuary system, and that such advancements can and will be used in the research;

III. the Secretary finds that the research is essential to address an important public health need; and

IV. the design of the research involves minimal pain and physical harm to the chimpanzee, and otherwise minimizes mental harm, distress, and disturbance to the chimpanzee and the social group in which the chimpanzee lives (including with respect to removal of the chimpanzee from the sanctuary facility involved).

And there will be many of us prepared to fight such a decision on a chimp-by-chimp basis.

Fortunately many chimpanzees have already been rescued from their tiny (and very cruel) cages when, for one reason or another, laboratories have been closed. (Jim Mahoney, a veterinarian who had worked for the University of New York at their Laboratory for Experimental Medicine and Surgery in Primates (LEMSIP), traveled the length and breadth of North America seeking retirement places for the 300 or so chimpanzees that were suddenly surplus goods. Many were taken in by various zoos and rescue centers.)

Richard Allen and Gloria Grow of the Fauna Foundation built a sanctuary for 15 chimpanzees near Montreal in Canada. It was the first sanctuary of its kind, built to house chimpanzees infected with AIDS and hepatitis, as well as "clean" individuals. It served as a precedent, inspiring others to make the same commitment. The next sanctuary for ex-laboratory chimps was built by Carole Noon in Florida. The first group to be housed there comprised 21 of the so-called Air Force chimpanzees from the Holloman Air Force Base in New Mexico. Some of these are descendents of the original group that was captured in the wild for the space research that culminated with sending the first astronauts to the moon. There are heart-warming and intensely moving descriptions of the abused LEMSIP and Air Force chimpanzees as they moved, for the first time, outside their cages, looked up at the clouds, felt the rain, and enjoyed foods other than the endless, boring monkey chow that is the lab staple.

Third generation baby chimpanzees from the original space research program inside a nursery at the Coulston Research Center.

© Getty Images.

In 2002, Carole Noon took on the biggest chimpanzee rescue in history when the Coulston laboratory in New Mexico was finally closed down (after logging up countless violations of the animal welfare act for years), and the whole facility was bought with an incredibly generous grant from the Arcus Foundation. Her first task was to make immediate improvements to the existing facility (the chimpanzees were sleeping on concrete or metal flooring with no blankets or straw, were given only one piece of fruit each per week, and many had no contact with each other (even visually). Gradually the Florida facility will be enlarged and more and more of these chimpanzees will move to the relative freedom of grassy islands with shade and climbing structures—and be cared for by humans who understand and love them.

One other sanctuary for ex-laboratory chimpanzees is almost complete in Austria. It will house about 50 chimpanzees who were used in hepatitis and HIV research by the infamous IMMUNO A.G. laboratory. The company was bought by Baxter A.G., which had no use for the chimpanzees, and they will be retired to this new sanctuary. The only other European laboratory using chimpanzees for medical research was the European Chimpanzee Research Facility at the Dutch Biomedical Primate Research Centre (BPRC). Many people protested the conditions in which the 50 or so chimpanzees were kept, and in 2001 the British nonprofit organization, People Against Chimpanzee Experimentation (PACE), under the leadership of Janie Reynolds, organized a coalition of pressure groups that began intensive lobbying for the Dutch government to ban the use of great apes in medical research. This finally happened in May 2003. Most of the chimpanzees will be rehabilitated at a sanctuary for exotic animals, Stichting AAP, then moved to a large sanctuary, Primadomus, being built for their retirement in southern Spain. Those chimpanzees that are infected with HIV, hepatitis, and other diseases will remain at the BPRC a little longer, but eventually move to a specially constructed sanctuary under the control of Stichting AAP.

Wherever a sanctuary is located, the chimpanzees rescued from abuse have so much to teach us. Many have lived for years alone, deprived of everything that a chimpanzee needs to enjoy life. Often they have acquired psychotic behaviors—such as rocking from side to side, banging their heads on the wall, mutilating themselves, showing sudden violent outbursts of rage, or huddling alone in a corner for hours on end. Many can never fully recover psychologically. But it is inspirational to watch how they gradually manage to lose some of their abnormal behavior and learn to live in chimpanzee society. And there is much they can teach us. Not so long ago psychiatrists and psychologists used to raise chimpanzees in conditions designed to replicate the abnormal early experiences of psychologically disturbed humans. It was argued that this would be helpful to scientists seeking to better understand mental illness in people and thus help human patients. Now there are hundreds of chimpanzees who have been exposed to all manner of abnormal conditions. It is important that sanctuaries open their doors to scientific observation of a strictly noninvasive, nondisruptive nature. Surely there are lessons we can learn from the rehabilitation of our closest relatives that will benefit the thousands of humans who, like the chimpanzees, carry deep psychological scarring from past traumatic experiences.

We rescued La Vielle (who was not really old) from an animal-holding facility in Pointe Noire where she and a few other chimpanzees had been more or less abandoned. When we first met La Vielle in 1990, the door of her cage was broken so that she could have escaped, but she never ventured out. We took her to our Tchimpounga Sanctuary in 1995. It was another 5 years before she dared leave her concrete floored sleeping quarters onto the sandy floor of the main indoor enclosure. I watched as again and again she tried to work up courage to leave her room, swaying from side to side with bristling hair, then charging the open door—and again and again stopping without going through. Something so horrible must have happened to her when, years ago, she walked out through a cage door. But eventually she overcame her phobia. Seeing her now, sitting on the grass in the sunlight in the little garden we have built for her and 60-year-old Gregoire, rescued from the now-defunct Brazzaville Zoo, makes it all worthwhile.

See also Reproductive Behavior—*Captive Breeding*
   Welfare, Well-Being, and Pain—*Back Scratching*
    and Enrichment in Sea Turtles
   Welfare, Well-Being, and Pain—*Carnivores in Captivity*
   Welfare, Well-Being, and Pain—*Enrichment for*
    Monkeys
   Welfare, Well-Being, and Pain—*Enrichment for*
    New World Monkeys
   Welfare, Well-Being, and Pain—*Experiments in*
    Enriching Captive Snakes
   Welfare, Well-Being, and Pain—*Primate Rescue*
    Groups
   Welfare, Well-Being, and Pain—*Psychological*
    Well-Being

Further Resources

Web Sites

 Editor's Note: For general information about animal sanctuaries see:

http://www.panafricanprimates.org
 The Pan African Sanctuary Alliance, or PASA, is an alliance of 16 primate sanctuaries from all over Africa. The Web site lists the sanctuaries and provides information on most them. It also provides an

extensive list of resources about animals, endangered species and conservation, sanctuaries, animal behavior and related topics.

http://www.taosanctuaries.org/sanctuaries/species.htm

The Association of Sanctuaries, TAOS, was founded in 1992 as a not-for-profit organization to assist sanctuaries in providing rescue and care for displaced animals. It accredits sanctuaries for wild, farmed, and companion animals. Provides a worldwide list of accredited sanctuaries.

http://www.cwu.edu/~cwuchci/index.html

The Chimpanzee and Human Communication Institute at Central Washington University in Ellensburg, Washington, provides sanctuary to a group of adult chimpanzees who communicate with humans and each other using American Sign Language (ASL).

Jane Goodall

■ Welfare, Well-Being, and Pain
Stress in Dolphins

Assessing stress in dolphins presents distinctive challenges, yet it is a critical component in the management of populations as well as individuals. The behavior of dolphins is often the primary, and one of the only available indicators of their physiological and psychological condition. Physical measurements of *stress* are particularly challenging to obtain in dolphins, largely because of the their unique anatomy and the aquatic environment in which they live. Yet over the past few decades, dolphins have been subjected to various forms of human impact to which most other animals are typically not exposed. Consequently, measurement of behavioral indicators of stress in dolphins has been increasingly important to their *welfare* and conservation.

Stress is a normal physiological and behavioral process that is designed to improve an animal's survival during challenging situations. Stress has been defined (by Broom and Johnson 2000, p. 72) as "an environmental effect on an individual which overtaxes its control systems and reduces its fitness or appears likely to do so." It may also be described as a condition in which abnormal or extreme adjustments to behavior, psychology, or physiology are necessary to cope with the environment. Psychological stressors can produce as much, or even more, of a physiological stress response in dolphins than can physical stressors. The physiological changes that are referred to as stress may be accompanied by subjective states of suffering as well.

Sources of Stress

Encountering sources of stress is natural for all wild animals. As noted by David St. Aubin and Leslie Dierauf in 2001, such stressors for dolphins include disease and illness, encountering predators, *agonistic* (ranging from submissive to aggressive) encounters with other dolphins, demanding geophysical conditions (such as storms), and biological sources of stress that challenge *homeostasis* such as environmental toxins and lack of food. However, anthropogenic (caused by humans) sources of stress are of growing concern because of increased human activity in the world's oceans and rivers. Examples include activities such as oil, gas, and mineral exploration, icebreaking, military activities such as explosions and active sonar transmissions, boat and aircraft noise, boat collisions and harassment, release of contaminants (such as oil spills), invasive research techniques such as tagging with transmitters, capture from the wild, and entanglement or encirclement in fishing gear. Lack of control is one of the most stressful events that an organism can encounter. Providing dolphins control over their environment has been recommended as an important method of improving their welfare in captivity as well as for animals in the wild who regularly encounter human activity.

In many parts of the world, dolphins encounter stressors that are fairly unique. Many people demonstrate an uncommonly strong fascination with dolphins and attempt not only to view them, but to also interact with them in captivity and in the wild. Encounters with dolphins from vessels, land, and in the water have become a very popular form of human recreation and tourism in many parts of the world. For example, tens of thousands of people attempt to swim with dolphins from commercial boats every year in New Zealand.

Some free-ranging dolphins are somewhat distinctive among wild animals in that they will exhibit attraction to humans on boats and in the water, even without the provisioning of food. However, more frequently, dolphins exhibit varying degrees of *habituation* (neutral response), *tolerance*, and *disturbance* to human activity. Accidental and intentional harassment of dolphins has become a serious problem around the world and has caused severe injuries and even mortalities in dolphins.

Various forms of regulatory protection have been implemented to address these situations in several countries, although few, if any, have been adequate to address the severity of the problems. Statutes like the Marine Mammal Protection Act in the United States include provisions that prohibit harassment of dolphins and other marine mammals. Yet just as there are challenges in defining the stress response in dolphins, providing a precise definition of the term "harassment" has been difficult. Consequently, a precautionary and conservative definition of this term is often warranted.

Facilities providing the public with an opportunity to watch and interact with dolphins in captivity have also become widespread over the past two decades. "Swim-with-the-dolphin programs" and "petting/feeding" programs use various species of dolphins, many of which are captured from the wild. The species used in these programs are typically bottlenose dolphins (*Tursiops truncatus*), but also include pilot whales (*Globicephala macrorhynchus*) and orcas (also called killer whales) (*Orcinus orca*), also considered dolphins, as well as beluga whales (*Delphinapterus leucas*). Laws governing the capture and maintenance of dolphins in captivity vary widely around the world, from being nonexistent to detailed. Some of these regulations include measures that are intended to respond to the psychological as well as the physical needs of the dolphins (such as requiring them to be housed with other dolphins). However, there are typically no special provisions enforced to protect dolphins from the additional sources of stress that may be encountered in interactive programs.

Measuring Stress

Because of the difficulty of obtaining and interpreting physiological data in dolphins beyond visible physical characteristics (e.g., perceptible wounds or emaciation), behavior is typically relied upon to assess their welfare. Recognition of stress-related behaviors, including submission (e.g., retreat and avoidance) to other dolphins or humans, loss of hunger, or agitation (e.g., erratically swimming back and forth) is critical in the identification of prepathological or pathological conditions which can occur even if the animal appears to be in otherwise optimal physical condition. As in the case of physical measures, a systematic and proven methodology for identifying and interpreting behavioral indices of stress in dolphins is lacking. Further, because their physical form and behavioral signals are so different from those of terrestrial animals, stress-related behaviors are easily overlooked and often misinterpreted. Regardless, even when physiological measurements are obtained, they are often only useful when observed in conjunction with the dolphin's behavior.

As with other animals, various characteristics of a potential stressor (such as a loud noise) affect the quality and consistency of the dolphins' response to a stimulus. These characteristics include controllability and predictability of the activity, duration, intensity, frequency, and location and distance from the animal. Also, individual variation in dolphins can produce vastly different responses to the same stressor. Such disparity may be related to variation in species, previous experience with the same or similar stimuli, age, gender, reproductive status, social and environmental conditions, group size and composition, physiological condition, psychological "temperament," and time of day and season.

For example, some captive dolphins may become reasonably acclimated to being transported in a stretcher, exhibiting a reduction in signs of stress. Conversely, other dolphins, perhaps different species, or individuals having more negative experiences being transported, may become sensitized to transport and respond with increasing *distress* over time. Similarly, introducing a novel stimulus into a dolphins' captive environment may provide an important source of environmental enrichment for some individuals while resulting in a stress response for others. Also, a single stressor can have various impacts on a dolphin. For example, a very loud noise can frighten or agitate dolphins, cause pain, physical damage, and interfere with important behaviors such as rest, acoustic exploration (echolocation), and communication.

Behavioral Indicators

Behavioral events directly and immediately associated with stimuli are generally regarded as short-term responses. Dolphins may exhibit such responses through changes in respiratory, surfacing and dive patterns, cohesion of groups (e.g., tightly bunching together), orientation and speed of swimming (e.g., avoidance, and retreat), communicative displays (e.g., threat displays, abrupt slaps, and splashes), postures (e.g., "hanging" in the water), defecation, and vocalizations. Short-term stress responses in dolphins can result in long-term, chronic behavioral patterns including *stereotypies* (e.g., repetitive circling), apathy and unresponsiveness, self-destructive patterns (e.g., ramming the head against a wall), and excessive aggressiveness (toward other dolphins or people). Such long-term responses are characterized by impacts to physical condition, cumulative or ongoing behavioral state (e.g., *attraction, habituation, desensitization, tolerance*, and *sensitization*), reproduction and survival (of the individual and population), and long-term habitat use and distribution. However, a lack of observable responses to a stimulus is not necessarily confirmation that a negative impact has not occurred. For example, dolphins in unusually stressful situations have been observed to be passive and this has been interpreted as being "calm." However, such behavior may indicate disorientation or fear as much as an ability to cope with the situation.

Studies of dolphins in the wild tend to focus on group behaviors while captive studies tend to focus on individuals, and sometimes more detailed or subtle actions. Many of the same behavioral manifestations of stress in dolphins can be observed in the wild and in captivity. Some behaviors are characterized by abruptness and include jaw snaps; "chuffing" (loud and sharp exhalations); head, fluke (tail), and body slaps and slams; head shakes; and retreat. These actions are often used as measures of agitation, submission, alarm, threat, or frustration in dolphins. However, abruptness has also been associated with play and excitement in other situations. Other behaviors, such as sustained avoidance, some postures (e.g., "sinking" in the water), and some open-jaw displays and "s-shaped" postures are exhibited more gradually. Variations in vocal rate, quality, and type (e.g., whistles, squeaks, or squawks) have been observed in a variety of situations (such as entanglement in nets, transport, "swim-with-the-dolphin" programs, and separation from other dolphins).

Agonistic behavior, ranging from submission to aggression, is observable in captivity and in the wild. Although it is not uncommon for dolphins to be aggressive toward one another, excessive aggressiveness is often considered to be a sign of stress. Such aggression often occurs in challenging situations such as forced, close confinement with others in captivity and in nets during tuna purse-seine fishing operations. Aggression toward humans in captivity and in the wild is also not uncommon, and has included broken bones, bruised and ruptured internal organs and bites, and rarely, human mortalities (three reported incidents; two in captivity involving one or more orcas and one free-ranging bottlenose dolphin). Such incidents have typically been related to inappropriate or abusive human behavior and the dolphin's response of frustration or self-defense. Sexual and aggressive behaviors are also observed concurrently with aggression and sometimes directed toward humans and other dolphins.

Free-Ranging Dolphins

A direct relationship between stressors and long-term responses can be particularly hard to determine in the wild, largely because many of these impacts are cumulative or delayed in expression. Also, pre-impact baseline data for most dolphin populations are lacking. However, it is in the wild where the impacts to welfare, reproduction, and survival may affect entire populations and species, thus stress assessment can be an important aspect of dolphin conservation.

Perhaps the most comprehensive studies on dolphin stress are those that have been conducted on dolphins chased and encircled (temporarily captured in nets) during tuna purse-seine fishing operations. This methods of fishing in the eastern tropical Pacific ocean threaten three depleted populations: northeastern offshore spotted dolphins (*Stenella attenuata attenuata*), eastern spinner dolphins (*Stenella longirostris orientalis*), and to a lesser extent, coastal spotted dolphins (*Stenella attenuata graffmani*). In addition to immediately observed mortalities (due to drowning or physical impact), indirect effects and delayed mortalities include the death of calves following separation from their mothers, failed pregnancies, injuries from boats and nets, organ damage as a result of cumulative stress, and increased susceptibility to predation following release. The shock of psychological (fear) and physiological (overexertion and overheating) responses to capture can manifest in metabolic processes that cause lethal damage to the kidneys and heart (as a heart attack) and other biological systems. This process, referred to as *capture myopathy*, results in immediate or delayed (up to months later) death.

Research has found that dolphins experience short- and long-term impacts in response

A fisherman prepares to free a dolphin that has been snagged in his net off the coast of France. Fishing nets are just one of many human intrusions that add stress to the life of a dolphin.
© *Guigo Constant / Corbis Sygma.*

to the activity of boats and human swimmers, although these responses seem to vary widely. For instance, a review that I conducted found that the observed impact of numerous swimmers was determined to be minimal in some environments while the impact of a single swimmer in other environments was severe enough to result in repeated disruption of rest, feeding, and other important activities. In some regions, boats and swimmers appeared to exert negative impacts on dolphins, such as disturbance and sensitization. For instance, Rochelle Constantine (Constantine & Yin 2003) has found that over time, the bottlenose dolphins in the Bay of Islands in New Zealand have been interacting with swimmers less and avoiding them more. Anna Forest suggested that spinner dolphins (*Stenella longirostris*) in Kealakekua Bay, Hawaii, may be avoiding important areas for resting, nursing, and mating because of the increased presence of swimmers and boaters. Intensive whale-watching activity from boats appears to regularly disrupt important activities like resting, feeding, and mating in seriously depleted populations of orcas in Washington State and British Columbia.

Some free-ranging dolphins who are regularly provisioned with food by people have also exhibited serious long-term responses such as alteration of natural foraging patterns and mother–calf interactions, reduced vigilance for predators, changes in distribution and habitat use, and increased calf mortality. Death from contaminated fish and serious injury and death due to human abuse have also resulted from these activities. Even when dolphins do not exhibit signs of stress in response to human interaction, closer proximity to humans place them at higher risk of being deliberately or accidentally harmed by people.

Captive Dolphins

Capture from the wild, transport, and confinement in captivity are stressful for dolphins. The capture of even a single animal can harm or kill multiple dolphins, since capture typically involves harassment, and sometimes temporary capture, of more animals than those targeted. Mortality is not uncommon following capture and transport, even in North America where facility standards generally exceed those in many other regions. Small and DeMaster (1995a) found that mortality rates of captured bottlenose dolphins increased six-fold immediately after capture—and did not decrease to "normal" captive levels for up to 35–45 days. Two of the most recent studies (by Small and DeMaster in 1995b and Woodley et al. in 1997) found that survivorship rates in bottlenose dolphins and orcas through the mid-1990s did not exceed those of their free-ranging counterparts. Considering that veterinary care, consistent food availability, and protection from predators would be expected to result in longer life spans, the effects of stress may play a significant role in the mortality of captive dolphins.

Captive environments have been associated with a variety of behavioral abnormalities such as stereotyped behavior, consumption of foreign objects (dolphins known to be "nervous" seem to do this more), self-inflicted trauma (e.g., obsessive cribbing/biting on objects, sometimes to the point where teeth are worn down), stress-induced vomiting, and excessive sexual behavior and aggressiveness. Submissive behaviors such as retreat and avoidance have been associated with intimidation by other dolphins, especially in captivity where dolphins are forced to remain in contact with more dominant and aggressive animals. Veterinarians have frequently found such submissive behavior to be associated with psychological and physiological distress resulting in the onset of serious health problems, sometimes leading to death.

Interactive programs such as "swim-with-the-dolphin" programs and petting/feeding pools present additional stressors to captive dolphins. Dolphins are typically afforded little or no control over the presence of human "visitors" in their enclosures. Increased noise, environmental stimuli, and disruption of rest, as well as greater risk of infection, disease, and harassment are potential sources of stress in these programs. The few studies conducted of these programs observed various stress-related behaviors indicating that these programs may have both short- and long-term negative effects on the participating dolphins. For example, dolphins frequently behaved submissively toward people in these programs, even when the people did not behave aggressively.

Physiological Indicators

Physiological indices of stress may be difficult or unfeasible to obtain because the very act of obtaining these data may be stressful, especially when they require restraint, removal from the water, or sedation (which is particularly dangerous for dolphins). Even when biological samples are obtained, a standardized, quantitative methodology for analyzing them is lacking. However, physiological measures of captive dolphins and dolphins involved in tuna purse-seine fishing operations, in particular, have yielded a valuable and growing body of information. Analysis of blood, tissue, fecal, and urine constituents includes endocrinological (stress-related hormones such as catecholamines, cortisol, corticosterone, and thyroid hormones), neurological (neurotransmitter activity), and immunological (such as changes in leukocyte counts related to susceptibility to disease) parameters. Physical measurements reveal changes in cardiovascular and respiratory parameters, body temperature, skin proteins, muscle, organ (e.g., adrenal, heart, and kidney lesions or enlargement), and other physical characteristics (e.g., gastrointestinal ulcers). Promising techniques for detecting stress levels using noninvasive techniques include measurement of glucocorticoid metabolites in fecal samples such as Samuel Wasser and his colleagues have been developing. This technique has tremendous potential since it is a noninvasive technique that measures adrenal responsiveness to a variety of stressors.

Summary

The impacts of human activity include an unknown number of unpredictable responses in dolphins, some of which are long-term and life threatening to individuals as well as to populations. Despite the need for a more objective and standardized methodology to identify stress in dolphins, behavioral measures—used alone, and in conjunction with physiological measures—are an increasingly critical component in their conservation and welfare.

See also Dolphins—*Dolphin Behavior and Communication*
       Welfare, Well-Being, and Pain—*Behavioral Assessment of Animal Pain*
       Welfare, Well-Being, and Pain—*Behavioral Correlates of Animal Pain and Distress*
       Welfare, Well-Being, and Pain—*Psychological Well-Being*

Further Resources

Broom, D. M. & Johnson, K. G. 1993. *Stress and Animal Welfare*. 2nd edn. London: Chapman and Hill.
Constantine, R. & Yin, S. 2003. *Swimming with dolphins in New Zealand*. In: *Between Species: A Celebration of the Dolphin–Human Bond* (Ed. by T. Frohoff & B. Peterson), pp. 257–263. San Francisco: Sierra Club Books.

Pryor, K. & Norris, K. S. (Eds.) 1991. *Dolphin Societies: Discoveries and Puzzles.* Berkeley: University of California Press.

Small, R. & DeMaster, D. P. 1995a. *Survival of five species of captive marine mammals.* Marine Mammal Science, 11, 209–226.

Small, R. & DeMaster, D. P. 1995b. *Acclimation to captivity: A quantitative estimate based on survival of bottlenose dolphins and California sea lions.* Marine Mammal Science, 11, 510–519.

Southwest Fisheries Science Center, National Marine Fisheries, National Oceanic and Atmospheric Administration. 2002. *Report of the Scientific Research Program under the International Dolphin Conservation Program Act.* http://swfsc.nmfs.noaa.gov/idcpa/tunadol_rep/

St. Aubin, D. J. & Dierauf, L. A. 2001. *Stress in marine mammals.* In: *CRC Handbook of Marine Mammal Medicine* (Ed. By L. A. Dierauf & F. M. D. Gulland), pp. 253–260. 2nd edn. New York: CRC Press.

Sweeney, J. C. 1990. *Marine mammal behavioral diagnostics.* In: *CRC Handbook of Marine Mammal Medicine: Health, Disease, and Rehabilitation.* 1st edn. (Ed. By L. A. Dierauf), pp. 53–72. Boston: CRC Press.

Woodley, T. H., Hannah, J. L., & Lavigne, D. M. 1997. *A comparison of survival rates for captive and free-ranging bottlenose dolphins* (Tursiops truncates), *killer whales* (Orcinus orca) *and beluga whales* (Delphinapterus leucas). IMMA Technical Report No. 97–02.

Toni G. Frohoff

■ Welfare, Well-Being, and Pain
Veterinary Ethics and Behavior

Like human medicine, veterinary medicine in the twentieth century became subordinated to science. The veterinarian's oath, for example, commits a veterinary practitioner to "use my scientific knowledge and skills for the benefit of society," and virtually all veterinary school faculty must do scientific research. Although basing medicine on science provided a solid, empirical foundation for medicine and also provided myriad new tools and medications to practitioners, the commitment to science also generated some problems. In particular, medicine—human and veterinary—inherited from science most of science's fundamental assumptions and philosophical commitments. Among these assumptions was the belief that science was aphilosophical and did not examine its basic assumptions. Thus dubious foundational assumptions tended to be preserved and unquestioned, and to harden into an ideology immune from criticism.

Such debatable assumptions, presuppositional to science, yet taught as gospel along with the facts relevant to scientific disciplines, I have called *scientific ideology* or *scientific common sense*, for they are as unquestionable to scientists as ordinary common sense is to daily life.

For our purposes, three aspects of scientific ideology are relevant. One is a *reductionist bias*. Whereas Aristotle believed that the best science explains the world that we perceive, that maintains the qualitative distinctions given in common experience between hot and cold, pretty and ugly, wet and dry, and so forth, post-Newtonian science tends to favor atomic or molecular or at least biochemical explanations. Physics is perceived as the master science; all phenomena can ideally be explained by the laws of physics. Thus the key commonsensical qualitative distinction between living and nonliving matter loses its importance, and biology becomes a subspecies of physics and chemistry.

The rise of reductionism in science is clear in the study of biology. It is far easier to get funded if one is doing molecular biological research than if one is doing ecology or ethology. Many universities actually have two separate biology departments marking this distinction. At the University of Colorado (Boulder), for example, one department is called Ecology and Molecular, Cellular and Developmental Biology, and is very well-funded. The second, not so well-funded, is called Evolutionary Biology. Fifty years ago, one could study biology with a minimum amount of chemistry. This is no longer possible.

Historically, reductionism has had the effect of slighting the whole organism in favor of its components. Explanations in terms of functions, purposes, or even evolution are discarded in favor of mechanistic explanations. For this reason, reductionism ignores an animal's behavior until such time as behavior can be reduced to neurophysiology and thence to neurochemistry and molecular biology. Insofar as veterinary medicine represented applied mainstream biology, it also tended to brush aside behavior as a legitimate object of study. Most veterinary schools did not have full-fledged behavior classes until relatively recently. Sometimes when a school did have a behavior course, it was taught by people whose specialty was not behavior, very much analogous to courses in ethics, which were also often seen as something anyone who was interested could teach. When such courses were established, they tended to be perceived as something pressed on veterinary medicine by social change, rather than as integral to the proper practice of medicine.

A second component of scientific ideology that militated against taking behavior seriously was scientific agnosticism about the legitimacy of talking about subjective mental states in animals or, indeed, in humans. This was a very strange development in United States science in the early twentieth century, in light of the fact that virtually all biologists were Darwinian. Darwin himself had been unequivocal in his belief that mental traits—thoughts and beliefs—were as much a matter of evolution and phylogenetic continuity as were physical traits. Darwin in fact wrote an entire book about evolutionary continuity of the psychological, called *The Expression of Emotion in Man and Animals*, and studied in a serious experimental way the intelligence of earth worms!

While Darwin believed that mental states in animals were scientifically studiable, both by ingenious experimental design and by judicious use of anecdotal information, his position was essentially overruled by excisive or exclusionary approaches to science known as positivism and behaviorism, which dominated early twentieth century thought. *Positivism* aimed at separating the domain of scientific inquiry from the nonscientific by affirming that all legitimate scientific concepts had to be cashed out empirically. Since we cannot experience the mental states of humans or animals, they are outside of the realm of science. In psychology (human and animal), this component of scientific ideology appeared as *behaviorism*, a movement begun by psychologist J. B. Watson. Watson argued that psychology would make progress only if it studied learning, since animals (and humans) were blank slates upon which conditioning wrote. This movement dismissed both evolutionary bases of behavior and mental states from science. It was widely acceptable to Americans because it was highly compatible with American political ideology, which in turn believed that "Yankee knowhow" could improve both the world and human behavior even as Pavlovian psychology was congenial to Soviet Marxist ideology for similar reasons. Watson, and later B. F. Skinner, promised that behaviorism could rehabilitate criminals, improve education and thus the intellectual ability of children, and redeem a broad range of social misfits. Thus humans and animals became seen as behavioral machines, with thought, feelings, and emotions deemed irrelevant to scientific psychology.

One might be inclined to believe that ethical reflection on science might have tempered the agnosticism about thought and feeling we referred to. For example, the issue of hurting animals for human benefit might have forced thinking about—and study of—animal pain, suffering, and distress. It might have, but in fact did not, because of a third component of scientific ideology, namely the dogma that science was "value-free" in general and ethics-free in particular. This component was a logical consequence of positivism's emphasis on verifiability and data. If the only concepts and propositions allowed in science were those that were empirically verifiable; ethical notions were forced out by implication. As Ludwig Wittgenstein, an Austrian philosopher, once remarked, if one takes an inventory of all the facts in the universe, one does not find it a fact that killing is wrong.

The above constellation of scientific ideological dicta sufficed to eliminate behavior from scientific veterinary medicine. Reductionism focused on the neurophysiological, behaviorism on learning, and the claim that science was value-free militated against ethics focusing on morally relevant behavior. For example, the study of pain for most of the twentieth century was the study of nociception or painful stimuli and the "plumbing" or neurochemistry and circuitry of pain, not of felt pain evidenced by behavior.

The final piece of the puzzle regarding neglect of behavior by veterinary medicine also has to do with ethics. Though ordinary common sense certainly was not agnostic about animal thought and feeling the way science was, it had little stake in worrying about it. Animal value to society tended to be exhausted by animals' economic value. Veterinarians overwhelmingly served the agricultural industry. The value of agricultural animals was economic—they needed to work, reproduce, gain weight, or produce milk or eggs. Well over 90% veterinarians at mid-century served agriculture. Disease was what limited productivity, and thus veterinary medicine was limited to states relevant to profit and production. Behavior—even pain behavior—was largely irrelevant to animals' economic purposes and was ignored. Indeed, as late as 1972, when the first United States textbook of veterinary anesthesia was published, control of felt pain wasn't listed as a reason for anesthesia. Anesthesia was employed to protect the practitioner from animal thrashing, to protect the animal's limbs, to allow ready manipulation of the animal. Even today, some older practitioners still equate anesthesia with "chemical restraint," and some will do surgical procedures using paralytic drugs or "bruticane" (i.e., physical restraint).

Given the science we have described, and a world in which animal value is economic, why worry about animal behavior? Conceivably to train animals, but such training methods were long established, albeit often very brutal. Perhaps because behavior impaired the animals' function, for example, as a guard dog? In such cases, it was easier, cheaper, and more predictable to replace the animal than to modify its behavior: "I need a hunting dog—it does not have to be *this* hunting dog!" Treating or understanding behavior rarely came up, was thus outside the purview of veterinary medicine, and was not seen as relevant to animal health.

In my view, veterinary medicine began to concern itself with animal behavior as a result of significant social ethical change, which served to highlight concern for animals in themselves, not merely as tools. This was, in turn, caused by two distinct but related social movements. The first was the major rise in status of companion animals roughly in the 1960s. As mentioned earlier, until the 1960s, veterinary medicine largely served agriculture, working to keep food animals healthy and productive. Until that time, roughly 85–90% of veterinarians were in rural food animal practice. If we were to take a time machine to the beginning of the twentieth century, whether to urban or rural areas, and administer a word association test to the person in the street wherein we said "animal" and they were to announce what comes into their mind, the response would surely be "horse," "cow," "food," "work." Animals were largely perceived in their agricultural role. If one were to run a similar word association test today, the overwhelming majority of people would say "dog," "cat," or "pet." Indeed, according to repeated studies conducted at veterinary schools, well over 90% of the public consistently declare their pets to be "members of the family."

The reasons for this shift are not difficult to unearth. In the early 1900s, more than half the United States population was engaged in producing food for the whole population, including animal products. Thus the view of animals was largely utilitarian and economic. But as the twentieth century progressed, the number of people making their living directly from animals declined precipitously, with roughly only 1.6% of the public today engaged in production agriculture, and no more than half of those engaged in animal agriculture.

After World War II and the Dustbowl, there was a massive exodus from farm areas to the cities. Economic dependence on animals declined, and their major social role became that of companion. Life in the cities, as opposed to life in farm and rural communities became a matter of *gesellschaft*, not *gemeinschaft*. People didn't know their neighbors and didn't care to. One could live for years in an apartment building and never speak to—let alone get to know—the people in neighboring apartments, let alone neighboring buildings. Privacy lay in lack of social intercourse. If one saw someone fall on the street, one stepped over or around them. "Don't get involved," became the urban dwellers' mantra. Don't talk to strangers; don't even make eye contact, lest it be taken as a challenge. At the same time, the security of the nuclear family began to crumble with more than half of the marriages recorded ending in divorce.

In such a world, it is not easy to make friends, nor is it possible to seek succor from the natural world, for the little nature that exists in massive cities is contrived, not natural. I recall meeting a child from New York City, perhaps 10 years old, who firmly believed that concrete grew wild and that grass and trees had to be artificially placed. If one believes that people need both friends who provide love and companionship, and some contact with nonhuman nature, one can readily see why dogs and cats would start to fill that lacuna in the human soul. Companion animals became family, nature, excuses to exercise, protectors, and friends.

Even more dramatically, companion animals, particularly dogs, became lubricants for social interaction. An entire culture made up of "dog people" taking their dogs for exercise to the parks sprung up, with people interacting in virtue of their common interest in their pets. Oftentimes people met daily in virtue of dog-walking and struck up relationships with other "dog people" without even knowing their names, identifying them only as "Fifi's mom" or "Red's dad." Even more bizarrely, they began to care for each other through their animals. When Red's owner went into the hospital for major surgery, we all took turns walking Red, a huge German shepherd, passing the key around. When I was devastated by asthma, and threatened with a long hospital stay, Red's owner brought me an envelope in the park. "What's that?" I asked. "The key to my cabin in Thunder Bay and a map showing how to get there. Go there for a few weeks and breathe some clean air."

My own dog, a 160-pound Great Dane, loved all people. Children climbed on her back; old people hugged her. As a graduate student, I kept odd hours, often walking the dog late at night for miles. Most moving, I remember the prostitutes in the theater district running over to hug and kiss the dog, buying her donuts, she the only living creature they could shower with genuine love.

New York City may represent an extreme, but we have lost *gemeinschaft* in most communities, and our animals are sources of

The rise of concern for animal behavior by veterinarians has a lot to do with the popularity of companion animals that's been on the rise since World War II. Dr. Short created resuscidogs and resuscicats on which veterinary students could practice CPR.
© Owen Franken / Corbis.

unconditional love. They play with and protect children and, as Dr. Leo Bustad recognized when he forced through the United States Congress Federal legislation allowing elderly people to keep animals in federally funded housing projects, they give old people a reason to keep living. Divorce lawyers have told me of divorces where the major issue is who gets custody of the family pet.

In such a world, people care more about their animals' behavior, not only intrinsically, but pragmatically. "I will not be able to keep my dog in my apartment if he barks." "I cannot have him tearing up the apartment during the day or snapping at strangers on the street." Euthanasia for behavior problems became the single largest cause of death for pet animals.

On the veterinary side, pet practice eclipsed all other areas of veterinary medicine. People love their animals, and will pay a great deal to keep them well and happy. As early as 1980, the Colorado State University oncology service had people bringing in their animals from thousands of miles away, and spending six figures on cancer treatment. Economics was no longer a constraint on spending for animal health, and people with no animal experience (and many with animal experience) needed help with their pets' behavior and expected their vets to be knowledgeable. To some extent, the reluctance to teach behavior in veterinary schools lost a portion of a very lucrative market to animal behaviorists who hung out a shingle, often without quality control. Eventually, veterinary schools were forced to start teaching behavior, and to start treating behavior problems, for the sake of both owners and animals. One of my ex-veterinary students has literally become a millionaire by setting up a boarding kennel with a staff of behaviorists to deal with training and behavior while owners are away. Tellingly, one of the first animal behavior services was set up at the prestigious Animal Medical Center in New York. Inevitably, owners' concerns forced veterinarians to look not only at behavior for corrective purposes, but also as signs of mental states like pain, fear, boredom, and anxiety.

Roughly contemporaneous with the rise of social concern for companion animals was the rise of a new and highly vigorous social concern for the welfare of all animals used by human society for human benefit. This is reflected not only in the rise of animal ethics works in the 1970s, arguing that animals are morally entitled to far better treatment than we give them, but in social dismay at the practices prevalent in such animal uses of "factory farming"; wildlife management; animal research and testing; animal shows, rodeos and circuses; hunting, fishing and trapping; animal training; sports; and science teaching and training. Numerous legislative initiatives across the world have addressed these practices with virtually every civilized country in the Western world now having, for example, laws constraining the use of animals in research. In 1988, Sweden banned factory farming in a series of timed phaseouts, and the rest of Europe is following suit. The management of wildlife by wildlife professionals accustomed to serving hunters, fishermen, and trappers has been seen as so out of phase with public concern about animal treatment that a wildlife specialist recently warned of "management of referendum" if the wildlife managers don't become more attuned to social ethics. (Steel-jawed traps, mountain lion hunting, and spring bear hunts where there is a danger of shooting lactating mothers and thereby orphaning cubs who starve to death, have all been abandoned in many jurisdictions by virtue of referenda.) Brutal horse training has been curtailed, and dog and horse racing, highly abusive in many ways, are being abandoned. Zoos as prisons have all but vanished.

Many animal users derisively dismiss this tide of social concern as anthropomorphic nonsense that people attribute to their pets and thence to all animals, wild and domestic. In my view, this is far too simplistic an explanation, but it is certainly correct that society now has no patience for scientific ideology's denial of the legitimacy of talking about animal consciousness, and is particularly focused on animals suffering pain, distress, fear, boredom, loneliness, anxiety, and so on. While society seems to accept the continued use of animals for human benefit in research and agriculture, it will not accept the suffering of these animals. The public wishes to see animals get the best possible treatment commensurate with their use, and seems willing to abandon certain uses if they cannot be shorn of that suffering.

The social concern for animal treatment briefly chronicled is accompanied by an insatiable public demand for knowledge of animal behavior and animal thought. (The public does not distinguish.) In 2002, a *New York Times* reporter told me that the issue occupying the most time on New York cable television channels was animal behavior and animal thought. And these subjects are not only valued for their intrinsic interest, but for their role in promoting animal welfare. The 1985, laboratory animal laws mandating control of pain and distress in research animals and accommodations for primates that "enhance their psychological well-being" both attest to social concern with animal behavior as a key to animal feeling, as does the growing international demand for farm animal facilities that allow these animals to fulfill the needs, physical and psychological, dictated by their natures (what I have called their *telos*).

All of this has (slowly to be sure) led or will lead veterinary medicine to study and master animal behavior as a way of understanding how we meet animals' psychological as well as physical needs, and alleviate the noxious states lumped together as "distress." For 15 years, the USDA, which enforces the United States laboratory animal laws, focused on control of pain, until that ideological barrier was breached, and veterinary medicine has a reasonable purchase on recognizing and alleviating animal pain. Now the USDA is turning its attention to distress, sensing that the veterinary and research communities have, under public ethical pressure, backed off from their resistance to attributing mental states to animals, states which are accessed largely by way of the animals' behavior. Françoise Wemelsfelder's masterful treatment of animal boredom in my book on laboratory animal care, *The Experimental Animal in Biomedical Research*, Volume I, is an exemplar for the scientific community as to how scientists can access mental states in animals by way of behavior understood through a carefully constructed set of categories. The same point is true regarding farm animals, and the systems respecting the animals' physical and behavioral needs that society is now increasingly demanding.

Thus, in my view, socioethical and valuational changes are driving the relevance of behavior and animal mentation (mental activity) for veterinary medicine. The rise of pets as objects of love and valued members of families has forced veterinary medicine to worry about aberrant behavior as a clinical issue, and also about behavior as signs of morally relevant subjective states such as pain, fear, and anxiety. No single paradigm for treating behavior in animals is yet prevelant in veterinary medicine—some veterinarians use learning, others use psychoactive drugs in many cases drawn from human medicine, and most use combinations of modalities.

Second, social concern for animal treatment has forced veterinarians to study behavior so that they can judge the animals' experiential life that is the basis for positive or negative animal welfare. If, for example, farmers must rid themselves of small sow stalls, 7 ft × 2 ft × 3 ft (2.1 m × .6 m × .9 m), as a way of producing swine, and if veterinarians are to help their clients, they must become knowledgeable in farm animal behavior and psychology. If we are to handle animals without traditional brutal methods such as whips, hotshots, and harsh bits, we must again know animal behavior and mentation. (Popular cultural works like *The Horse Whisperer* have soundly belied the need for superior force in managing large animals.) If we are to treat pain and suffering, we must know their often highly subtle and species-specific behavioral manifestations.

As I wrote more than 20 years ago in *Animal Rights and Human Morality*, ethics and behavior must work hand in hand to guide us in eliminating animal pain and suffering *and* in attempting to achieve animal happiness through the satisfaction of animal physical and psychological needs. And nowhere is this requirement more evident daily than in veterinary medicine, the applied science that has as its aim fostering the health and well-being of the animals we depend on for our own health and well-being.

See also Welfare, Well-Being, and Pain—*Behavioral Assessment of Animal Pain*

Welfare, Well-Being, and Pain—*Behavioral Correlates of Animal Pain and Distress*

Welfare, Well-Being, and Pain—*Psychological Well-Being*

Further Resources

Rollin, B. E. 1992. *Animal Rights and Human Morality*. 2nd edn. Amherst, NY: Prometheus Books.
Rollin, B. E. 1995. *The Unheeded Cry: Animal Consciousness, Animal Pain and Science*. Rev. edn. Ames: Iowa State University Press.

Bernard E. Rollin

■ Welfare, Well-Being, and Pain
Wildlife Trapping, Behavior, and Welfare

More animals are trapped in the United States than in any other nation. Roughly 3-5 million animals are trapped and killed in the United States annually by commercial and recreational trappers (Fox 2004a). Millions more are trapped and killed by wildlife damage and predator control trappers, researchers, and wildlife managers. Animals are also trapped and released as part of reintroduction and relocation programs. Our understanding of the full impact of trapping on individual animals, population dynamics, or ecosystem health, however, is strikingly deficient. Notably, there are few comprehensive assessments of the effects that trapping has on animal behavior and physiology, and wildlife population dynamics.

A survey of the literature on trapping and its effects on animal behavior follows, with an accompanying list of the relevant literature.

The paucity of research on the effects of trapping on animal behavior and welfare reflects fundamental flaws and political bias in current trap testing programs and the development of national and international mammal trap standards (Fox 2004b). For example, the United States government is currently conducting a national "Best Management Practices" (BMP) trap-testing program to test leghold traps and other restraining traps. The BMP program was implemented as a result of pressure from the European Union to prohibit use of leghold traps in those countries that still allow their use. Instead of banning leghold traps, however, the United States government agreed to conduct a national trap-testing program of traditional restraining traps. According to Tom Krause, editor of *American Trapper* magazine, one of the stated goals of the BMP program is to "maintain public acceptance" of trapping. However, while injury rates, capture efficiency, and selectivity are part of the testing protocols, behavioral and overall physiological analyses are not. Previous studies that have considered the behavioral and physiological responses of animals caught in traps have shown that the trauma of being caught in a trap can alter the behavior of released animals and reduce survival rates (Redig 1981; Bortolotti 1984; Proulx 1999) and disrupt the social dynamics of territorial species (Hornocker & Hash 1981; Banci & Proulx 1999). That behavioral and physiological assessments are not part of the BMP trap testing protocols suggests that trapping proponents are unwilling to conduct comprehensive evaluations of traditional trapping devices for fear that such information could challenge the status quo and require that wildlife management agencies question the appropriateness of certain trap types and trapping practices.

The Impacts of Trapping

To date, research on trapping in the United States has focused primarily on trap injury rates, selectivity, and efficiency. In an effort to standardize the assessment of the injuries caused by body-gripping traps, several injury or "trauma" scales have been developed to quantify trap-induced injuries in restraining traps and time to unconsciousness and death in killing traps. B. F. Tullar (1984) and G. H. Olsen et al. (1998) evaluated injuries to the limbs of animals caught in traps. D. K. Onderka & colleagues (1990), and G. F. Hubert & colleagues (1996, 1997) attempted to consider all physical trap-related injuries by conducting whole-body necropsies (autopsies). Hubert et al. (1997) found whole-body scores were approximately 15% higher than leg injury scores and noted that previous "studies fail[ed] to tabulate all trap-related injuries that should be assigned to an individual [animal]." They recommended "whole-body necropsies should be conducted to insure no type of trap-related physical trauma is overlooked." Physical trauma, however, is not the only measurement of trap impact. Psychological distress (i.e., fear), physiological stress, and pain can also be observed and assessed in trapped animals through behavioral analyses and stress-related hormonal and blood-cell measurements. However, at present no scoring system for restraining traps integrates physical injuries with behavioral and physiological responses (Proulx 1999). Without such analyses, no comprehensive evaluation can be made of the full impact of trapping on animals or the dynamics of wildlife populations.

Leghold Traps

By intentionally underestimating the adverse effects of traps on animals, use of inhumane traps can be more easily justified by those with a vested interest in ensuring their continued use. The steel-jaw leghold trap—a device condemned as inhumane by the American Veterinary Medical Association, the American Animal Hospital Association, the World Veterinary Association, and the National Animal Control Association—is still the most widely used trap in the United States today (IAFWA 1993). Leghold traps can cause severe swelling, lacerations, joint dislocations, fractures, damage to teeth and gums, limb amputation, and death (Proulx 1999, Papouchis 2004). Many injuries result from the animal's struggle to escape, while others are incurred from the clamping force of the trap's metal jaws onto the animal's limb. Steel (unpadded) leghold traps have been shown to cause significant injuries for a number of commonly trapped species and generally fail to meet basic trap standards with regard to injuries, according to Dr. Gilbert Proulx of Alpha Wildlife Research and Management Ltd. These traps are widely used by the United States government in its federal predator control programs (Fox & Papouchis 2004). While more than 80 countries have banned the controversial device, leghold traps remain legal in most U.S. states and public land systems. In 1995, the European

Union banned the use of leghold traps in member states and sought to bar the import of furs from countries still using these traps. The United States—one of the world's largest fur producing and consuming nation—continues to defend commercial fur trapping and the use of the leghold trap, and even threatened the European Union with a trade war if it prohibited the importation of fur from countries allowing the use of leghold traps (Fox 2004b).

Kill-Traps

In addition to leghold devices, kill-traps are commonly used by wildlife managers and commercial fur trappers throughout North America. Kill-traps (also called rotating-jaw traps) have been shown to cause extreme trauma, pain, and stress to trapped animals (Gilbert 1981; Waller 1981; Proulx 1999). Conibear traps and other common models of kill-traps may not cause instant death because of the numerous variables needed to produce a killing blow to the neck or head (i.e., correct sized animal entering the trap at the correct angle and speed) (Gilbert 1981; Waller 1981; Proulx 1999). Proulx reviewed the most popular models of kill traps and concluded that standard rotating-jaw traps do not meet basic quick-kill trap standard criteria, which he defined as traps that result in equal to or more than 70% of animals being irreversibly unconscious within less than or equal to 3 minutes (with a 95% confidence level).

Drowning Traps

Although few studies of kill-traps conducted in the United States have included comprehensive trap impact assessments, several studies conducted outside of the U.S. have analyzed the behavioral, physiological, pathological, and/or clinical responses of trapped semi-aquatic mammals in drowning sets. Most of these studies have been conducted in Canada and other countries where trap researchers are often more independent and less influenced by political lobbies than in the United States. Using killing traps underwater reduces their efficiency so that when the strike is of insufficient strength or improperly placed to kill the animal, they act as restraint devices and death is caused by drowning. Leghold and submarine traps act by restraining the animals underwater until they drown. Most semi-aquatic animals, including mink, muskrat, and beaver, are adapted to diving by means of special oxygen conservation mechanisms. The experience of drowning in a trap must be extremely terrifying: Animals have displayed intense and violent struggling and were found to take up to 4 minutes for mink (Gilbert and Gofton 1982), 9 minutes for muskrat, and 10–13 minutes for beaver (Gilbert and Gofton 1982) to die. Mink have been shown to struggle frantically prior to loss of consciousness, an indication of extreme trauma. Because most animals trapped in aquatic sets struggle for more than 3 minutes before losing consciousness, Gilbert Proulx (1999) concluded that they did not meet basic trap standards and could therefore not be considered humane.

Minimizing Impacts of Trapping in Wildlife Research

Whether trapping animals for scientific research, relocation, or reintroduction programs, wildlife researchers and managers require state-of-the-art, humane, live traps. They need to know, for example, if a particular trap type may negatively alter an animal's behavior after they are released. Roger A. Powell and Gilbert Proulx (2003) have argued that researchers should choose traps that minimize pain, stress, and discomfort if for no other reason than to minimize the effect on the behavior and survival of animals, which ultimately affect research results.

Until completely unobtrusive techniques can be developed to study wildlife, the ability to capture and safely release animals will remain critical. Animals captured in box and cage traps generally experienced fewer traumas than those captured in limb-holding traps (White et al. 1991; Powell and Proulx 2003; Seddon et al. 1999, Warburton et al. 1999, Kolbe et al. 2003). Seddon et al. (1999) found that only 8% of Ruppell's foxes (*Vulpes rueppellii*, Saudi Arabia) captured in padded leghold traps were alive 6 months later, compared to 48% of foxes caught in cage traps. The authors concluded that:

> [E]ven apparently minor injuries assessed at the time of release may result in lameness which, directly through increase risk of predation, or indirectly through reduced ability to find food in combination with possible stress of capture, could significantly reduce the likelihood of survival. (Seddon et al. 1999, p.76)

Nontarget animals trapped in leghold traps and then released may be so severely injured that they are unable to survive in the wild. P. Redig (1981) reported that 21% of the bald eagles admitted to the University of Minnesota Raptor Research and Rehabilitation Program over an 8-year period had been caught in leghold traps (Redig 1981). Of these, 64% had sustained injuries that proved fatal. Oftentimes, trap-related injuries may be internal and therefore less readily apparent. Furthermore, the somatic and psychological stress that can result from trapping wild animals can suppress their immune systems and significantly compromise their post release recovery (Jordan 2001).

Studies have shown that the trauma of being caught in a trap can alter the behavior of released animals, reduce survival rates, and disrupt the social dynamics of territorial species. This red fox was caught as part of a relocation program for living too close to humans.
© Joel Sartore / National Geographic Image Collection.

Trapping can alter the behavior of territorial species with well-developed social systems. In their study of wolverines in Montana, M. G. Hornocker and H. S. Hash (1981) suggested that intensive trapping contributed to behavioral instability and home range overlap among resident adults. Research is needed to determine the impacts of trapping on the behavior of other sensitive and socially developed species, including wolves, fisher, marten, and lynx.

As animal ethologists and ethicists continue to demonstrate the cognitive, emotional, and behavioral similarities between humans and other animals (Fox 2001), it will become increasingly difficult to justify continued testing and use of traps known to inflict fear, pain, and suffering on wildlife. Ideally, in the field of wildlife research, trapping will be replaced with less invasive methods that preclude the need for trapping. Track plates, hair traps, remotely triggered cameras, and DNA hair testing offer noninvasive alternatives to trapping. When trapping is necessary, researchers should ensure that traps minimize physical injury as well as behavioral and physiological stress. Researchers must also be aware that when they conduct what appears to be benign, least-invasive research that involves trapping, there may be post-release impacts that affect the individual animal(s), and ultimately their research results (Powell & Proulx 2003).

Wildlife Trapping: Resources

Banci, V. & G. Proulx. 1999. *Resiliency of Furbearers to Trapping in Canada*. In: *Mammal Trapping*. (Ed. by G. Proulx), pp. 1–46. Sherwood Park, Alberta, Canada: Alpha Wildlife Research and Management Ltd.

Bortolotti, G. R. 1984. *Trap and poison mortality of golden and bald eagles*. Journal of Wildlife Management, 48, 1173–1179.

Colleran, E., Papouchis, C., Hofve, J. & Fox, C. 2004. *The Use of Injury Scales in the Assessment of Trap-Related Injuries*. In: *Cull of the Wild: A Contemporary Analysis of Wildlife Trapping in the United States* (Ed. by C. H. Fox & C. M. Papouchis), pp. 55–59. Sacramento, CA: Animal Protection Institute.

Fox, C. H. 2004a. *Trapping in North America: A Historical Overview*. In: *Cull of the Wild: A Contemporary Analysis of Wildlife Trapping in the United States* (Ed. by C H. Fox & C. M. Papouchis), pp. 1–12. Sacramento, CA: Animal Protection Institute.

Fox, C. H. 2004b. *The Development of International Trapping Standards*. In: *Cull of the Wild: A Contemporary Analysis of Wildlife Trapping in the United States* (Ed. by C. H. Fox & C. M. Papouchis), pp. 61–69. Sacramento, CA: Animal Protection Institute.

Fox, C. H. & Papouchis, C. M. 2004. *Coyotes in our Midst: Coexisting with an Adaptable and Resilient Carnivore*. Sacramento, CA: Animal Protection Institute.

Fox, M. W. 2001. *Bringing Life to Ethics: Global Bioethics for a Humane Society*. Albany, NY: New York State University Press.

Gilbert, F. F. 1981. *Maximizing the humane potential of traps—the Vital and the Conibear 120*. In: *Proceedings of the Worldwide Furbearer Conference* (Ed. by J. A. Chapman & D. Pursley), pp. 1630–1646. Frostburg: University of Maryland.

Gilbert, F. F. & Gofton, N. 1982. *Terminal dives in mink, muskrat and beaver*. Physiology & Behavior, 28, 835–840.

Hornocker, M. G., & Hash, H. S. 1981. *Ecology of the wolverine in northwestern Montana*. Canadian Journal of Zoology, 59, 1286–1301.

Hubert, G. F., Hungerford, L. L. & Bluett, R. D. 1997. *Injuries to coyotes captured in modified foothold traps*. Wildlife Society Bulletin, 25, 858–863.

Hubert, G. F., Hungerford, L. L., Proulx, G., Bluett, R. D. & Bowman, L. 1996. *Evaluation of two restraining traps to capture raccoons*. Wildlife Society Bulletin, 24, 699–708.

International Association of Fish and Wildlife Agencies Fur Resources Technical Committee. 1993. *Ownership and Use of Traps by Trappers in the United States in 1992*. Fur Resources Washington, D.C: Technical Committee of the International Fish and Wildlife Agencies and the Gallup Organization.

Jordan, W. J. 2001. *Veterinary and Conservation Aspects of Trapping*. From: www.infurmation.com

Kolbe, J. A., Squires, J. R. & Parker, T. W. 2003. *An effective box trap for capturing lynx*. Wildlife Society Bulletin, 31, 980–985.

Olsen, G. H., Linhart, S. B., Holmes, R. A. Dasch, G. J. & Male, C. B. 1986. *Injuries to coyotes caught in padded and unpadded steel foothold traps*. Wildlife Society Bulletin, 14, 219–223.

Olsen, G. H., Linscombe, R. G., Wright, V. L. & Holmes, R. A. 1988. *Reducing injuries to terrestrial furbearers by using padded foothold traps*. Wildlife Society Bulletin, 16, 303–307.

Onderka, D. K., et al. 1990. *Injuries to Coyotes and Other Species Caused by Four Models of Footholding Devices*. Wildlife Society Bulletin, 18, 175–181.

Papouchis, C. M. 2004. *Trapping: A Review of the Scientific Literature*. In: *Cull of the Wild: A Contemporary Analysis of Wildlife Trapping in the United States* (Ed. by C. H. Fox & C. M. Papouchis), pp. 41–45. Sacramento, CA: Animal Protection Institute.

Powell, R. A., & Proulx, G. 2003. *Trapping and Marking Terrestrial Mammals for Research: Integrating Ethics, Performance Criteria, Techniques, and Common Sense*. ILAR Journal, 44, 259–276.

Proulx, G. 1999. *Review of current mammal trap technology in North America*. In: *Mammal Trapping* (Ed. by G. Proulx), pp. 1–46. Sherwood Park, Alberta, Canada: Alpha Wildlife Research and Management Ltd.

Redig, P. 1981. *Significance of trap-induced injuries to bald eagles*. In: *Eagle Valley Environmental Technical Report BED 81*, pp. 45–53. St. Paul: University of Minnesota.

Seddon, P. J., VanHeezik, Y. & Maloney, R. F. 1999. *Short- and medium-term evaluation of foothold trap injuries in two species of fox in Saudi Arabia*. In: *Mammal Trapping* (Ed. by G. Proulx), pp. 67–78. Sherwood Park, Alberta, Canada: Alpha Wildlife Research and Management Ltd.

Tullar, B. F. 1984. *Evaluation of a Padded Leg-Hold Trap for Capturing Foxes and Raccoons*. New York Fish and Game Journal, 31, 97–103.

Waller, D. J. 1981. *Effectiveness of kill-type traps versus leg-hold traps utilizing dirt-hole sets*. Proceedings of the Annual Conference of the Southeastern Association Fish & Wildlife Agencies, 35, 256–260.

Warburton, B., Gregory, N. & Nunce, M. 1999. *Stress response of Australian brushtail possums captured in foothold and cage traps*. In: *Mammal Trapping* (Ed. by G. Proulx), pp. 53–66. Sherwood Park, Alberta, Canada: Alpha Wildlife Research and Management Ltd.

White, P. J., Kreeger, T. J., Seal, U. S., & Tester, J. R. 1991. *Pathological response of red foxes to capture in box traps*. Journal of Wildlife Management, 55, 75–80.

The Future of Trapping and Wildlife Research

For an activity that affects millions of wild animals each year, it is astounding that so little is known about the full impact of trapping on individual animals, wildlife populations, and ecosystem health. Political forces and powerful lobbies have greatly influenced trapping research, especially in the United States where commercial fur trapping and predator control trapping are considered "sacred cows." Trap researchers and wildlife management agencies should establish research protocols that ensure that behavioral and welfare parameters are included in trap research, and standards should be developed that adequately measure all trapping related impacts. Traps that fail to meet these standards should be immediately prohibited. By resisting and undermining efforts to reduce adverse effects of trapping, wildlife management agencies and trap researchers open themselves to public and scientific criticism and will undoubtedly face increasing pressure to address these issues. Ultimately, as society places greater value on wildlife and the humane treatment of all animals, use of traps and other management methods known to harm individual animals, wildlife populations, and ecosystem health will no longer be condoned.

See also Human (Anthropogenic) Effects—*Human (Anthropogenic) Effects on Animal Behavior*
Welfare, Well-Being, and Pain—*Behavioral Assessment of Animal Pain*

Camilla H. Fox

■ Whitman, Charles Otis
(1842–1910)

Charles Otis Whitman (1842–1910) was a distinguished American biologist who believed that the study of an animal species should include everything about the animal's life, not just, for example, its anatomy, its physiology, or its genetics. His approach is epitomized by the claim that his students greeted each other not with the question, "What is your special field?," but rather, "What is your beast?"

Whitman headed the Division of Biology at the newly founded University of Chicago. He also directed the Marine Biological Laboratory at Woods Hole, Massachusetts. He served in these influential capacities at a time when the experimental laboratory was coming to be the very symbol of modern biology. He did not feel, however, that the laboratory was a sufficiently ample setting for promoting biology as he conceived of it. He envisioned biology as "experimental natural history," and to this end he recommended establishing special experiment stations—"biological farms," as he called them—where "the study of life histories, habits, instincts and intelligence" could be conducted in conjunction with "the experimental investigation of heredity, variation, and evolution."

Whitman had several different sorts of "beasts" that he favored in his own research over the course of his career. In the last decade and a half of his life, he concentrated primarily on pigeons. He sought to reconstruct the evolutionary history of pigeons through a painstaking study of their heredity, variation, and development. He expected this work would simultaneously shed light on the mechanisms of heredity, variation, development, and evolution. One part of his pigeon studies was the comparative study of the birds' calls and instinctive behavior patterns.

In his work on the calls and instincts of pigeons, Whitman was putting into practice the idea for which he is best known today among ethologists. This is the idea that animal instincts can be used just like the study of animal structures to reconstruct the evolutionary histories of related species. He initially set this claim forth in a lecture entitled "Animal Behavior" that he delivered at Woods Hole in the summer of 1898. There he stated: "Instinct and structure are to be studied from the common standpoint of phyletic descent."

Whitman intended to write a monograph on the behavior of pigeons after he wrote up his analyses of the birds' heredity and evolution. He died, unfortunately, before completing any of these projects. He died from pneumonia, which he developed after spending a December afternoon outside in the cold, transferring his beloved pigeons to their winter quarters.

Whitman's comparative approach to the study of animal behavior was taken up by his American pupil, Wallace Craig, and later, most famously, by the Austrian ethologist Konrad Lorenz. The latter liked to call Whitman's comparative study of instincts the "Archimedean point" from which the modern science of ethology began.

See also Craig, Wallace (1876–1954)
History—*History of Animal Behavior Studies*
Lorenz, Konrad Z. (1903–1989)

Further Resources

Burkhardt, R. W. Jr. 2005. *Patterns of Behavior: Konrad Lorenz, Niko Tinbergen, and the Founding of Ethology*. Chicago: University of Chicago Press.

Richard W. Burkhardt, Jr.

■|Wildlife Management and Behavior

Wildlife management is the set of coordinated actions taken to control the size, composition, distribution, and behavior of wildlife populations. The behavior of animals is linked to wildlife management in three critical ways. First, the behavior of a species determines its ability to survive and thrive in an environment that is increasingly dominated by human activities. Second, wildlife behavior determines whether management is needed or not. Finally, understanding wildlife behavior is the essential ingredient for wildlife management success—whether attempting to reduce overabundant wildlife populations through hunting, or discouraging a family of squirrels from invading the attic.

For much of the twentieth century, wildlife management in North America was dedicated to the restoration of the many game species that had nearly been annihilated by unrestricted hunting and trapping. Massive killing of wildlife in the 1800s was driven mainly by economic demand—for feathers and beaver pelts for hats; skins for clothing, blankets, and rugs; and meat for restaurants in the city and railroad workers on the plains. Beginning around 1900, the United States banned the sale of most wildlife products, established strict limits on hunting seasons and the number of animals that could be taken, and aggressively transplanted wildlife from successful populations into unoccupied areas. Most of these efforts were funded directly or indirectly by sport hunters, linking wildlife management for the purposes of population restoration to wildlife management for the purposes of sport hunting.

However, the success of wildlife restoration efforts brought about unexpected new conflicts between some wildlife species and human needs. These conflicts also grew because of fundamental changes in land use. Notably, the regrowth of forests on abandoned farmlands and the spread of suburbs with their networks of parks and other greenspaces have brought wildlife and people together and into conflict.

At the same time that changes in land use were helping fill the human environment with species such as deer, pigeons, and beaver, other species became scarce. Thus, modern wildlife management remains deeply concerned with the restoration of imperiled species.

Whether a species is rare or abundant is a matter of environment, ecology, human attitudes, and animal behavior. From a behavioral standpoint, for example, species that range over large areas of undisturbed habitat (such as grizzly bears and other large carnivores) often become threatened or endangered when wild habitat is transformed by agriculture or development.

Perhaps most importantly, wild animals that have become abundant in human-dominated environments show great flexibility in their behavior, and an enormous ability to learn. The ever-present raccoon is equally at home in a hollow tree, under a porch, or in a chimney, and feeds on whatever he can catch or pry open. The stealthy, but widespread, great horned owl nests in hollow trees, old crows' nests, in caves, on tree stumps, or on the ground, in forest, parkland, steppe, or swamps.

On the other hand, behavioral inflexibility, the inability to learn new behaviors or to utilize a variety of food sources or habitat types, can lead to scarcity or extinction. The Kirtland's warbler, a small blue-gray and yellow migratory songbird, chooses to nest only on the ground under low-hanging branches of 5–12-year-old jack pines. Whereas such stands of young jack pines were widespread in the upper Midwest 200 years ago, now only intensive forest management, consisting of repeated cutting, burning, and replanting large pine stands, protects this highly endangered bird from extinction.

As the examples above indicate, however, not all abundant wild species cause conflicts with humans, becoming nuisances or pests. Animals are not nuisances; it is their behavior that can make them nuisances. Almost everyone likes beaver, but the sophisticated dam-building behavior for which they are often admired may anger suburban residents when the resulting dams flood suburban yards and basements. A few crows cawing overhead or picking at a road-killed squirrel are not a nuisance. However, aggregations of hundreds of crows feeding together in a cornfield, or thousands roosting together in a suburban woodlot, may well draw human displeasure. Naturally, animals that tend to aggregate in large flocks or herds can exert impacts on human activities that most people will not tolerate, whereas species that prefer a solitary life or regulate their own numbers (for example, through defense of territories from which nonrelatives are excluded) likewise limit their own impacts.

Examining Canada geese is especially instructive in discovering the relationship between behavior and the creation of conflicts between humans and wildlife. First, geese show the behavioral flexibility that allows wild species to thrive in the human environment. The favored foraging habitats of geese are broad, open, short-grass meadows in which are embedded ponds or lakes—which perfectly describes the campuses, industrial parks, recreational areas, and open spaces that humans create for themselves through much of the world. Although some populations of geese migrate, some do not; where they can find suitable habitat year-round, many birds may breed and winter in the same area. They quickly identify and utilize new feeding areas and food sources, and quickly learn to approach people who offer them bread or other foods. They nest on islands, in courtyards, and on rooftops. And they aggregate in flocks of hundreds or even thousands, dropping enormous amounts of fecal material in the landscapes so favored by humans.

Three approaches are taken when building comprehensive solutions to wildlife conflict management. First is deterring or discouraging animals from engaging in the behaviors that cause conflicts, or at least modifying these behaviors such that conflicts become tolerable. Second is regulating animal numbers; fewer animals can (but do not necessarily) lead to fewer conflicts. The last, and possibly most important, is encouraging changes in human behavior that reduce the potential for conflict, largely by increasing tolerance for (and interest in!) wildlife through education about wildlife behavior.

Eliminating or modifying animal behaviors that cause conflicts usually requires a fairly sophisticated understanding of those behaviors. Dam-building by beaver is a good example. Dams can be knocked down, or beaver trapped and removed, but neither method works for long. Beaver quickly repair damaged dams, and suitable habitat from which they are removed will be recolonized by dispersing beaver within a few months to a few years. The most successful devices for resolving conflicts with beaver are water flow devices, pipes that are passed through the dam that lower water levels to acceptable levels. Since beaver are highly sensitive to the sound and feel of flowing water, however, a simple pipe will not do: They will simply dam the pipe if they can. The upstream end of the pipe must be fenced off or multiple holes drilled along the length of the pipe.

Most birds and mammals quickly habituate to (learn to ignore) otherwise harmless stimuli intended to scare them off. Scarecrows do not scare crows, nor are pigeons and starlings fooled by plastic owls positioned on rooftops. Noisemakers do not frighten away deer for long. (Deer carelessly graze on the grass between runways of major airports!) Thus, discouraging wild animals from entering places they are not wanted requires knowledge of animal locomotion patterns, perception, antipredator behavior, and feeding behavior. Deer, for example, are great leapers but have poor depth perception, so they are reluctant to jump a fence which is erected on a slant. Because deer will return repeatedly to preferred food sources, chemical repellents work best if applied to a shrub before the deer has had a chance to sample and develop a taste for it. Geese are best discouraged by changing landscaping. Planting shrubs interferes with their ability to see potential predators and blocks escape routes to water; allowing grass to grow taller also blocks their vision, and makes the grass itself less appealing as a food source.

The control of animal numbers as a wildlife management objective has become increasingly controversial in recent years. Such control is usually carried out by some form of regulated public hunting which, as noted above, was the principal management tool and the economic foundation for game conservation efforts for most of the twentieth century. However, some people now object to public hunting on ethical grounds; others feel it is unsafe or inappropriate, especially when carried out near urban or suburban residences or in parks heavily used by the public.

A general knowledge of the behavior of game animals has always lay at the heart of game management, and many hunters (among them, Theodore Roosevelt) have written detailed descriptions of the behavior of their prey animals. The success and safety of the hunter has always depended on his knowledge of the habits of game animals, their daily activity patterns, movements, ability to detect danger, and responses to danger. Hunters lure game into range with a variety of decoys, calls, scents and other tricks that manipulate the social, communication, and mating behavior of their target species.

Regulation of hunting also requires knowledge of behavior. Hunting seasons are generally set to avoid interfering with the birth and rearing of young. For waterfowl, deer, and indeed most game animals in North America, most public hunting is carried out in autumn and early winter, when the young born the previous spring have matured enough to survive on their own. Among waterfowl, the autumn hunt also focuses mortality on first-time migrants, which may in any case be less likely to survive the winter.

For many years, hunting of deer and elk has targeted males, both for cultural reasons (antlers are desirable trophies) and because this gender bias encourages population growth. In these polygynous species, males are not involved in raising offspring and each male can mate with many females. Consequently, the number of offspring produced depends largely on the number of females reproducing.

Another implication of a polygynous mating system, however, is that any wildlife management action intended to reduce populations must focus on removing females. In areas where deer have been perceived as overabundant, this shift in management tactics is now occurring, although slowly.

The work of Louis Verme, John Ozoga, Karl Miller, and others is also demonstrating that the practice of hunting can change behavior and social organization in wildlife.

In white-tailed deer, for example, hunting that focuses on males may extend breeding seasons. This effect has been attributed both to the scarcity of males, and to the scarcity of the male visual and olfactory signals that stimulate females to become sexually receptive. Generally, intensive hunting pressure on both sexes may permit or stimulate reproduction at early ages.

Research by William Porter, Nancy Mathews, and others has shown that, in the absence of hunting, young female white-tailed deer establish permanent home ranges overlapping with or adjacent to those of their mothers. Consequently, female relatives commonly maintain long-term spatial and social associations. (This phenomenon may be hidden in heavily hunted populations, where females rarely live long enough to encounter their granddaughters.) Interestingly, when such matrilocal groups of deer were experimentally removed, their home ranges were not recolonized by other female deer for several years.

If widespread, this phenomenon has profound implications for deer management. First, it implies that deer numbers can be reduced locally. This may be especially helpful in managing conflicts with deer populations in suburbs, where females typically display very small home ranges (1.5 km^2 or .6 mi^2 or less). The other side of the local management coin, however, is that removing deer from a suburban park will not help reduce the number of deer–vehicle collisions on a highway 3–4 km (2–2.5 mi) away.

In recent years, there has been much interest in the use of birth control as an alternative to killing for controlling the size of wild populations, especially on islands and in cities, towns, and suburbs. For monogamous wildlife species, where both parents are involved in all stages of reproduction (such as most geese, pigeons, and some wild canids), such efforts might successfully target either sex. However, for polygynous species (including most deer, antelope, wild horses, and ducks), birth control will almost certainly have to be targeted at females, since females only have to find one uncontracepted male to reproduce successfully.

Jay Kirkpatrick, John Turner, and other researchers who have done extensive work in the field have argued that an acceptable wildlife contraceptive will minimize effects on natural behavior. Some tests of steroid hormone contraceptives in wildlife have failed for this reason. For example, estrogen implants in female African elephants produce abnormal, continuous estrus and elevated activity levels, disrupting cow–calf bonds and destabilizing female social groups. Steroids have not been useful as wildlife contraceptives for other reasons as well, including difficulty in administering treatments, harmful side effects, and the ability of steroids to pass through the food chain.

Current wildlife contraceptive research focuses on what are called immunocontraceptives. Immunocontraceptive vaccines stimulate the immune system to block some essential reproductive process. The most promising and widely tested immunocontraceptive is called PZP (porcine zona pellucida), a naturally occurring pig protein. When injected (by hand or dart) into a female, PZP stimulates the production of antibodies that prevent sperm from attaching to eggs. PZP vaccines have been successfully tested on wild horses, white-tailed deer, elk, African elephants, grey seals, and a wide variety of other species in zoos. PZP has so far displayed only minor effects on behavior, and no serious health effects. Recent tests have demonstrated not only that PZP significantly reduces births among treated individuals in a variety of species, but that localized deer populations can be stabilized and reduced using contraceptive vaccines. As with other techniques, managing local deer populations with contraception relies on the site fidelity of individual adult females.

See also Conservation and Behavior—*Wildlife Behavior as a Management Tool in United States National Parks*

Further Resources

Hadidian, J., Hodge, G. R., & Grandy, J. W. 1997. *Wild Neighbors*. Golden, CO: Fulcrum Publishing.
Leopold, A. 1949. *A Sand County Almanac*. New York: Oxford University Press.

Allen T. Rutberg

■ Wolf Behavior
Learning to Live in Life or Death Situations

The behavior of wolves is what makes them remarkable. It also determines what we think about them, how we manage them, how we conserve them, why we like them and why we hate them. We can think of them as blood-thirsty killers who kill for the fun of it, or we can think of them as primary agents in structuring ecosystems as a symbol of wildness. Which view we hold depends on what we think of how they behave.

Their behavior is one reason why wolves have done well in the northern Rockies, the program with which I am involved, but not so well in the southern Rockies and North Carolina. As leader for the Yellowstone Wolf Project for the National Park Service, I know that understanding wolf behavior is important, because it influences our management and conservation of them.

I was recently radio tracking wolves in Yellowstone National Park. About a third of the wolves in Yellowstone National Park are radio collared. Radio collars emit a radio signal that allows us to follow individual wolves, look in on their lives, and learn about the wolf population. On this particular day, the radio signal led me to a dead wolf. I was flying at the time in a small airplane we use to track the wolves. The radio signal was beeping at 120 beeps/minute. Had the wolf been alive it would have been beeping at 55 beeps/minute. I knew right away that female wolf #251 was dead.

The next day I hiked in with my crew to find #251 and to learn why she had died. When we found her, it was apparent that she had been eaten by a grizzly bear. Grizzlies leave tell-tale signs that are hard to mistake. Once recognized, it also puts the hair up on the back of your neck, but that is another story. None of us could imagine a wolf being caught and killed by a grizzly bear. We had seen wolves and grizzlies interact numerous times, and in every case the wolves were too quick for the grizzly to catch. How could #251 let a grizzly catch her?

The next time I was flying, the pilot, naturally curious, asked me how #251 had died. I had to say I didn't know, because I still could not imagine a grizzly catching and killing her. I told this to the pilot, but he was not so sure that this was not accurate. Our pilot was the classic bush pilot; he had been flying over Yellowstone since 1952, and he had seen about everything in the wilds there is to see from an airplane. He had seen animals make mistakes before. His next comment to me was, "Yeah, unless she made a mistake; if she did, then the bear could catch her." I was struck by this comment. Ultimately we will never know how #251 died, but I had dismissed the possibility of a grizzly killing the much more agile wolf. Why did I think this way; had I subconsciously categorized wolf and bear behavior? I had not considered how behavior can vary from one individual to the next and how this might account for this unusual event. I had thought of wolf behavior as of one kind, yet no two people I know behave alike; why should wolves? Or bears for that matter? That bush pilot had taught me a valuable lesson.

This is important because it is a human tendency to think of animal behavior generically: Wolves behave this way, grizzly bears that way, all behave specific to their species. But we know from our own life experiences that this is not true. Since we do not expect consistent behavior from people, why should we from wolves? Indeed, #251 might have made a mistake and gotten too close to that bear, slipped, or misjudged, and fell prey to the grizzly who partially ate her. We do not know what happened, but can I say—just because every time I have seen grizzlies and wolves interact, wolves have not been killed because they are quicker—that this is always true?

Nunamiut Eskimo do not share this view, and they have tremendous experience and exposure to wolves in the Arctic. Virtually all Nunamiut wolf hunters think of wolves as individuals. Their view, as characterized by Alaskan wolf biologist Robert Stephenson, clearly indicates nongeneric, individual behavior: "The Nunamiut recognize broad variation in the physical and behavioral attributes of wolves. Some variation is correlated with sex and age and environmental conditions, but a good deal is attributed to inherent individual differences in morphology and temperament." Stephenson goes on to discuss the temperamental differences these aboriginal peoples have noted between individual wolves. This perspective on wolf behavior is a very different one than you would find in a textbook on animal behavior. They look at each wolf with no preconceived notions on how it should behave. I would like to talk with them about wolf #251. They have probably seen more wolf–grizzly interactions than I have.

Scientists are gradually and incrementally learning how individual wolves are different from each other, and how varied wolf behavior can be in similar situations. The times that I refer to seeing wolves outmaneuver grizzly bears often occur by wolf den sites. When a bear is by a wolf den, where there are young vulnerable pups that could be killed by the bear, the wolves are very nervous. What the wolves do is basically harass the bear away. They cannot physically evict the bear from the area because the bear is larger and stronger, but they can act like a bunch of mosquitoes flying around and annoying the bear until the bear leaves. But the most interesting phenomenon is that each wolf has a role to play in harassing the grizzly bear. A couple of wolves wander around the bear, staying between the bear and the pups, but never getting too close to the bear. They operate in a kind of buffer zone between bear and pups. These wolves tend (I use this word because there is variation, again warning against labeling wolf behavior as "generic") to be older, and often involve the mother wolf. Closer to the bear, near the action, a couple of wolves, sometimes three, directly harass the bear. Typically these are younger wolves, yearlings and two-year-olds, and they seem to enjoy the job of pressing the bear. They dart in and out, coming close to biting the bear but rarely do so; they play bow with tails wagging, they change position front to back, and generally keep the bear off-guard. They are the mosquitoes. You can see the bear's concern, even agitation. On a couple of occasions, I actually saw a wolf bite the bear in the butt when the bear turned to leave—the final insult! What I noticed though, is that individual wolves had different roles, and these roles tended to be consistent, but were not always (I did not always have marked animals to observe; some of them were uncollared). Different wolves had different roles probably based on experience and learning, and even possibly enthusiasm (young wolves are often more active than older wolves). Perhaps this is where wolf #251 made her mistake?

Another life or death behavior that effects wolf survival is how they kill their prey. Wolves kill animals that are 2–20 times as large as they are, and all of their potential prey can kill them. A friend of mine who studies wolf–deer relationships has recorded two instances where deer have killed wolves. One time he saw a deer running full speed right at a wolf, right at the enemy. What did the wolf do? Run for its life. If wolves are not careful, if they do not behave correctly while trying to bring down moose (1000 lbs), or buffalo (2000 lbs), they will be killed. In Yellowstone, we have had eight wolves killed by prey: six by elk, one by a bison, and one by a moose. As a result, wolves must be exceedingly careful in how they go about killing; learning and experience matter, and some wolves are better at it than others. Often older wolves are much better killers for all the above stated reasons.

For example, we are beginning to learn that the outcome of an encounter with an elk may be determined by which wolves in the pack are involved (not all wolves participate in all hunts). If the alphas, the most dominant animals in the pack, are involved, the probability of a kill being made goes up. Yearlings participate in hunts more than any other age class, but their success rate is the lowest. One year later, as 2-year-olds, participation in hunts declines, but success rate in making kills increases. This seems counterintuitive but shows vividly that experience matters. Older wolves are becoming more selective about the prey they choose to attack, only attacking prey that are weaker and more vulnerable, increasing the chances that a kill will be made. This is an amazing discovery!

This can be understood further by examining some of the details of what happens when wolf meets prey. Prey that stand their ground, or elk that do not flee when approached by wolves, pose great difficulty to kill. In general, wolves cannot kill them. In 45 years of wolf–moose research on Isle Royale, an island in Lake Superior, no one has ever seen a wolf kill a moose that stood its ground. All of the moose killed were running away from the wolves. The same can be said about elk in Yellowstone National Park. When elk stand their ground they are too dangerous for wolves to kill, and wolves know this—so much so, that it is almost a prerequisite for prey to flee for wolves to be able to kill them.

Typically, when wolves attack larger prey they approach from the rear of the running animal. The front is the more dangerous end because of antlered or horned males, or sharp front hooves. The rear is a safer approach for the wolf, a risk sensitive behavior for a careful predator. One time I saw a wolf pass up attacking a bull elk on the rear. Instead, the wolf ran ahead and grabbed the elk by the neck. I had seen wolves do this on smaller prey, like cow elk, but not on the much more dangerous bull elk. Perhaps this wolf was trying to kill bulls the same way it had had success killing cows. The wolf was varying its behavior in other circumstances. The wolf got a hold below the bull elk's neck, hanging by powerfully evolved canine teeth. This had no effect on the elk. It kept running with the wolf dangling off its neck, getting hit by the churning front legs. The elk then came to a log that it leapt over slamming the dangling wolf's

Aerial shot (winter 2000) of the Chief Joseph wolf pack testing a bull elk that stands its ground in Yellowstone National Park. The elk was not killed.
Courtesy of Douglas Smith / NPS.

mid-section into the log. The wolf let go and fell back from the chase leaving the elk to other wolves, which they eventually killed by attacking from behind after a 25-minute battle. We have found, as has other published research, that most wolves *necropsied* (autopsied) have had or currently have several broken bones. Killing for a living has its risks, and wolves have to know how to do it or they will be killed themselves.

Knowing how to kill has conservation implications. Wolves were extirpated from most of the contiguous United States by the 1950s. Recent conservation efforts have attempted to restore wolves to some of the locations where they originally existed. Because of wolf–human conflicts, these areas are typically wild and sparsely populated by humans. Three of these locations are North Carolina (red wolves—*Canis rufus*), the northern Rocky Mountains (gray wolves—*Canis lupus*) and the southern Rocky Mountains (Mexican gray wolves, a subspecies of the gray wolf—*Canis lupus baileyi*). Each program attempted to restore a population of wolves by reintroducing them into the area.

Wolves aren't the only one attracted to a fresh kill.
Courtesy of Douglas Smith / NPS.

In the northern Rockies, wild wolves were obtained from Canada, where large populations of wild gray wolves still existed, but for North Carolina and the southern Rockies, the species (red wolf) or subspecies (Mexican gray wolf) was extinct in the wild, so captive stock had to be used. This difference in animals—wild versus captive—has had tremendous bearing on the success of the three different programs.

Wild wolves knew how to kill prey. They also knew how to do other things like avoid people. Wild wolves were versed in the ways of the wild at the moment they were liberated from their acclimation pens. Few died post-release. There are now (as of 2003) roughly 300 wolves in the Yellowstone Recovery area and about 700 in the northern Rockies. Most (some wolves existed naturally in northern Montana) of these wolves came from only 31 reintroduced to Yellowstone plus 35 more to Idaho.

Conversely, the red wolf project, which began releasing wolves in 1986, has only 85 wolves in the wild, and the Mexican wolf project has about 20 wolves (releases began in 1997) in the wild. The difference in success of the three programs is not this simple; there are other reasons for the disparity among the programs, but experience in the wild, appropriate behavior, knowledge of killing prey, and guarding dens from other carnivores certainly has had bearing. When we released the wolves in Yellowstone from their pens, I found a elk killed by wolves a half mile away. They had immediately upon release killed prey because they knew how. The crew in both the Mexican wolf and red wolf projects had to feed the wolves supplementally immediately after release to help them get a foothold in the wild. One red wolf choked on a raccoon kidney. I know of no death to a wild wolf from choking on a kidney.

Another aspect of wolf behavior that has bearing on their management and conservation is the tremendous proclivity of wolves to move. Wolves are built to roam. Watch one at a zoo and you will notice that all they do is pace. In the wild, this transfers into hundreds of miles across mountainous terrain should they choose to do so. I have seen wolves travel 30 mi, (48 km) in a night or travel in winter over mountain passes with 6 ft (1.8 m) of snow. As I tracked them, it seemed as if they didn't flinch from this kind of travel. This wandering behavior has tremendous bearing on how we manage them and how they fit in with our lifestyle.

The Mollies in wolf pack traveling in YNP, 2002.
Courtesy of Douglas Smith / NPS.

Wolves run up against us because they basically outwalk every place we set up for them to live, except maybe the remote reaches of the far north. I know that Yellowstone, one of the largest, wildest areas in the lower 48 states, is small to a wolf. I tracked one crossing the entire breadth of the Yellowstone ecosystem, which from east to west is about 150 mi (240 km), in 4 days. And the wolf took his time. Then he headed back. No big deal. This exploratory behavior has kept them at odds with humans. We don't have country available like that anymore; we have used it all. Conservation of wolves is confounded by small wild areas and big traveling needs. They are built to roam, and roaming is difficult with parking lots and plazas.

In the end, our desire to think of animals as part of a population, the mindset for most modern wildlife biology, is useful only to a point. Eventually we will have to think of animals as individuals and manage them as such. We started in Yellowstone with 14 wolves (later we brought in 17 more). It would have been hard to restore wolves to the world's first national park had we not considered a perspective a least a little bit like that of the Nunamiut Eskimos.

See also Coyotes—*Clever Tricksters, Protean Predators*

Further Resources

Busch, R. H. 1995. *The Wolf Almanac*. New York: Lyons & Burford, Publishers.

Carbyn, L. N., Fritts, S. H. & Seip, D. R. *Ecology and Conservation of Wolves in a Changing World*. Edmonton: Canadian Circumpolar Institute, University of Alberta.

Halfpenny, J. C. 2003. *Yellowstone Wolves in the Wild*. Helena, MT: Riverbend Publishing.

Mech, L. D. & Boitani, L. 2003. *Wolves: Behavior, Ecology, and Conservation*. Chicago: University of Chicago Press.

Smith, D. W., Peterson, R. O. & Houston, D. B. 2003. *Yellowstone after wolves*. BioScience, 53, 330–340.

Douglas W. Smith

Zoos and Aquariums
Animal Behavior Research in Zoos

In zoos and aquariums, questions about biology and behavior can be addressed in ways that would be impossible in any other setting. This is due to the facts that animals in zoos are highly accessible (which bypasses the considerable logistical and financial burdens of field research); already acclimated to the presence of human beings (which obviates the need for long periods of habituation); and available 24 hours per day (which avoids the sporadic observability problem that often occurs in the field). It is not surprising, therefore, that many animal behaviorists focus their scholarly research efforts upon exotic animals held in captivity in zoos.

There is, however, a major pitfall inherent in zoo-based research, and this is the fact that captivity itself undoubtedly affects the behavior of the animals in substantial ways. After all, one can not learn very much about the natural social structure of a given species by studying the behavior of animals that live in groups that have been artificially arranged by zookeepers and who are often obliged by their enclosures to remain closer to each other than they might otherwise be inclined to do.

Consider the following example. In a recent killer whale birth at Marineland of Canada, the expectant mother swam away from her group during the end of her labor and positioned herself in a partially separate pool at a distance of about 40 m (132 ft) from the other whales. On the one hand, the circumstances of captivity afforded a truly priceless opportunity to observe an event that has never been witnessed in the wild, and the researchers present were able to document important behavioral details in a killer whale birth. The mother's behavior seemed to suggest that it is a natural tendency for killer whale mothers to move away from her group and give birth alone (as it is in many species of terrestrial mammals).

On the other hand, this single observation leaves many questions unanswered. For one thing, we surely can not know if the distance of 40 m (132 ft) was what the mother would have chosen if she had more room to maneuver. In this case, she moved herself to the opposite end of the pool complex and 40 m was the maximum distance that was available to her. We can guess that she might have separated herself even further if she had the opportunity. But we don't know.

Perhaps even more significantly, it is important to note that this particular whale had separated herself from a group that consisted of an adult male

Studies of killer whales at Marineland of Canada show that mother–infant pair bonds are maintained via frequent touch.

© Amos Nachoum / Corbis.

and some adolescents. Perhaps she would have behaved differently if she had been in the presence of other adult females. We only know what she did under the particular social conditions in which she found herself.

Uncertainties like these do not mean that nothing useful can be derived from such studies. After all, we do now know what she did do, and such observations have the potential to tell us a great deal about the needs and motivations of animals. But they do mean that we should be cautious about the questions we attempt to address, and we need to be careful not to over interpret findings derived from captive animal studies. It is a matter of balancing the opportunities that zoos provide with conservative scientific reasoning so that valid insights are reached.

Some of the best animal behavior studies conducted in zoos are focused upon the animal's reactions to captivity itself, and these are often inspired by concern for the welfare of the animals being studied. One important focus in this respect is on what scientists refer to as "behavioral budgets." The reasoning is that one way to assess the welfare of an animal in captivity is to compare its lifestyle to that of its counterparts in the wild. Consider an example in which an animal in captivity is shown to spend 80% of its time resting or sleeping. If the animal under consideration were a chimpanzee, it would probably be reasonable to judge the environment to be inadequate because wild chimpanzees ordinarily spend a much greater portion of their time actively exploring their environment. However, one might reasonably draw a very different conclusion if the animal in question were a lion, since that species ordinarily spends a majority of its time in rest, even when presented with all of the sources of stimulation that the natural world provides in the wild. Every good zoo tries to provide its animals with appropriate habitat and sources of enrichment, but it is a constant challenge to get these things right. Carefully conducted assessments of the behavioral budgets of the animals in the zoo can help immensely in making such judgments.

Another very significant point of focus in zoo-based research pertains to their efforts to breed their captive animals. Consider, for example, the management of a breeding program for giraffes. These animals give birth standing up and this means that the newborn falls approximately 2 m (6.5 ft) to the ground when it emerges. Unfortunately, it is not unusual for the calf to be injured in the process, particularly when it lands on the concrete floor that is typical of giraffe barns in zoos. A preventive measure that zookeepers can take is to fill the maternal pen with a thick layer of straw so that the baby's fall is cushioned somewhat. However, since very thick straw is not practical in day-to-day husbandry, the challenge is to anticipate the night of the birth, and that is not at all easy to do. To address this problem, a recent study carefully tracked the behavior of expectant giraffes and found that mothers do not typically drink during the last day leading up to delivery. Since giraffes do not drink very often anyway, this is something that zookeepers had never noticed on their own. By providing this new insight, this behavioral study gave the keepers a useful behavioral index for anticipating a birth.

Although the zoo community has made great strides in maintaining viable populations of many endangered species via Species Survival Plans (SSPs), there are some very puzzling exceptions. For example, although many individuals of Asian elephant experience apparent good health in zoos, their rate of reproduction in captivity is much lower than that shown by their wild counterparts. In this species, it evidently not enough to simply pair a male and female of reproductive age. Even when a female is experiencing apparently normal hormonal cycles and a male has a healthy sperm count, mating too often fails to lead to a successful birth. Scientists are therefore increasingly convinced that behavioral/social variables such as herd size, dominance relationships, and long-term pheromonal signaling must play essential roles when elephants reproduce. Consequently, there is a large research effort now underway attempting to identify the conditions that are essential. This is a major collaborative effort involving the combination of data from across the zoo community.

Of course, the range of behavioral studies at zoo is not limited to welfare and reproductive topics. They include studies of social communication, animal cognition, and many other topics. Often the questions addressed relate to the animal's natural history and the investigations complement those conducted in the field. Whatever the specific topic may be, the 200 modern zoos and aquariums that operate worldwide present a truly priceless opportunity for research, and animal behaviorists are often among the zoo visitors, taking data that will increase our understanding of animals in the future.

Efforts to breed captive Asian Elephants have proven to be surprisingly complicated, raising questions about the breeding habits of their wild counterparts. Kandula celebrates her first birthday at the National Zoo in Washington.
© Getty Images.

See also Welfare, Well-Being, and Pain—*Carnivores in Captivity*
Zoos and Aquariums—Studying Animal Behavior
in Zoos and Aquariums

Further Resources

Gibbons, E., Durrant, B. & Demarest, J. (Eds.). 1995. *Conservation of Endangered Species in Captivity.* Albany: State University of New York Press.
Kleiman, D., Allen, M., Thompson, K. & Lumpkin, S. (Eds.). 1996. *Wild Mammals in Captivity.* Chicago: University of Chicago Press.

Michael Noonan

■ Zoos and Aquariums
Giant Pandas in Captivity

The giant panda (*Ailuropoda melanoluca*) is an endangered species native to China and part of the heritage of the world's natural history. Its lovely face and funny postures have attracted much attention from both inside and outside of China. Morris and Morris (1996) explained that its popularity is probably the result of many different character traits that appeal

This work was supported by National Natural Science Foundation of China, NSFC (Grant No. 30070107 and 30170169).

independently to different people and diverse age groups. The giant panda used to flourish in ancient China. Fossil evidence showed its historical distribution stretched from the Jianghuai Plain in the east to the Yungui Plateau and north Burma, Vietnam and Thailand in the west, from Guangdong in the south, to Beijing in the north. Unfortunately, the wild populations of giant pandas sharply decreased from the 1980s because of the shrinking of their habitat, resulting in a decrease in the the regular flowering of their unique food, bamboo. Currently, bamboo only survives in six small fragmented areas, the Qinling, Minshan, Qionglai, Big Xiangling, Small Xiangling, and Liangshan Mountains. For many years, great efforts have been made by the Chinese government and other organizations, for example the World Wildlife Federation, to rescue and protect the giant pandas. In 1992, the former Ministry of Forestry began to implement China's National Conservation Project for the Giant Panda and Its Habitat. This project enforces giant panda conservation in China.

One important measure was to enforce the protection of their natural habitat by establishing more panda nature reserves in areas in which panda live. At present, there are 34 nature reserves established principally to protect giant pandas and their habitats. There are also 17 corridors being built to link those reserves. The updated estimation of wild panda populations in 2000 is around 1,100. Another important measure was to conduct captive breeding in 1983. There are four large institutions that conduct captive breeding: the Beijing Zoo, the China Conservation and Research Center for the Giant Panda (CCRCGP) at Wolong, the Chengdu Panda Base/Chengdu Zoo, and the Fuzhou Zoo. Besides captive breeding, CCRCGP also plays an important role in rescuing wounded, starved, and weak pandas from the wild. After they have lived in captivity, those who have recovered have been released into the wild again; others, including those individuals too weak or sick to survive in the wild are used in captive breeding. The ultimate goal of captive breeding is to reintroduce captive-born individuals into the wild. To realize this goal, a large captive population of pandas is necessary. Yet, it is hard work, and there is a much work to be done. Scientists first need to solve many problems existing in captive breeding with a top priority of solving behavioral problems of captive giant pandas.

A panda mother and her young baby.
Courtesy of Dingzhen Liu.

Behavioral Problems Associated with Captive Breeding

There are primarily three difficult problems for captive breeding: poor sexual performance of males, difficulties in detecting a female's pregnancy, and hand-raising twin infants or those abandoned by the mother. During the past 10 years, scientists have been successful in assisting reproductive techniques for giant pandas, and have completely solved the third problem in captive breeding. The 2001 *International Studbook for Giant Pandas* shows that there were 157 litters and 205 infants born in captivity, and a captive population of 142 individuals by December 2001. Although the first successful captive breeding that happened in Beijing Zoo in 1963 was a successful natural mating between a male and female, most of the breeding that happened in captivity depended greatly on artificial insemination. Because only one third of the adult males can naturally mate with females; nearly two thirds of the females do not show typical estrus behavior and are without an obvious estrus peak. Most of the captive pandas have lost their species-typical behavior.

The second difficulty of captive breeding is the dection of pregnancy in females. The newborn infant weighs only 1/900 of it mother's weight, and the gestation period ranges from 97 to 163 days, averaging 135 days. Moreover, like black bears and brown bears, giant pandas have delayed implantation of the fertilized egg, and the time lag of implantation is affected by both physiological and environmental factors. Thus, it is difficult for scientists to figure out effective approaches to pregnancy detection.

To solve these problems, scientists both from China and abroad are focusing on behavioral studies of captive giant pandas, which include the behavioral characteristics, mechanisms of chemical communication between male and female pandas, environmental enrichment, and behavioral development of young pandas. It is necessary to develop a standard ethogram of captive giant pandas so that panda scientists from all over the world can share the results and facilitate intensive research on the basis of the whole species. Based on a 9-year study by the panda research group lead by the author, and combining the behavioral definitions of terms by George B. Schaller, Devra G. Kleiman, and Ronald R. Swaisgood, a comprehensive ethogram of captive giant pandas has been developed. This ethogram can serve as a guideline for giant panda behavioral study in the future.

Behavioral Categories

The following behavioral categories are included in the ethogram for giant pandas:

- *anogenital marking*: rubbing the anogenital area around or up and down on the surface of an object or on the wall;
- *eating*: handling and eating steamed bread, apples and grass, and drinking water and milk;
- *exploring*; investigating an object with a distance of 0.1 m or more between the nose and the object;
- *sniffing*: investigating an object with a distance of less than 0.1 m between the nose and the object accompanied by; a flehmen response;
- *grooming*: scratching and licking of the pelage;
- *resting*: inactivity, including sleeping and nonsleeping;
- *urine marking*: urinating in a squat, leg-cock, handstand, or standing posture on a wall or the ground;
- *locomotion*: rapidly pacing and moving around an enclosure without placing the feet in the same position each time and without following the same path;
- *playing*: rolling and somersaulting with manipulation of objects, such as food dishes, bamboo stalks, tree branches or toys provided by the keeper; and

A male adult panda scent marks on the wall.
Photographed by Rongping Wei.

An artificial wooden frame for young pandas to prevent them from engaging in stereotyped behavior.
Courtesy of Dingzhen Liu.

- *stereotyped behavior*: pacing (continuous walking back and forth following the same path), circling (walking following a defined route placing feet in the same position each time), self-mutilation (self-inflicted physical harm, such as biting or chewing tail or leg, or hitting head against a wall), head bobbing (standing in one place and continuously moving head up and down), and standing bi-pedally at a window, a door, or a fence, often seemingly in expectation of food delivery.

- *vocalizations*: usually can be discriminated by using a with a sound spectrograph Common behavior includes: bleating, chirping, chomping, huffing, whistling, and honking.

Sampling and Recording Methods

Behavioral studies of giant pandas in captivity will help us learn more about the nature of these fascinating animals. This information is essential in our efforts to conserve this engangered and charismatic mammal.

References

Lehner, P. N. 1996. *Handbook of Ethological Methods*. New York: Garland STPM Press.

Martin, P. & Bateson, P. 1986. *Measuring Behaviour*. Cambridge: Cambridge Univesity Press.

Morris, R. & Morris, D. 1966. *Men and Pandas*. New York: McGraw-Hill.

Peters, G. 1982. *A note on the vocal behavior of the giant panda (Ailuropoda melanoleuca, David, 1869)*. Zeitschrift für Saugetierkunde, 47(4), 236–246.

Zhao, C. N. & Wang, P. Y. 1988. *The sound spectrum analysis of calls in the baby giant panda (Ailuropoda melanoleuca)*. Exploration of Nature, 7(2), 99–101.

Zhao, C. N., Wang, P. Y. & Wang, A. Q. 1988. *Exploring of making use of calls to infer peak oestrus and promote oestrus in giant panda (Ailuropoda melanoleuca)*. Exploration of Nature, 7(2), 93–99.

Zhu, J. & Meng, Z. B. 1987. *On the vocal behavior during the estrous period of the giant panda, Ailuropoda melanoleuca*. Acta Zoologica Sinica, 33(3), 285–292.

Dingzhen Liu

■ Zoos and Aquariums
Studying Animal Behavior in Zoos and Aquariums

Studying animals in zoos and aquariums provides the observer with a rare opportunity to be able to find the animal in a predetermined space, under locations determined by other people, rather than by the activity of the animal itself. This helps in the planning of studies, but it also presents some complicated issues that need to be resolved before a study can be done. First, it is necessary to obtain permission from the institution in which the animals are kept. This permission usually is circumscribed by the fact that the primary mission of the zoo and aquarium is to make it possible for people to see the animals in the best of all circumstances—in settings that are as much like their natural settings as possible (where they may have evolved or developed), that enable the animal to be active in ways that are typical of the animal in its natural setting, that protect the animals from the public viewers, that maintain them in health and promote their physiological and psychological well being.

Frequently, observations of the animals in their exhibit areas are most easily permitted during the hours when they are in those enclosures. Marine and water animals and many terrestrial animals usually are kept in the same environment throughout the day and night. In the case where animals are kept in enclosures that need to be cleaned of feces, urine, food and other objects which the animals would not ordinarily have in their environment, or where the animals avoid feces, the animals are removed from the exhibit areas and kept in night locations. It is most difficult to study animals in the nonexhibit areas, or night cage areas, because it is necessary to protect the animals from possible infection from people, and to protect the humans who care for the animals. These requirements lead to using special instruments to record the activity of the animals without the presence of people. In some zoos, behavioral scientists are permitted to be in the night cage area if they have been examined by a physician who can attest to their freeedom from disease-causing symbionts. A *symbiont* is an organism living with or in a host and is usually smaller than the host (e.g., bacterium, virus).

The basic method of studying the behavior (activity) of animals in zoos and aquariums is observation by a human: A human looks at the nonhuman animal and attempts to become familiar with its activity in the place in which it is living. In the zoos and aquariums not only do students of behavior become familiar with the animals, but also the caregivers, the maintenance people, the docents, and the veterinarians all spend time looking at the animals. They may have interesting and sometimes very significant information that the behavioral investigator could use. To obtain this information, observers may have the caregivers fill out questionnaires which then can be analyzed. In getting this information, it is important to recognize that it may not be as systematic as the behavioral observer would wish. The scientific observer needs not only to record what is being told by the ancillary observers described above, but needs to verify the material and ascertain its reliability and accuracy. But, such observers frequently have valuable clues for the scientific observer, and they should be part of the project's data base.

Most observers start with hunches (these are hypotheses when hunches become formulated carefully and integrated in a research plan) or questions about what they have observed or what they have been told about the animals.

What does observation mean? To observe is to watch carefully, to inspect, to come to know especially through consideration of noted facts. The behavioral scientist not only looks at the whole animal, how it moves, where it is in space, and how it is oriented, (i.e., in terms of the position and direction of the head of the animal), but also looks at the changes that the animal makes in the place in which it is: feces, urine, deposition of material on surfaces of rocks, plants, objects in the environment (rubbing glands in the skin), feeding on leaves or other substances, dropping material from its mouth, manipulating objects, with or without changing their locations, and so on. The sounds produced by the animal, as for example, vocalizations by dolphins, can be recorded by special microphones that do not interfere with the animal's activity, or are not in danger of being abused by the animal, or the environment. All of this activity may offer clues to the relationship of the animal to its environment and to other animals in its environment, including the human being who is doing the observation.

It is also necessary to consider the animal's use of the space in which it is living. To understand this aspect of its activity, a map of the area in which it is living, with information about the dimensions of the enclosure, the type of fences or walls used, and so on, and the relation of the daytime housing to the nighttime, is important. Additionally, materials that the animal may manipulate which are provided to the animal need to be described and recorded. This is especially necessary with animals that make nests or are offered special enrichment programs (objects that give the animal satisfactory manipulating experience).

Inanimate, as well as animate aspects of the environment also require observation and recording: What are the auditory, visual, and olfactory changes in the environment which will be effective in causing the animal's activity to remain the same or change? Are these products of living organisms (members of the same species, e.g., kin, parents, members of a group of which the observed animal is a part), predators, other organisms, such as plants, or casual passersby, or are they produced by inanimate processes or substances (sunlight, minerals, soil, rain, etc.)? To describe and record such information will require special instruments that are the responsibility of the observer and are a reflection of the questions the observer is asking about the animal: Does the animal see, hear, smell other organisms?

The history of the animal must also be recorded: sex, its age, its time in the zoo or aquarium, its origin (born there or elsewhere), whether it was hand reared or parent reared, reproductive history, and more. In addition to recording information about the setting in which the organism is being observed, such as the date, time of observation, weather conditions, and temperature in the enclosure; the presence of conspecifics in the same enclosure or within sensory input to the animal; or whether heterospecifics, particularly known possible predators were within sensory input of the animal, number of visitors viewing the exhibit, and if possible, the age of the visitors (children?) should also be recorded. Incidents of interruption or activities by the visitors which bring about a change in the animal's activity need also to be recorded. Such information may be found to be irrelevant to the observation, but if it is not recorded, there is no way to ascertain whether it is or is not relevant.

To obtain the proper climatic information, the observer may record temperature, light, and humidity. In addition, governmental and other public sources of information are sometimes available.

Observations are only useful if they are recorded properly. The above items are necessarily recorded for each observation. The observer also has to decide whether the observation of the animal is continuous, that is, starting at some time and recording everything the animal does for a given period of time (5 seconds, 5 minutes, 5 hours), or whether the activity of the animal is to be sampled, that is, recording what it is doing (every 5 seconds, every 5 minutes, every 5 hours) at the precise time when the sampling is to take place. In continuous observations, one is able to record the duration of an activity, that is, starting at 5 seconds and ending at 12 seconds equals 7-second duration. In the sample technique, one records if an animal did something at 5 seconds, 10 seconds, 15 seconds, and so on. A count is then made of the number of times an activity was seen. Both these techniques are useful, and the choice of one or the other depends on the questions being asked, the conditions of the observation, and instruments available.

For both types of recording, certain instruments will be required: a timing device which gives a signal at the desired interval (every 5 seconds or 5 minutes); a device for holding this so that it can be read, or heard without being sensed by the animal, leaving the hands of the observer free to record the description of the activity; something on which the observation can be recorded—paper and pencil, keyboard connected to a computer, a motion recording or still camera (video or otherwise), a voice activated instrument (tape recorder) on which the observer describes the activity in speech; and a sound recorder on

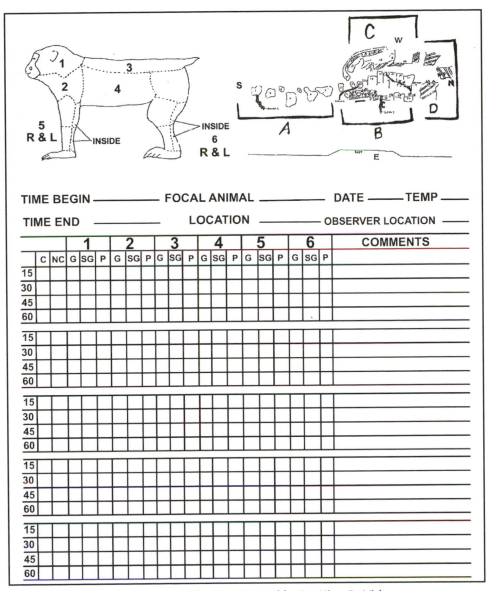

A typical data sheet to record behavior. Created by Dr. Nino DeVidze at the Central Park Zoo.
Courtesy of Ethel Tobach.

which the sounds made by the animal may be recorded. In using cameras and video equipment, the observer does not need to be present at all times. Each of these instruments has advantages and disadvantages, and the choice is made by the observer.

The observer needs to define the patterns of activity and devise a code by which the activity can be recorded. In other words, if the animal is swimming in water, the code may use an "S" for swimming; another letter for type of swimming, for example, in the case of dolphins, swimming on side, and thus might be coded as SS, or swimming on ventrum as SV, or on back, SD. Such codes would be devised for each behavior or activity that the observer wished to record. The use of a camera and a tape recorder to which the observer speaks also requires the observer to analyze the visual or auditory recording to be analyzed to yield the desired information.

In the case of a computer with a keyboard, the timing of an activity is automatically recorded because the recording device is continuously moving. In the case of a paper and pencil protocol where the activity at given times is recorded, as in continuously recording or in a time sampling technique, the data are recorded on paper that is formated by columns and lines so that the observer records the activity in the appropriate column and on the proper time line.

Depending on conditions of observation, a check list of behavior items can be created, and each item seen is scored by the observers whenever possible.

In addition to recording by camera, it is possible to record the activity of an animal by a monitoring system which yields, for instance, the location of an animal, the geolocation (depth of animal in water, for example), the temperature, or the time spent in different locations. Again, such data are continuous but also require analysis of the recorded changes in the animal's activity.

Working with animals in zoos and aquariums is very satisfying. It is, however, demanding of the observer's flexibility in designing the studies to be done. The first consideration is the well-being of the animals. This may sometimes mean that the animals are not on display and therefore are not available for observation. This means that the investigator needs to be flexible and creative.

See also Zoos and Aquariums—*Animal Behavior
Research in Zoos*
Welfare, Well-Being, and Pain—*Carnivores
in Captivity*

Further Resources

Cain, N. W. 1995. *Animal Behavior Science Projects*. New York: John Wiley & Sons, Inc.

Hediger, H. 1950. *Wild Animals in Captivity*. London: Butterworths Scientific Publications.

Hediger, H. 1970. *Man and Animal in the Zoo: Zoo Biology*. London: Routledge & K. Paul.

Mallapur, A. & Chellam, R. 2002. *Environmental influences on stereotypy and the activity budget of Indian leopards* (Panthera pardus) *in four zoos in southern India*. Zoo Biology, 21, 585–595.

McCowan, B. & Reiss, D. 2001. *The fallacy of 'signature whistles' in bottlenose dolphins: A comparative perspective of 'signature information' in animal vocalizations*. Animal Behaviour, 62(6), 1151–1162.

Mc Farland, D. (Ed.). 1987. *The Oxford Companion to Animal Behavior*. New York: Oxford University Press.

McGeehan, L., Li, X., Jackintell, L., Huang, S., Wang, A. & Czekala, N. M. 2002. *Hormonal behavioral correlates of estrus in captive giant pandas*. Zoo Biology, 21, 449–466.

McPhee, M. E. 2002. *Intact carcasses as enrichment for large felids: effects on on-and off-exhibit behaviors*. Zoo Biology, 21, 37–47.

Simeone, A., Wilson, R. P., Knauf, G. Knauf, W. & Schutzendube, J. 2002. *Effects of attached data-loggers on the activity budgets of captive Humbolt penguins*. Zoo Biology, 21, 365–373.

Ethel Tobach

■ Organizations and Related Web Sites

The following list offers organizations and Web sites that provide information on numerous aspects of animal behavior. Clearly, there are many different disciplines that are interested in this broad and exciting field, including anthropology, biology, law, medical researchers, neuroscience, psychology, veterinary medicine, and more.

Contributors to this encyclopedia suggested many of these organizations. There also are many other organizations that are not listed here that will be useful, and I suggest that you follow the numerous links that are provided or conduct searches on specific topics to discover what else is available. The two professional organizations that are most important for those seriously interested in animal behavior are the **Animal Behavior Society (ABS)** and the **Association for the Study of Animal Behaviour (ASAB).** Together the organizations sponsor the monthly journal, *Animal Behaviour.*

For convenience, the organizations and Web sites are first listed alphabetically and then by specific topics and geographical locations (for those not in the United States).

Alphabetical Listings

American Anthropological Association

http://www.aaanet.org/

The American Anthropological Association (AAA), the primary professional society of anthropologists in the United States since its founding in 1902, is the world's largest professional organization of individuals interested in anthropology. AAA seeks to advance anthropology as the science that studies humankind in all its aspects, through archeological, biological, ethnological, and linguistic research, and to further the professional interests of American anthropologists, including the dissemination of anthropological knowledge and its use to solve human problems.

American Anti-Vivisection Society

www.aavs.org

The American Anti-Vivisection Society (AAVS) is a nonprofit animal advocacy and educational organization dedicated to ending experimentation on animals in research, testing, and education. AAVS also opposes and works to end forms of cruelty to animals.

American Malacological Society

http://erato.acnatsci.org/ams/

The American Malacological Society (AMS) is an international association of professional malacologists (scientists who study mollusks). AMS promotes the conservation and study of mollusks, encourages students, recognizes the contribution of serious molluskan enthusiasts who often work with professional malacologists, and holds annual meetings.

American Ornithologists' Union

www.aou.org

Founded in 1883, the American Ornithologists' Union (AOU) is the oldest and largest organization in the New World devoted to the scientific study of birds. Although AOU is primarily a professional organization, its membership of about 4,000 includes many amateurs dedicated to the advancement of ornithological science. The major activity of AOU is the publication of scientific information relating to birds.

American Psychological Society

http://www.psychologicalscience.org/

The American Psychological Society seeks to promote, protect, and advance the interests of scientifically oriented psychology in research, application, teaching, and the improvement of human welfare.

The American Society for the Prevention of Cruelty to Animals

www.aspca.org

Founded in 1866, the American Society for the Prevention of Cruelty to Animals (ASPCA) is the oldest humane society in the western hemisphere. ASPCA offers national programs in humane education, public awareness, government advocacy, shelter support, animal medical services, and placement as well as a wide variety of behavior programs and information for public and professional audiences. The ASPCA Animal Behavior Center operates a behavior clinic for companion animals, offers dog training programs, provides postdoctoral fellowships in applied animal behavior, and consults with animal shelters across the country on behavior evaluations and enrichment programs for companion animals.

American Society of Mammalogists

www.mammalsociety.org/

The American Society of Mammalogists (ASM) was established in 1919 to promote interest in the study of mammals worldwide and to foster communication among scientists. ASM also provides information to the public about mammals that may be used for education and to inform public policy, resources management, and conservation. The society hosts annual meetings and maintains several publications.

American Society of Primatologists

www.asp.org

The American Society of Primatologists (ASP) is an educational and scientific organization that aims to understand, conserve, and inform about nonhuman primates. ASP promotes and encourages the discovery and exchange of information regarding primates and offers membership to anyone engaged in scientific primatology.

American Veterinary Medical Association

www.avma.org

The American Veterinary Medical Association (AVMA), established in 1863, is a nonprofit association representing more than 69,000 veterinarians. AVMA seeks to advance the science and art of veterinary medicine, including its relationship to public health, biological science, and agriculture.

American Zoo and Aquarium Association

www.aza.org

The American Zoo and Aquarium Association (AZA) is the largest zoo and aquarium organization in the world with more than 5,500 members. AZA is a nonprofit organization dedicated to the advancement of zoos and aquariums in the areas of conservation, education, science, and recreation. AZA's Web site includes information on its organization, issues related to zoos and aquariums, and links to more than 214 zoos and aquariums across the United States and Canada that have been accredited by the AZA.

Animal Behavior Society

http://www.animalbehavior.org/

The Animal Behavior Society (ABS) is the primary American nonprofit scientific society founded to encourage and promote the study of animal behavior. Membership is open to all those interested in the study of animal behavior. The organization and Web site provide information on careers, links to other organizations, a directory of United States and Canadian graduate programs in animal behavior, and education projects for elementary and high school teachers. Current members' research activities span the invertebrates and vertebrates, both in the field and in the laboratory, and include experimental psychology, behavioral ecology, neuroscience, zoology, biology, applied ethology, and human ethology, as well as many other specialized areas.

Animal Behaviour Web Sites

http://www.societies.ncl.ac.uk/asab/websites.html

The British-based Animal Behaviour Web Sites contains links to more than 70 organizations and resources worldwide, which are concerned with various aspects of animal behavior, behavioral ecology, applied and clinical ethology, neuroscience and conservation.

Animal-Info Net (or Animal Information Network)

http://www.animal-info.net

Animal-info.net is an international network composed of a number of individuals working in several organizations who share the goal of providing information on issues about animals and promoting their welfare. Affiliated with Cambridge University's Animal Welfare Information Centre among others, the network provides access to education, information on pets, jobs, funding, animal shelters, animal issues, and animal welfare. They provide such services as resources, education, glossaries, bibliographies, news, campaigns, and more.

Animal Sciences Group of Wageningen University and Research Centre

http://www.asg.wur.nl/english/index.asp

The Animal Sciences Group of Wageningen University and Research Centre in the Netherlands offers research, education, and expertise on livestock, fish, and companion animals. The group's work is concerned with sustainable methods for food production; sound ecological management of landscape, nature, and sea; and the social functions that animals fulfill.

Applied Ethology on the World Wide Web

http://www.usask.ca/wcvm/herdmed/applied-ethology/

Applied Ethology is dedicated to the field of applied animal behavior, including the behavior of domestic animals and other animals kept in captivity. This Web site provides students and scientists with information on animal behavior and includes articles, information on animal behavior conferences, e-mail forums, and links to other animal behavior websites.

The Association for the Study of Animal Behaviour

http://www.societies.ncl.ac.uk/asab/

The Association for the Study of Animal Behaviour (ASAB) was founded in 1936 to promote the study of animal behaviour. ASAB members primarily hail from Britain and Europe and include professional biologists who work in universities, research institutes, or schools. ASAB publishes a scientific journal, holds conferences, and provides grants for animal behavior. ASAB also promotes the ethical treatment and conservation of animals and encourages the teaching of animal behavior in schools.

Atlantic Salmon Federation

www.asf.ca

The Atlantic Salmon Federation (ASF) is an international nonprofit organization, which promotes the conservation and wise management of the wild Atlantic salmon and its environment.

BioAcoustics Team

http://www.cb.u-psud.fr/cb/index.html

The BioAcoustics Team is a research unit of the Centre National de la Recherche Scientifique (CNRS) and has a lab at the University of Paris XI-Orsay and a field lab at the University Jean Monnet of Saint-Étienne in France. The BioAcoustics Team works to understand how animals can vocally communicate in spite of environmental constraints. Currently, the team is conducting research on birds (tropical passerines, gulls, penguins, zebra finches, and others), mammals (fur seals and sheep), and insects (flies, ants, and cicadas) with ethological and neuroethological approaches.

The Borror Laboratory of Bioacoustics

http://blb.biosci.ohio-state.edu

The Borror Laboratory of Bioacoustics (BLB) is part of the Department of Evolution, Ecology, and Organismal Biology at the Ohio State University. The BLB houses one of the largest collections of recorded animal sounds in the world, with approximately 28,000 recordings of more than 1000 animal species.

BugBios

www.insects.org

This site provides basic information, including common and scientific names and descriptions of hundreds of insects and links to many other sources. All insects featured have detailed color photos.

Center for the Integrative Study of Animal Behavior

http://www.indiana.edu/~animal/

The Center for the Integrative Study of Animal Behavior (CISAB) fosters the integrative study of animal behavior by facilitating interaction among animal behavior researchers and educators at Indiana University and elsewhere. CISAB's program incorporates various disciplines including biology, psychology, neural science, anthropology, and other disciplines. CISAB hosts a seminar series, workshops, visiting scholars, an EthoInformatics Center, and an extensive Web site providing information to the general public about animal behavior.

Centre for Neuroscience and Animal Behaviour

http://sciences.une.edu.au/bbms/centre_neuroscience_animal_behaviour.asp

The Centre for Neuroscience and Animal Behaviour at the University of New England in Armidale, Australia, conducts research on a range of species, including marmosets, chicks, toads, and Australian birds. The Centre interests include social behavior and cognition, the development of behavior, aging and behavior, lateralization of brain and behavior, spatial cognition, vocal communication (especially referential signaling in the Australian magpie), and animal welfare.

Colorado Bat Society

www.coloradobats.org

The Colorado Bat Society provides educational information about bats and works for the conservation of bats and their habitats in Colorado. The Colorado Bat Society's Web site provides educational information about bats, including bat behavior and identification, as well as suggested reading and links to other Web sites.

Comparative Cognition Society

http://www.pigeon.psy.tufts.edu/ccs/default.htm

The Comparative Cognition Society is dedicated to gaining a broad scientific understanding of the nature and evolution of cognition in human and nonhuman animals. The Comparative Cognition Society is a nonprofit scientific society with no doctrine or philosophy, except the scientific method, as it is commonly understood in all natural sciences. Anyone who studies perception, learning, memory, or any other cognitive or representational process in animals is welcome to join.

Cooper Ornithological Society

www.cooper.org

The Cooper Ornithological Society has an international membership of professional and amateur ornithologists. Its purpose is to advance our knowledge of birds and their habitats.

Cornell Lab of Ornithology

http://birds.cornell.edu

The Cornell Lab of Ornithology is a nonprofit membership institution whose mission is to interpret and conserve the earth's biological diversity through research, education, and citizen science focused on birds. The Cornell Lab of Ornithology believes that bird enthusiasts of all ages and skill levels can and do make a difference, and their programs work with citizen scientists, government, and nongovernment agencies across North America and beyond.

Dolphin Communication Project

www.dolphincommunicationproject.org

The Dolphin Communication Project (D.C.P.), located at the Mystic Aquarium & Institute for Exploration, looks at how dolphins communicate and attempts to shed more light on the meaning of their interactions. With research ongoing since 1991, D.C.P. focuses on signal exchange (communication) in Bahamas' Atlantic spotted dolphins and Indo-Pacific bottlenose dolphins at Mikura Island, Japan, with comparisons between species and geographies. In addition to research, D.C.P. works to share information about their findings on dolphin communication with children and adults through a variety of workshops, seminars, and symposia.

The Dolphins of Monkey Mia Research Foundation

http://www.monkeymiadolphins.org/

The Dolphins of Monkey Mia Research Foundation promotes research on the bottlenose dolphins in Monkey Mia off the coast of Western Australia. Their research is done without tagging or capturing the dolphins and includes research on behavior, ecology, genetics, development, communication, social structures, and predator–prey relationships. The research conducted at Monkey Mia has provided basic information on the effects of provisioning on wild animals and has informed local and international management policies pertaining to dolphin-focused tourism.

Eco-Ethology Research Unit

http://www.ispa.pt/ispa_frame_00.asp?url=ispa_cinvest_01.asp%3Fsubseccao%3D4

The Eco-Ethology Research Unit of the Institute of Applied Psychology in Lisbon, Portugal focuses on animal behavior. The first studies were conducted on fish and included behavior, reproductive biology, and life history. These studies were further expanded to include marine biology, marine ecology, behavior, and conservation. Research is also performed on the behavioral endocrinology of fishes, communication and social ethology, and behavior and bioacoustics of cetaceans. The Eco-Ethology Research Unit is recognized as a Research & Development Research Unit by the Portuguese Science and Technology Foundation.

The Electronic Zoo

http://netvet.wustl.edu/e-zoo.htm

The Electronic Zoo provides educational information and links to hundreds of other educational Web sites for a variety of topics, including veterinary care information and species-specific information. For example, you can click on a particular animal and the electronic zoo will provide basic information such as health and genetics, as well as links to other educational websites.

Ethologists for the Ethical Treatment of Animals/Citizens for Responsible Animal Behavior Studies

www.ethologicalethics.org

Founded by Marc Bekoff and Jane Goodall, Ethologists for the Ethical Treatment of Animals/ Citizens for Responsible Animal Behavior Studies (EETA/CRABS) is devoted to developing the highest of ethical standards in comparative ethological research. EETA/CRABS also promotes the use of the latest development in cognitive ethology and animal sentience to inform public policies governing the use of animals. Membership is open, and scientists, nonscientists, teachers, and students are invited to join.

Hamburgische Gesellschaf für Verhaltensbiologie

http://www.rrz.uni-hamburg.de/HGV/

The Hamburgische Gesellschaft für Verhaltensbiologie promotes the behavioral sciences by funding research projects and organizing conferences. In German.

The Humane Society of the United States

www.hsus.org

The Humane Society of the United States (HSUS) is the nation's largest animal protection organization with more than 8 million members and constituents. HSUS is a mainstream voice for animals, with active programs in companion animals and equine protection, wildlife and habitat protection, animals in research, and farm animals and sustainable agriculture. For nearly 50 years, HSUS has protected animals through legislation, litigation, investigation, education, advocacy, and fieldwork. The HSUS envisions a world in which people meet the physical and emotional needs of domestic animals; protect wild animals and their environments; and change their interaction with other animals, evolving from exploitation and harm to respect and compassion.

Institute for the Conservation of Tropical Environments

http://icte.bio.sunysb.edu

The Institute for the Conservation of Tropical Environments (ICTE) promotes scientific research, training, and conservation in the tropics. ICTE coordinates and catalogs the work of over 300 natural and social scientists at Ranomafana National Park and other parts of Madagascar, organizes and conducts biodiversity surveys and ecological assessments of tropical ecosystems, and trains scientists at all levels through field-based courses, collaborations, and academic exchanges. ICTE maintains offices and staff at SUNY Stony Brook and in Antananarivo, Madagascar.

Instituto de Investigación en Recursos Cinegéticos

http://www.uclm.es/IREC/

The Instituto de Investigación en Recursos Cinegéticos (Spanish National Institute of Game Research) is focused on all aspects of game research, particularly behavioral ecology. In Spanish.

International Primatological Society

http://www.primate.wisc.edu/pin/ips.html

The International Primatological Society was created to encourage all areas of nonhuman primatological scientific research, to facilitate cooperation among scientists of all nationalities engaged in primate research, and to promote the conservation of all primate species. The society is organized exclusively for scientific, educational, and charitable purposes and includes an international directory of primatology, which provides information about programs in primate research, education, and conservation.

International Society for Anthrozoology

http://www.vetmed.ucdavis.edu/CCAB/ISAZ.htm

Formed in 1991, the International Society of Anthrozoology (ISAZ) is an academic organization for the scientific and scholarly study of human–animal interactions and relationships. ISAZ members represent a broad range of fields, including anthropology, archaeology, animal welfare science, art and literature, education, ethology, folklore studies, human and veterinary medicine, psychology, and sociology.

International Society for Applied Ethology

http://www.applied-ethology.org/

The International Society for Applied Ethology (ISAE) covers all applied aspects of ethology and other behavioral sciences, including farming, wildlife management, companion animals, laboratory animals, and pest control.

International Society for Behavioral Ecology

http://web.unbc.ca/isbe/

Founded in 1988, the International Society for Behavioral Ecology publishes a journal and holds international meetings every other year on a wide variety of topics relating behavioral ecology to ethology, population ecology, and evolutionary biology.

International Society for Comparative Psychology

http://www.comparativepsychology.org/

The International Society for Comparative Psychology (ISCP) promotes the comparative study of behavior in human and nonhuman animals, sponsors a biannual meeting, and publishes a scientific journal, the *International Journal of Comparative Psychology*, which publishes studies on the evolution and development of human and animal behavior.

International Society for Developmental Psychobiology

http://www.oswego.edu/isdp/

The International Society for Developmental Psychobiology (ISDP) seeks to promote and encourage research on the development of behavior in all organisms including man, with special attention to the effects of biological factors operating at any level of organization. This research may be descriptive or experimental and may take place in natural or controlled conditions. ISDP hold conferences and publishes a journal, which includes articles on human and animal behavior.

International Society for Neuroethology

http://www.neuroethology.org

The International Society for Neuroethology (ISN) is a scholarly society devoted to neuroethology, which is a branch of science that seeks to understand the neural basis of natural behavior in animals. ISN seeks to facilitate dialogue among researchers and provide students and the curious public with information about neuroethology.

International Society on Biotelemetry

http://www.biotelemetry.org/

The International Society on Biotelemetry (ISOB) addresses various subspecialties in the field of biotelemetry, including wildlife telemetry and telemetry in birds and small animals. ISOB holds a conference every other year and its Web site provides links to research, commercial vendors, literature, and other societies in biotelemetry, biomedical engineering, and related fields.

International Union for the Study of Social Insects

http://www.iussi.org

The International Union for the Study of Social Insects (IUSSI) was formed to facilitate communication among social insect researchers worldwide.

Jane Goodall Institute

www.janegoodall.org

Founded by Jane Goodall, the Jane Goodall Institute (JGI) is a global nonprofit which seeks to empower people to make a difference for all living things through healthy ecosystems, promoting sustainable livelihoods, and nurturing new generations of committed, active citizens. JGI works to increase primate habitat conservation; increase awareness of issues related to human's relationships with each other, the environment, and other animals; expand noninvasive research programs on primates; and promote activities that ensure the well being of primates and animal welfare. The Web site provides information on the JGI institute, as well as information on chimpanzee behavior and conservation issues.

J. P. Scott Center for Neuroscience, Mind and Behavior

http://caspar.bgu.edu/%7eneuro/

The J. P. Scott Center for Neuroscience, Mind and Behavior (NM&B) at Bowling Green State University offers an interdisciplinary neuroscience program, which includes strong programs in behavioral pharmacology, classical ethology, behavioral neuroscience, and cognitive science.

Living Links Center

http://www.emory.edu/LIVING_LINKS/

The Living Links Center was established in 1997 to conduct primate studies that shed light on human behavioral evolution. The research goals are to reconstruct human evolution, pinpoint differences and similarities between humans and nonhuman primates, educate the public about primates, and promote primate well-being and conservation. It is an integrated part of the Yerkes Primate Center, the United States's oldest and largest primate center.

Marine Biological Laboratory

www.mbl.edu

The Marine Biological Laboratory (MBL), located in Woods Hole, Massachusetts, is an international center for research, education, and training in biology, biomedicine, and ecology.

Marine Mammal Behavior and Cognition

http://www.usm.edu/psy-kuczaj/

The Marine Mammal Behavior and Cognition program is housed in the department of psychology at the University of Southern Mississippi. The research team includes scientists from around the world and both graduate and undergraduate students. Research efforts are focused on social behavior and cognition in a variety of marine mammal species, with particular attention to the role of the environment and the social group in social and cognitive development.

National Center for Animal Law

http://www.lclark.edu/org/ncal

The National Center for Animal Law seeks to encourage those who wish to practice animal law, further the field of animal law, and promote legal protection for animals. The Center encourages animal law curriculum development, educational opportunities in animal law for students, and financial support for animal law students and attorneys. The center also publishes the journal *Animal Law*.

National Institute of Advanced Studies

www.iisc.ernet.in/nias/

The National Institute of Advanced Studies (NAIS), located in Bangalore, India, is a multi-disciplinary research institute in the natural sciences, social sciences, and humanities. NAIS has an active unit dedicated to research in the areas of cognition, communication, and consciousness in primates; behavioral ecology and conservation biology; genetics; evolutionary biology; zoo biology; and animal welfare.

Noldus Information Technology

http://www.noldus.com/resources/zoo_links.html

The Noldus Information Technology Web site contains links to professional organizations worldwide, Web sites, and news groups, which are interested in all aspects of animal behavior, including ethology, behavioral ecology, and animal welfare, as well as entomology and bioacoustics.

Orca Network

http://orcanetwork.org/

The Orca Network seeks to connect whales and people in the Pacific Northwest and provides information on an extended clan of orcas, which socialize and forage in the inland waters of Washington State and British Columbia. This Web site provides information on orca sightings in this area, orca behavior, and problems facing orcas, including habitat degradation, industrial poisons, and the impacts of human pollution.

Paleoanthropology Society

http://www.paleoanthro.org/default.htm

The Paleoanthropology Society is a group of scholars interested in human evolution and includes physical anthropologists, archaeologists, paleontologists, geologists, and others whose work has the potential to shed light on hominid behavioral biological evolution. One of the pillars of paleoanthropological research is the use of comparative evidence, particularly the use of nonhuman primate behavior, to develop models for early hominid behavior.

Pelagic Shark Research Foundation

www.pelagic.org

The Pelagic Shark Research Foundation (PSFR) funds research projects on shark ethology and behavior and the conservation of sharks, which are particularly vulnerable to habitat destruction. PSFR also seeks to improve shark protection through public education and outreach efforts, which dispel myths about these animals and promote increased awareness of the problems surrounding shark fisheries.

People for the Ethical Treatment of Animals

www.peta-online.org

People for the Ethical Treatment of Animals (PETA) is an international animal advocacy organization. PETA's agenda is interdisciplinary and broad-based and is concerned with the treatment of wild, captive, and domestic animals in such activities as education, research, entertainment, clothing, and factory farming.

PetPlace.com
www.petplace.com

PetPlace.com provides information on pet care, including dogs, cats, birds, horses, small mammals, reptiles, or fish. This is a commercial site that provides, among services and products, short articles on various aspects of living with a pet, including feeding and nutrition, exercise, well being of pets, bonding, behavioral problems, first aid, and dealing with the loss of a pet.

Physicians Committee for Responsible Medicine
www.pcrm.org

The Physicians Committee for Responsible Medicine (PCRM) was founded in 1985 to promote preventative medicine, conduct clinical research, and encourage higher standards for ethics and effectiveness in research. PCRM promotes discussion about the role of animal models in education and studying human disease, and advocates for humane alternatives to medical research using animals, including ending gruesome experiments and promoting nonanimal research methods.

PRBO Conservation Science
http://prbo.org

PRBO Conservation Science is dedicated to conserving birds, other wildlife, and ecosystems through innovative scientific research and outreach. Founded in 1965 at Point Reyes Bird Observatory, the more than 120 staff and seasonal biologists study birds to protect and enhance biodiversity in marine, terrestrial, and wetland systems in western North America.

Primate Research Institute, Kyoto University
http://www.pri.kyoto-u.ac.jp

The Primate Research Institute at Kyoto University is Japan's only national center for the study of nonhuman primates. The institute houses 20 species and around 800 individuals and also has several field stations in Africa and East Asia. The institute seeks to understand the biological, behavioral, and socioecological aspects of primates, and the origin and evolution of man.

Psychologists for the Ethical Treatment of Animals
www.psyeta.org

The Psychologists for the Ethical Treatment of Animals (PSYETA) is a national nonprofit organization of social scientists and mental health providers working to reduce the exploitation and suffering of nonhuman animals through education, research, and applied psychology.

The Research School of Biological Sciences, Australian National University
http://www.rsbs.anu.edu.au/

The Research School of Biological Sciences conducts research in biology and biotechnology, including research in agriculture, environment, health, and technology. Their Centre for Visual Sciences specializes in insect vision, perception, navigation, memory, and cognition.

School of Botany and Zoology, Australian National University

http://www.anu.edu.au/BoZo/

The School of Botany and Zoology at the Australian National University is concentrated on behavioral ecology and evolution, especially in the areas of mating systems, sexual selection and communication in birds, reptiles, frogs, and invertebrates. The school is involved in research and offers undergraduate and graduate programs.

Shark Research Institute

http://www.sharks.org/main_menu.html

The Shark Research Institute (SRI) was created in 1991 at Princeton to sponsor and conduct research on sharks and promote their conservation. SRI works with the scientific community, individuals, and organizations concerned about the health of our marine ecosystem. SRI seeks to correct misperceptions about sharks and end the slaughter of 100 million sharks annually. Among their research projects, SRI has collected and analyzed shark attacks and studied the behavior pattern and body language of sharks when they are in proximity to humans in order to determine the cause of attacks and reduce future risks.

Società Italiana di Etologia

http://w3.uniroma1.it/sie/

The Società Italiana di Etologia (Italian Society for the Study of Animal Behavior) is a professional scientific society devoted to the study of animal behavior.

Society for Conservation Biology

http://conbio.net/

The Society for Conservation Biology (SCB) is dedicated to promoting the scientific study of the phenomena that affect the maintenance, loss, and restoration of biological diversity. Members comprise a wide range of people interested in the conservation and study of biological diversity, including resource managers, educators, government and private conservation workers, and students. SCB was formed to develop the scientific and technical means for the protection, maintenance, and restoration of life on Earth, including species, ecosystems, and the processes that sustain them.

Society for Integrative and Comparative Biology

www.sicb.org/

The Society of Integrative and Comparative Biology (SICB) is dedicated to promoting the pursuit and public dissemination of important information relating to all aspects of biology, including animal behavior. SICB holds an annual meeting and publishes a bimonthly journal.

The Society for Marine Mammalogy

http://www.marinemammalogy.org/

The Society for Marine Mammalogy is a professional society dedicated to understanding the biology, behavior, and conservation of marine mammals.

Society for Neuroscience

http://web.sfn.org/

The Society for Neuroscience is a nonprofit membership organization of scientists and physicians who study the brain and nervous system. Neuroscience includes the study of brain development, sensation and perception, learning and memory, movement, sleep, stress, aging, and neurological and psychiatric disorders. It also includes the molecules, cells, and genes responsible for nervous system functioning. Recognizing the tremendous potential for the study of the brain and nervous system as a separate field, the society was formed in 1970. It has grown from 500 members to more than 34,000 and is the world's largest organization of scientists devoted to the study of the brain.

Society for Quantitative Analyses of Behavior

http://sqab.psychology.org/

The Society for Quantitative Analyses of Behavior (SQAB) was founded in 1978 to present symposia and publish material, which bring a quantitative analysis to bear on the understanding of animal behavior. SQAB holds an annual meeting where talks focus on the development and use of mathematical formulations to characterize one or more dimensions of an obtained data set, derive predictions to be compared with data, and generate novel data analyses.

TerraMar Research

www.TerraMarResearch.org

TerraMar Research is dedicated to the study and protection of marine mammals. Research activities include human/cetacean interactions; dolphin behavior and communication; behavioral indicators of stress, dolphin emotion, cognition, and awareness; and marine mammal protection and conservation.

Texas Marine Mammal Stranding Network

http://www.tmmsn.org/

Texas Marine Mammal Stranding Network (TMMSN) is dedicated to the conservation of marine mammals through rescue and rehabilitation, research, and education. TMMSN rescues and rehabilitates marine mammals that strand along the Texas Coast, providing food, medical treatment, and 24-hour-a-day observation. TMMSN also collects data from both live animals and those found dead, providing insight into the basic biology, causes of illness and death, feeding, and migration habits of marine mammals.

University of Houston–Clear Lake Dolphin Research Team

http://members.aol.com/rpweeks/uhcl-drt.htm

The University of Houston–Clear Lake Dolphin Research Team (UHCL-DRT) Web site provides information about behavioral observations of free-ranging bottlenose dolphins in the area of Galveston Island, Texas. This site includes photo identification of individual dolphins, population surveys, and behavioral observations, including how dolphins spend their time, the structure of dolphin groups, interactions between individuals and groups, and behavioral adaptability.

University of Plymouth

www.plymouth.ac.uk

The University of Plymouth's Web site provides information about the university's programs, which include animal behavior research in both the biology and psychology departments, on a wide range of species, from invertebrates to livestock and companion animals to zoo animals.

What You Need to Know about Animals/Wildlife

http://animals.about.com/

Part of the "What You Need to Know about . . ." series, this commercial Web site offers news and basic information about animals and wildlife. Provides many links to other sites.

Wildlife Conervation Society and Living Landscape Program

http://www.wcs.org

The Wildlife Conservation Society (WCS) promotes conservation of wildlife and wild lands through science, international conservation, education, and the management of the world's largest system of urban wildlife parks, led by the flagship Bronx Zoo. WCS works to change individual attitudes toward nature and help people imagine wildlife and humans living in sustainable interactions on both a local and a global scale. For example, the WCS's Living Landscape Program (*http://www.wcslivinglandscapes.org*) realizes that animals do not recognize park boundaries and so seeks to involve local communities in protecting wildlife in areas surrounding parks.

The Wilson Ornithological Society

www.ummz.lsa.umich.edu/birds/wos.html

The Wilson Ornithological Society is an international organization composed of people who share a curiosity about birds including their behavior, ethology, and conservation. This society recognizes the importance of serious amateurs in ornithology.

Category Listings

General Organizations and Societies

American Psychological Society
http://www.psychologicalscience.org/

American Society of Mammalogists
www.mammalsociety.org/

Animal Behavior Society
http://www.animalbehavior.org/

Animal Behaviour Web Sites
http://www.societies.ncl.ac.uk/asab/websites.html

The Association for the Study of Animal Behaviour
http://www.societies.ncl.ac.uk/asab/

Center for the Integrative Study of Animal Behavior
http://www.indiana.edu/~animal/

Comparative Cognition Society
http://www.pigeon.psy.tufts.edu/ccs/default.htm

The Electronic Zoo
http://netvet.wustl.edu/e-zoo.htm

Hamburgische Gesellschaft für Verhaltensbiologie
http://www.rrz.uni-hamburg.de/HGV/

Instituto de Investigación en Recursos Cinegéticos
http://www.uclm.es/IREC/

International Society for Behavioral Ecology
http://web.unbc.ca/isbe/

International Society for Comparative Psychology
http://www.comparativepsychology.org/

International Society for Developmental Psychobiology
http://www.oswego.edu/isdp/

Marine Biological Laboratory
www.mbl.edu

National Institute of Advanced Studies
www.iisc.ernet.in/nias/

Paleoanthropology Society
http://www.paleoanthro.org/default.htm

Società Italiana di Etologia
http://w3.uniroma1.it/sie/

Society for Conservation Biology
http://conio.net/

Society of Integrative and Comparative Biology
www.sicb.org/

Applied Ethology

Applied Ethology on the World Wide Web
http://www.usask.ca/wcvm/herdmed/applied-ethology/

International Society for Anthrozoology
http://www.vetmed.ucdavis.edu/CCAB/ISAZ.htm

International Society for Applied Ethology
http://www.applied-ethology.org/

Acoustics

BioAcoustics Team
http://www.cb.u-psud.fr/cb/index.html

The Borror Laboratory of Bioacoustics
http://blb.biosci.ohio-state.edu

Bats

Colorado Bat Society
www.coloradobats.org

Birds

American Ornithologists' Union
www.aou.org

Cooper Ornithological Society
www.cooper.org

Cornell Lab of Ornithology
http://birds.cornell.edu

PRBO Conservation Science
http://prbo.org

The Wilson Ornithological Society
www.ummz.lsa.umich.edu/birds/wos.html

Conservation

American Zoo and Aquarium Association
www.aza.org

Institute for the Conservation of Tropical Environments
http://icte.bio.sunysb.edu

Jane Goodall Institute
www.janegoodall.org

Orca Network
http://orcanetwork.org/

PRBO Conservation Science
http://prbo.org

Society for Conservation Biology
http://conbio.net/SCB/Information/Mission/

The Society for Marine Mammalogy
http://www.marinemammalogy.org/

Wildlife Conservation Society
http://www.wcs.org

The Wilson Ornithological Society
www.ummz.lsa.umich.edu/birds/wos.html

Dolphins

Dolphin Communication Project
www.dolphincommunicationproject.org

The Dolphins of Monkey Mia Research Foundation
http://www.monkeymiadolphins.org/

University of Houston–Clear Lake Dolphin Research Team
http://members.aol.com/rpweeks/uhcl-drt.htm

Education

The American Society fo the Prevention of Cruelty to Animals
www.aspca.org

Animal Behavior Society
http://www.animalbehavior.org

American Zoo and Aquarium Association
www.aza.org

Iowa Primate Learning Sanctuary
www.iowagreatapes.com

Living Links Center
http://www.emory.edu/LIVING_LINKS/

Wildlife Conservation Society
http://www.wcs.org

Ethics and Animal Well-Being

American Anti-Vivisection Society
www.aavs.org

Animal Sciences Group of Wageningen University and Research Centre
http://www.asg.wur.nl/english/index.asp

The American Society for the Prevention of Cruelty to Animals
www.aspca.org

Ethologists for the Ethical Treatment of Animals/Citizens for Responsible Animal Behavior Studies
www.ethologicalethics.org

The Humane Society of the United States
www.hsus.org

People for the Ethical Treatment of Animals
www.peta-online.org

Physicians Committee for Responsible Medicine
www.pcrm.org

Psychologists for the Ethical Treatment of Animals
www.psyeta.org

Insects

BugBios
www.insects.org

International Union for the Study of Social Insects
http://www.iussi.org

Marine Mammals

Marine Biological Laboratory
www.mbl.edu

Marine Mammal Behavior and Cognition
http://www.usm.edu/psy-kuczaj/

Orca Network
http://orcanetwork.org/

The Society for Marine Mammalogy
http://www.marinemammalogy.org/

TerraMar Research
www.TerraMarResearch.org

Texas Marine Mammal Stranding Network
http://www.tmmsn.org/

Methods

International Society on Biotelemetry
http://www.biotelemetry.org/

Noldus Information Technology
http://www.noldus.com/resources/zoo_links.html

Society for Quantitative Analyses
http://sqab.psychology.org/

Neuroethology

International Society for Neuroethology
http://www.neuroethology.org

Society for Neuroscience
http://web.sfn.org/

Mollusks

American Malacological Society
http://erato.acnatsci.org/ams/

Primates

American Anthropological Association
http://www.aaanet.org/

American Society of Primatologists
www.asp.org

International Primatological Society
http://www.primate.wisc.edu/pin/ips.html

Iowa Primate Learning Sanctuary
www.iowagreatapes.com

Jane Goodall Institute
www.janegoodall.org

Living Links Center
http://www.emory.edu/LIVING_LINKS/

Primate Research Institute, Kyoto University
http://www.pri.kyoto-u.ac.jp

Salmon

Atlantic Salmon Federation
www.asf.ca

Sharks

Pelagic Shark Research Foundation
www.pelagic.org

Shark Research Institute
http://www.sharks.org/main_menu.html

Veterinary Medicine

American Veterinary Medical Association
www.avma.org

Animal Behavior Network
www.AnimalBehavior.Net

PetPlace.com
www.petplace.com

Zoos

American Zoo and Aquarium Association
www.aza.org

International Animal Behavior Organizations and Web Sites (Geographical Listings Outside of the U.S.A.)

Australia

Centre for Neuroscience and Animal Behaviour
http://sciences.une.edu.au/bbms/centre_neuroscience_animal_behaviour.asp

Research School of Biological Sciences, Australian National University
http://www.rsbs.anu.edu.au/

School of Botany and Zoology, Australian National University
http://www.anu.edu.au/BoZo/

France

BioAcoustics Team
http://www.cb.u-psud.fr/cb/index.html

Germany

Hamburgische Gesellschaft für Verhaltensbiologie
http://www.rrz.uni-hamburg.de/HGV/

India

National Institute of Advanced Studies
www.iisc.ernet.in/nias/

Italy

Società Italiana di Etologia
http://w3.uniroma1.it/sie/

Japan

Primate Research Institute, Kyoto University
http://www.pri.kyoto-u.ac.jp

Madagascar

Institute for the Conservation of Tropical Environments
http://icte.bio.sunysb.edu

The Netherlands

Animal Sciences Group of Wageningen University and Research Centre
http://www.asg.wur.nl/english/index.asp

Portugal

Eco-Ethology Research Unit
http://www.ispa.pt/ispa_frame_00.asp?url=ispa_cinvest_01.asp%3Fsubseccao%3D4

Spain

Instituto de Investigación en Recursos Cinegéticos
http://www.uclm.es/IREC/

Switzerland

Geneva International Centre for Humanitarian Demining
www.gichd.ch

United Kingdom

Animal Behaviour Web Sites
http://www.societies.ncl.ac.uk/asab/websites.html

Animal-Info Net (or Animal Information Network)

http://www.animal-info.net

The Association for the Study of Animal Behaviour
http://www.societies.ncl.ac.uk/asab/

University of Plymouth
www.plymouth.ac.uk

■ Suggested Resources in the Study of Animal Behavior

The "Further Resources" sections that accompany each essay along with the list of international organizations and Web sites at the end of this encyclopedia, when taken as a whole, provide a gold mine of information about the wonderful and challenging field of animal behavior. There also is much information on possible careers in the field of animal behavior. One career that is not mentioned is cartoon drawing—think of the work of Gary Larson ("The Far Side") and Patrick McDonnell ("Mutts").

Here are a few more general resources. While this list does not contain *all* available references, the compilation of sources in this encyclopedia and the references that are contained in each of the sources (books, monographs, journal articles, magazine essays, videos, and the lists of professional journals and magazines) is encyclopedic in and of itself, and you will be able to find information on just about any topic in the field of animal behavior. Note that some of these books have been published in 2004, so they are very current. An excellent series titled *Perspectives in Ethology* is published by Plenum Press (New York).

Journals and Magazines

Some professional journals and magazines that will be of interest include:
American Journal of Primatology; American Naturalist; Animal Behaviour; Animals and Society; Animal Cognition; Animal Conservation; Anthrozoös; Auk; BBC Wildlife; Behaviour; Behavioral and Brain Sciences; Behavioral Ecology; Behavioral Ecology and Sociobiology; Biological Conservation; Canadian Journal of Zoology; Condor; Conservation Biology; Copeia; Ecological Monographs; Ecology; Ethology; Ethology, Ecology, and Evolution; Evolution; Folia Primatologica; Journal of Ethology; Journal of Herpetology; Journal of Ichthyology; Journal of Theoretical Biology; Journal of Zoology (London); National Geographic; Nature; Natural History Magazine; New Scientist; Oecologia; Primates; Science; Science News; Scientific American; Smithsonian Magazine; Trends in Cognitive Science; Trends in Ecology and Evolution; Trends in Neuroscience; and Wilson Bulletin.

There also are many other journals that are published by professional societies that focus on a particular group of animals and by regional professional societies (see the list of International Animal Behavior Organizations and Web Sites).

Recommended Articles and Books

Alcock, J. 2001. *Animal Behavior: An Evolutionary Approach.* Sunderland, MA: Sinauer Associates, Inc.

Alcock, J. 2001. *The Triumph of Sociobiology.* New York: Oxford University Press.

Allen, C. & Bekoff, M. 1997. *Species of Mind: The Philosophy and Biology of Cognitive Ethology.* Cambridge, MA: MIT Press.

Altmann, S. A. 1998. *Forging for Survival: Yearling Baboons in Africa.* Chicago: University of Chicago Press.

Barash, D. P. 1982. *Sociobiology and Behavior.* New York: Elsevier.

Barnard, C. 2004. *Animal Behaviour: Mechanism, Development, Function, and Evolution.* Essex, UK: Pearson Educational Limited.

Barrows, E. M. 2001. *Animal Behavior Desk Reference: A Dictionary of Animal Behavior, Ecology, and Evolution.* Boca Raton, FL: CRC Press.

Bekoff, M. 1994. *Cognitive ethology and the treatment of non-human animals: How matters of mind inform matters of welfare.* Animal Welfare, 3, 75–96.

Bekoff, M. (Ed.) 1998. *Encyclopedia of Animal Rights and Animal Welfare.* Greenwood Publishing Group, Inc., Westport, Connecticut.

Bekoff, M. (Ed.) 2000. *The Smile of a Dolphin: Remarkable Accounts of Animal Emotions.* Washington, D. C.: Random House/Discovery Books.

Bekoff, M. 2000. *Strolling with our Kin: Speaking for and Respecting Voiceless Animals.* Jenkinstown, PA: AAVS; Distributed by: New York: Lantern Books.

Bekoff, M. 2002. *Minding Animals: Awareness, Emotions, and Heart.* New York: Oxford University Press.

Bekoff, M. 2005. *Consummate Companions: Why We Need Animals Now More than Ever.* Philadelphia: Temple University Press.

Bekoff, M., Allen, C. & Burghardt, G. M. (Eds.) 2002. *The Cognitive Animal: Empirical and Theoretical Perspectives on Animal Cognition.* Cambridge, MA: MIT Press.

Bekoff, M. & Jamieson, D. 1990. (Eds.) *Interpretation and Explanation in the Study of Animal Behavior. Volume I: Interpretation, intentionality, and communication. Volume II: Explanation, evolution, and adaptation.* Boulder, CO: Westview Press.

Birkhead, T. 2000. *An Evolutionary History of Sperm Competition.* London, UK: Faber and Faber Limited.

Boitani, L. & Fuller T. K. (Eds.) 2000. *Research Techniques in Animal Ecology: Controversies and Consequences.* New York: Columbia University Press.

Bolhuis, J. J. & Hogan, J. A. (Eds.) 1999. *The Development of Animal Behavior: A Reader.* Malden, MA: Blackwell Publishers.

Bradbury, J. W. & Vehrencamp, S. L. 1998. *Principles of Animal Communication.* Sunderland, MA: Sinauer Associates, Inc.

Brown, J. L. 1975. *The Evolution of Behavior.* New York: W. W. Norton & Company, Inc.

Burghardt, G. M. & Bekoff, M. (Eds.) 1978. *The Development of Behavior: Comparative and Evolutionary Aspects.* New York: Garland Publishers.

Burkhardt, R. W. Jr. 1981. *On the emergence of ethology as a scientific discipline.* Conspectus of History, 1, 62–81.

Burkhardt, R. W. Jr. 1990. *Theory and practice in naturalistic studies of behavior prior to ethology's establishment as a scientific discipline.* In: *Interpretation, and Explanation in the Study of Behavior, Vol 2. Explanation, Evolution, and Adaptation.* (Ed. by M. Bekoff & D. Jamieson), pp. 1–25. Boulder, CO: Westview Press.

Burkhardt, R. W., Jr. 2005. *Patterns of Behavior: Konrad Lorenz, Niko Tinbergen, and the Founding of Ethology.* Chicago: University of Chicago Press.

Caro, T. (Ed.) 1998. *Behavioral Ecology and Conservation Biology.* New York: Oxford University Press.

Darwin, C. 1872/1998. *The Expression of the Emotions in Man and Animals.* 3rd edn. New York: Oxford University Press. (with an introduction, afterword, and commentaries by Paul Ekman)

Dawkins, M. S. 1993. *Through Our Eyes Only? The Search For Animal Consciousness.* Oxford, UK: Blackwell.

Dawkins, M. S. 1995. *Unraveling Animal Behaviour.* Essex, UK: Longman Scientific & Technical.

Dawkins, R. 1976. *The Selfish Gene.* New York: Oxford University Press.

Drickamer, L. C., Vessey, S. H. & Jakob, E. M. 2002. *Animal Behavior: Mechanisms, Ecology, and Evolution.* New York: McGraw-Hill.

Dugatkin, L. A. 2004. *Principles of Animal Behavior.* New York: W. W. Norton & Company.

Eibl-Eibesfeldt, I. 1970. *Ethology.* New York: Holt, Rinehart, and Winston.

Eisner, T. 2003. *For Love of Insects.* Cambridge, MA: Harvard University Press.

Ewer, R. F. 1968. *Ethology of Mammals.* New York: Plenum Press.

Festa-Bianchet, M. & Apollonio, M. (Eds.) 2003. *Animal Behavior and Wildlife Conservation.* Washington, D. C.: Island Press.

Friend, T. 2004. *Animal Talk: Breaking the Codes of Animal Language*. New York: Free Press.

Goodall, J. 1986. *The Chimpanzees of Gombe*. Cambridge, MA: Harvard University Press.

Goodall, J. & Bekoff, M. 2002. *The Ten Trusts: What We Must Do to Care for the Animals We Love*. San Francisco: HarperCollins.

Goodenough, J., McGuire, B. & Wallace, R. 1993. *Perspectives on Animal Behavior*. New York: John Wiley & Sons.

Gosling, L. M. & Sutherland, W. J. (Eds.) 2000. *Behavior and Conservation*. New York: Cambridge University Press.

Gould, J. L. 1982. *Ethology: Mechanisms and Evolution of Behavior*. New York: W. W. Norton & Sons.

The Great Ape Project Census. 2003. *The Great Ape Project Census: Recognition for the Uncounted*. Portland, OR: Great Ape Project (GAP) Books.

Greenberg, G. & Haraway, M. M. (Eds.) 1998. *Comparative Psychology: A Handbook*. New York: Garland Publishers, Inc.

Greenwood, P. J., Harvey, P. H. & Slatkin, M. (Eds.). 1985. *Evolution: Essays in Honor of John Maynard Smith*. New York: Cambridge University Press.

Griffin, D. R. 2001. *Animal Minds*. Chicago: University of Chicago Press.

Grzimek, B. 1977. *Grzimek's Encyclopedia of Ethology* (Ed. by K. Immelman). New York: Van Nostrand Reinhold.

Hafez, E. S. E. (Ed.) 1975. *The Behaviour of Domestic Animals*. 3rd edn. London: Balliere, Tindall.

Hancocks, D. 2001. *A Different Nature: The Paradoxical World of Zoos and their Uncertain Future*. Berkeley: University of California Press.

Hauser, M. 2000. *Wild Minds*. New York: Henry Holt.

Hinde, R. A. 1970. *Animal Behavior: A Synthesis of Ethology and Comparative Psychology*. New York: McGraw-Hill.

Hölldobler, B. & Wilson, E. O. 1990. *The Ants*. Cambridge, MA: Harvard University Press.

Honoré, E. K. & Klopfer, P. H. 1990. *A Concise Survey of Animal Behavior*. New York: Academic Press.

Houck, L. D. & Drickamer, L. D. (Eds.) 1996. *Foundations of Animal Behavior: Classic Papers with Commentaries*. Chicago: University of Chicago Press.

Hrdy, S. B. 1999. *Mother Nature: A History of Mothers, Infants, and Natural Selection*. New York: Pantheon Books.

Immelman, K. & Beer, C. 1989. *A dictionary of Ethology*. Cambridge, MA: Harvard University Press.

Keller, E. F. & Lloyd, E. A. (Eds.). 1992. *Keywords in Evolutionary Biology*. Cambridge, MA: Harvard University Press.

Kemp, David D. 1998. *The Environment Dictionary*. New York and London: Routledge.

Klopfer, P. H. 1999. *Politics and People in Ethology*. Lewisburg, PA: Bucknell University Press.

Krebs, J. R. & Davies, N. B. 1993. *An Introduction to Behavioural Ecology*. Boston and London: Blackwell Scientific Publications.

Krebs, J. R. & Davies, N. B. (Eds.) 1997. *Behavioural Ecology: An Evolutionary Approach*. Boston and London: Blackwell Scientific Publications. (Earlier editions were published in 1978, 1984, and 1991.)

Kruuk, H. 2003. *Niko's Nature*. New York: Oxford University Press.

Lehner, P. N. 1996. *Handbook of Ethological Methods*. New York: Cambridge University Press.

Lockwood, R. & Ascione, F. R. (Eds.) 1998. *Cruelty to Animals and Interpersonal Violence*. West Lafayette, IN: Purdue University Press.

Lorenz, K. Z. 1965. *Evolution and Modification of Behavior*. Chicago: University of Chicago Press.

Lorenz, K. Z. 1966. *On Aggression*. New York: Harcourt, Brace and World.

Lorenz, K. Z. 1981. *The Foundations of Ethology*. New York: Springer-Verlag.

Lorenz, K. Z. 1988. *Here am I—Where Are You? The Behavior of the Greylag Goose*. New York: Harcourt Brace Jovanovich, Publishers.

Manning, A. & Dawkins, M. D. 1998. *An Introduction to Animal Behaviour*. New York: Cambridge University Press.

Marler, P. & Hamilton III, W. D. 1966. *Mechanisms of Animal Behavior*. New York: John Wiley & Sons.

Martin, P. & Bateson, P. 1986. *Measuring Behavior: An Introductory Guide*. New York: Cambridge University Press.

Maynard, C. 2001. *Bugs: A Close View of the Insect World*. New York: DK Publishing.

Maynard, T. 2000. *Working with Wildlife: A Guide to Careers in the Animal World*. New York: Franklin Watts.

McDonnell, P. 1996. *Mutts*. Kansas City, MO: Andrews and McNeel.

McFarland, D. (Ed.) 1982. *The Oxford Companion to Animal Behaviour*. New York: Oxford University Press.

McFarland, D. 1985. *Animal Behavior: Psychobiology, Ethology, and Evolution*. Reading, MA: Benjamin/Cummings Publishing Company, Inc.

Perspectives in Ethology (Series). New York: Plenum Press.

Sax, B. 2001. *The Mythical Zoo: Encyclopedia of Animals in World Myth, Legend, and Literature*. Denver, CO: ABC Clio.

Schiller, C. H. (Trans. and Ed.) 1957. *Instinctive Behavior: The Development of a Modern Concept*. New York: International Universities Press, Inc.

Schneider, S. H. & Root, T. L. (Eds.) 2002. *Wildlife Responses to Climate Change: North American Case Studies*. Washington, D. C.: Island Press.

Sherman, P. W. & Alcock, J. (Eds.) 1998. *Exploring Animal Behavior: Readings from American Scientist*. Sunderland, MA: Sinauer Associates, Inc.

Siiter, R. 1999. *Introduction to Animal Behavior*. Pacific Grove, CA: Brooks/Cole Publishing Company.

Slater, P. (Ed.) 1986 *The Encyclopedia of Animal Behavior*. London: William Collins Sons and Company Ltd.

Smuts, B. 1999. *Sex and Friendship in Baboons*. Cambridge, MA: Harvard University Press.

Thorpe, W. H. 1979. *The Origins and Rise of Ethology*. London: Heinemann Educational Books, Ltd.

Thorpe, W. H. & Zangwill, O. L. (Eds.). 1966. *Current Problems in Animal Behaviour*. New York: Cambridge University Press.

Tinbergen, N. 1951/1989. *The Study of Instinct*. New York: Oxford University Press.

Tinbergen, N. 1953. *The Herring Gull's World*. London: Collins.

Tinbergen, N. 1958. *Curious Naturalists*. London: Country Life.

Tinbergen, N. 1972. *The Animal in Its World: Field Studies*. Cambridge: MA: Harvard University Press.

Tinbergen, N. 1973. *The Animal in its World: Explorations of an Ethologist*. Cambridge, MA: Harvard University Press.

Trivers, R. 1985. *Social Evolution*. Reading, MA: Benjamin/Cummings Publishing Company, Inc.

Wilson, E. O. 1975. *Sociobiology: The New Synthesis*. Cambridge, MA: Harvard University Press.

Zuk, M. 2002. *Sexual Selections: What We Can and Can't Learn about Sex from Animals*. Berkeley: University of California Press.

◼ About the Editor and Contributors

The Editor

Marc Bekoff
Professor of Ecology and Evolutionary Biology, University of Colorado, Boulder, Colorado
  Cognition—*Cognitive Ethology: The Comparative Study of Animal Minds*
  Communication—Olfaction—*Dog Scents and "Yellow Snow"*
  Communication—Vocal—*Social Communication in Dogs: The Subtleties of Silent Language*
  Coyotes—*Clever Tricksters, Protean Predators*
  Human (Anthropogenic) Effects—*Human (Anthropogenic) Effects on Animal Behavior*
  Play—*Social Play Behavior and Social Morality*

The Contributors

Charles I. Abramson
Professor of Psychology, Oklahoma State University, Stillwater, Oklahoma
  Education—Classroom Actitivites—*"Petscope"*
  Education—Classroom Activities—*Planarians*
  Turner, Charles Henry (1867–1923)
  Washburn, Margaret Floy (1871–1939)

Rick A. Adams
Associate Professor of Biology, University of Northern Colorado, Greeley, Colorado
  Bats—*The Behavior of a Mysterious Mammal*

Elsa Addessi
Istituto di Scienze e Tecnologie della Cognizione, CNR Consiglio Nazionale delle Ricerche, Rome, Italy
  Feeding Behavior—*Social Learning: Food*

John Alcock
Professor of Biology, Arizona State University, Tempe, Arizona
  Sociobiology

Colin Allen
Professor, Department of History and Philosophy of Science and Program in Cognitive Science, Indiana University, Bloomington, Indiana
  Cognition—*Animal Consciousness*
  Cognition—*Cognitive Ethology: The Comparative Study of Animal Minds*

Vitor Almada
Professor of Biology, Eco-Ethology Research Unit, Instituto Superior de
  Psicologia Aplicada (ISPA), Lisboa, Portugal
  Caregiving—*Parental Care in Fish*
  Reproductive Behavior—*Adaptations and Exaptations in the Study of Behavior*
  Reproductive Behavior—*Alternative Male Reproductive Tactics in Fish*

Sofyan Alyan
Associate Professor of Biology, United Arab Emirates University, United Arab Eimrates.
  Navigation—*Spatial Navigation*

Roland C. Anderson
Biologist, The Seattle Aquarium, Seattle, Washington
  Cephalopods—*Roving Octopuses*
  Reproductive Behavior—*Mollusk Mating Behaviors*

Kenneth B. Armitage
Professor Emeritus of Ecology & Evolutionary Biology, University of Kansas, Lawrence, Kansas
  Social Organization—*Socioecology of Marmots*

Byron Arnason
Geophysicist, Austin, Texas
  Communication—Auditory—*Long Distance Calling, the Elephant's Way*

Pat Backwell
Zoologist, The Australian National University, Canberra, Australia
  Cognition—*Deception*

Jonathan Balcombe
Research Consultant, Physicians Committee for Responsible Medicine, York, England
  Emotions—*Pleasure*

Russell P. Balda
Regents' Professor of Biology, Northern Arizona University, Flagstaff, Arizona
  Cognition—*Caching Behavior*

Elizabeth Balko
Visiting Assistant Professor of Biology, SUNY—Environmental Science & Forestry, Syracuse,
  New York; and State University of New York—Oswego, Oswego, New York
  Lemurs—*Behavioral Ecology of Lemurs*

Jonathan R. Banks
School of Biological Sciences, University of Manchester, Manchester, United Kingdom
  Communication—*Electrocommunication*
  Electric Fish

Craig Barnett
Evolution and Behaviour Research Group, University of Newcastle upon Tyne, Newcastle upon
  Tyne, United Kingdom
  Behavioral Plasticity

Jeffery L. Beacham
Environmental Research and Service Institute for Public Service and Policy Research, University of South Carolina, Columbia, South Carolina
Dominance—*Development of Dominance Hierarchies*

Guy Beauchamp
Research Officer, Faculty of Veterinary Medicine, St-Hyacinthe, Quebéc, Canada
Feeding Behavior—*Social Foraging*

Andy Beck
White Horse Equine Ethology Project, Northland, Aotearoa, New Zealand
Horses

Peter A. Bednekoff
Associate Professor of Biology, Eastern Michigan University, Ypsilanti, Michigan
Antipredatory Behavior—*Finding Food While Avoiding Predators*
Antipredatory Behavior—*Sentinel Behavior*

Anne Bekoff
Professor of Integrative Physiology, University of Colorado, Boulder, Colorado
Development—*Embryo Behavior*

Joel Berger
Senior Field Biologist, Wildlife Conservation Society, Moose, Wyoming
Predatory Behavior—*Ghost Predators and their Prey*

Penny Bernstein
Associate Professor of Biological Sciences, Kent State University Stark Campus, Canton, Ohio
Communication—Vocal—*Jump-Yips of Black-Tailed Prairie Dogs*

Mark S. Blumberg
Professor of Psychology, University of Iowa, Iowa City, Iowa
Behavioral Physiology—*Thermoregulatory Behavior*

Daniel T. Blumstein
Assistant Professor of Ecology and Evolution, University of California Los Angeles, California
Antipredatory Behavior—*A Brief Overview*
Communication—Vocal—*Alarm Calls*

Gerald Borgia
Professor of Biology, University of Maryland, College Park, Maryland
Reproductive Behavior—*Bowerbirds and Sexual Displays*

Dalila Bovet
Associate Professor, Laboratoire d'Ethologie et Cognition Comparees, Université Paris X, Nantere, France
Cognition—*Categorization Processes in Animals*

Sarah T. Boysen
Professor of Psychology, The Ohio State University, Columbus, Ohio
Learning—*Spatial Learning*

Michael D. Breed
Professor of Ecology and Evolutionary Biology, University of Colorado,
   Boulder, Colorado
   Communication—*Honeybee Dance Language*
   Communication—Olfaction—*Chemical Communication*
   Education—Classroom Activities—*Insects in the Classroom*

Alyn R. Brereton
Instructor of International Baccalaureate Psychology/International Baccalaureate
   Anthropology, Modesto High School, Modesto, California
   Aggressive Behavior—*Dominance: Female Chums or Male Hired Guns*

William Bright
Professor of Linguistics, University of Colorado, Boulder, Colorado
   Animals in Myth and Lore—*Animals in Native American Lore*

Susan Brooks
Clinical Psychologist, Green Chimneys Children's Services, Brewster, New York
   Careers—*Animal-Assisted Psychotherapy*

Sarah F. Brosnan
Postdoctoral Fellow, Emory University, Atlanta, Georgia
   Cognition—*Fairness in Monkeys*

Joanna Burger
Distinguished Professor of Biology, Rutgers University, Piscataway, New Jersey
   Caregiving—*Incubation*
   Human (Anthropogenic) Effects—*Pollution and Behavior*

Gordon M. Burghardt
Professor of Psychology and Ecology & Evolutionary Biology, University of Tennessee,
   Knoxville, Tennessee
   Anthropomorphism
   Nature and Nurture—*Baldwin Effect*

Richard W. Burkhardt, Jr.
Professor of History, University of Illinois at Urbana-Champaign, Illinois
   Craig, Wallace (1876–1954)
   History—*Niko Tinbergen and the "Four Questions" of Ethology*
   Whitman, Charles Otis (1842–1910)

John A. Byers
Professor of Zoology, University of Idaho, Moscow, Idaho
   Antipredatory Behavior—*Hiding Behavior of Young Ungulates*

Jennifer Calkins
Research Scientist, University of Washington, Seattle, Washington
   Animal in Myth and Lore—*Mythology and Animal Behavior*
   Methods—*Molecular Techniques*

Frank Cézilly
GDR Ecologie Comportementale, Université de Bourgogne, Dijon, France
   Reproductive Behavior—*Assortative Mating*
   Reproductive Behavior—*Monogamy*

Arnold S. Chamove
Senior Lecturer, Massey University, Palmerston North, New Zealand
Careers—*Unconventional Uses of Animal Behavior*
Welfare, Well-Being, and Pain—*Enrichment for Monkeys*

Jackie Chappell
School of Biosciences, University of Birmingham, Birmingham, United Kingdom
Tools—*Tool Use and Manufacture by Birds*

Ivan D. Chase
Professor of Psychology, State University of New York, Stony Brook, New York
Dominance—*Development of Dominance Hierarchies*

Dorothy L. Cheney
Professor of Biology, University of Pennsylvania, Philadelphia, Pennsylvania
Cognition—*Social Cognition in Primates and Other Animals*

Mertice M. Clark
Department of Psychology, McMaster University, Hamilton, Ontario, Canada
Development—*Intrauterine Position Effect*

Rebecca Clark
School of Life Sciences, Arizona State University, Tempe, Arizona
Social Organization—*Eusociality*

N. S. Clayton
Department of Experimental Psychology, University of Cambridge, Cambridge, United Kingdom
Cognition—*Cache Robbing*

Michael Lamport Commons
Harvard Medical School, Boston, Massachusetts
Development—*Behavioral Stages*

Kathleen Conlee
Program Officer, Animal Research Issues, The Humane Society of the
United States, Washington, DC
Welfare, Well-Being, and Pain—*Behavioral Correlates of Animal Pain and Distress*

Richard Connor
Associate Professor of Biology, University of Massachusetts, Dartmouth, Massachusetts
Social Organization—*Alliances*

Robert G. Cook
Professor of Psychology, Tufts University, Medford, Massachusetts
Behavioral Physiology—*Visual Perception Mechanisms*
Cognition—*Concept Formation*

Matthew A. Cooper
Postdoctoral Fellow, Center for Behavioral Neuroscience, Georgia State
University, Atlanta, Georgia
Social Organization—*Reconciliation*

Alex Córdoba-Aguilar
Researcher, Centro de Investigaciones Biolgicas, Universidad Nacional
    Autónoma de México
    Reproductive Behavior—*Sperm Competition*

Mary Crowe
Director, Center for Undergraduate Research, Xavier University,
    New Orleans, Louisiana
    Behavioral Physiology—*Thermoregulation*
    Careers—*Careers in Animal Behavior Science*

Alexander Cruz
Professor of Ecology and Evolutionary Biology, University of Colorado, Boulder, Colorado
    Caregiving—*Brood Parasitism in Freshwater Fish*

Davide Csermely
Associate Professor of Vertebrate Zoology, University of Parma, Parma, Italy
    Caregiving—*Parental Care and Helping Behavior*
    Welfare, Well-Being, and Pain—*Rehabilitation of Raptors*

Sasha R. X. Dall
Research Associate, Department of Zoology, University of Cambridge, Cambridge,
    United Kingdom
    Social Evolution—*Optimization and Evolutionary Game Theory*

Rebekka L. Darner
Center for Research in Math and Science Education, San Diego State
    University, San Diego, California
    Antipredatory Behavior—*Finding Food While Avoiding Predators*

Chantel Dawson
Student, Baylor University, Waco, Texas
    Welfare, Well-Being, and Pain—*Back Scratching and Enrichment in Sea Turtles*

Tagide deCarvalho
Department of Biology, University of Maryland, College Park, Maryland
    Aggressive Behavior—*Ritualized Fighting*

V. J. DeGhett
Department of Psychology, State University of New York, Potsdam, New York
    Communication—Auditory—*Ultrasound in Small Rodents*

Ingrid C. de Jong
Animal Sciences Group, Wageningen University and Research Centre,
    Wageningen, The Netherlands
    Welfare, Well-Being, and Pain—*Feather Pecking in Birds*

Francien H. de Jonge
Assistant Professor, University of Utrecht, Utrecht, The Netherlands
    Welfare, Well-Being, and Pain—*Animal Welfare and Reward*

Robert H. Devlin
Research Scientist, Fisheries and Oceans Canada, West Vancouver, British Columbia, Canada
   Human (Anthropogenic) Effects—*Genetically Modified Fish*

Donald A. Dewsbury
Professor of Psychology, University of Florida, Gainesville, Florida
   Frisch, Karl von (1886–1982)
   History—*History of Animal Behavior Studies*
      Some Leaders in Animal Behavior Studies
   Lorenz, Konrad Z. (1903–1989)
   Tinbergen, Nikolaas (1907–1988)
      A Visit with Niko and Lies Tinbergen

Tony DiPasquale
Department of Clinical Psychology, University of New Mexico, Albuquerque, New Mexico
   Methods—*Deprivation Experiments*

F. Stephen Dobson
Professor of Biological Sciences, Auburn University, Auburn, Alabama
   Social Organization—*Social Dynamics in the Plateau Pika*

Nicholas Dodman
Professor, Tufts University School of Veterinary Medicine, North Grafton, Massachusetts
   Applied Animal Behavior—*Social Dynamics and Aggression in Dogs*

Judith Donath
Assistant Professor of Media Arts and Sciences, Massachusetts Institute of Technology,
      Cambridge, Massachusetts
   Robotics—*Artificial Pets*

C. M. Drea
Assistant Professor of Biological Anthropology & Anatomy/Assistant Professor of Biology,
      Duke University, Durham, North Carolina
   Communication—Olfaction—*Mammalian Olfactory Communication*
   Development—*Spotted Hyena Development*
   Learning—*Social Learning and Intelligence in Primates*

Kathleen M. Dudzinski
Director, Dolphin Communication Project, Mystic Aquarium & Institute for Exploration,
      Scientist-In-Residence, Mystic Aquarium & Institute for Exploration, Mystic, Connecticut
   Dolphins—*Dolphin Behavior and Communication*
      Dolphins

Lee Alan Dugatkin
Professor of Biology, University of Louisville, Louisville, Kentucky
   Cooperation
   Hamilton, William D. III (1936–2000)
   Maynard Smith, John (1920–2004)
   Memes

Reuven Dukas
Professor of Psychology, McMaster University, Hamilton, Canada
   Cognition—*Limited Attention and Animal Behavior*

Diane Dutton
Psychology Department, Liverpool Hope University College, Hope Park, Liverpool, United Kingdom.
Personality and Temperament—*Personality in Chimpanzees*

Robert A. Eckstein
Professor of Biology, Warren Wilson College, Asheville, North Carolina
Welfare, Well-Being, and Pain—*Obsessive–Compulsive Behaviors*

N. J. Emery
Royal Society University Research Fellow, University of Cambridge, United Kingdom
Cognition—*Cache Robbing*

Rachel Endicott
Independent Scholar, San Francisco, California
Caregiving—*Parents Desert Newborns after Break-In! Mother Goes for Food as Siblings Battle to the Death! Father Eats Babies!*

Linda Evans
Research Officer, Animal Behaviour Laboratory, Department of Psychology, Macquarie University, Sydney, Australia
Art and Animal Behavior

Dorit Urd Feddersen-Petersen
University of Kiel, Kiel, Germany
Communication—Vocal—*Communication in Wolves and Dogs*

Linda Marie Fedigan
Professor of Anthropology and Canada Research Chair, University of Calgary, Canada
Methods—*Zen in the Art of Monkey Watching*

John Fentress
Adjunct Professor in Psychology and Neuroscience, Dalhousie University, Halifax, Nova Scotia, Canada
Cognition—*Dogs Burying Bones: Unraveling a Rich Action Sequence*

Jennifer Fewell
Associate Professor, School of Life Sciences, Arizona State University, Tempe, Arizona
Social Organization—*Eusociality*

Tommy G. Finley
Graduate Student, Department of Biology, University of Arkansas at Little Rock, Little Rock, Arkansas
Communication—Vocal—*Vocalizations of Northern Grasshopper Mice*

Diana O. Fisher
Postdoctoral Fellow, Australian National University, Canberra, Australia
Caregiving—*Parental Behavior in Marsupials*

Nancy N. FitzSimmons
Molecular Ecologist, University of Canberra, Canberra, Australia
Reproductive Behavior—*Marine Turtle Mating and Evolution*

Deborah Fouts
Friends of Washoe, Chimpanzee and Human Communication Institute, Central Washington
University, Ellensburg, Washington
Cognition—*Talking Chimpanzees*

Roger Fouts
Friends of Washoe, Chimpanzee and Human Communication Institute, Central Washington
University, Ellensburg, Washington
Cognition—*Talking Chimpanzees*

Camilla H. Fox
Director of Wildlife Programs, Animal Protection Institute, Sacramento,
California
Welfare, Well-Being, and Pain—*Wildlife Trapping Behavior and Welfare*

Gabriel Francescoli
Ethology Section, Facultad de Ciencias, Universidad de la República,
Montevideo, Uruguay
Communication—Tactile—*Communication in Subterranean Animals*

Jo Fritz
Primate Foundation of Arizona, Mesa, Arizona
Education—Classroom Activities—*All About Chimpanzees!*

Toni G. Frohoff
TerraMar Research, Consultant and Director of Research, TerraMar Research,
Bainbridge Island, Washington
Dolphins—*Dolphin Behavior and Communication
Dolphins*
Welfare, Well-Being, and Pain—*Stress in Dolphins*

Mildred Sears Funk
Assistant Professor, Biology Department, Roosevelt University, Chicago,
Illinois
Play—*Birds at Play*

Bennett G. Galef, Jr.
Emeritus Professor of Psychology and Adjunct Professor of Biology, McMaster
University, Canada
Development—*Intrauterine Position Effect*
Learning—*Social Learning*

Howard E. Garrett
Independent Scholar, Greenbank, Washington
Culture
Springer's Homecoming

Stefanie K. Gazda
University of Massachusetts at Dartmouth, Massachusetts
Dolphins—*Cooperative Hunting among Dolphins*

Kathleen C. Gerbasi
Resource Coordinator, Psychologists for the Ethical Treatment of Animal (PsyETA)
    Lewiston, New York
    Animal Abuse—*Violence to Human and Nonhuman Animals*

Carl Gerhardt
Curators' Professor of Biological Sciences, University of Missouri, Columbia, Missouri
    Communication—Vocal—*Choruses in Frogs*

Geoffrey Gerstner
School of Dentistry, University of Michigan, Ann Arbor, Michigan
    Neuroethology

Asif A. Ghazanfar
Research Scientist, Max Planck Institute for Biological Cybernetics, Tübingen, Germany
    Communication—Auditory—*Audition*

John P. Gluck
Professor of Psychology, University of New Mexico, Albuquerque, New Mexico; The
    Kennedy Institute of Ethics, Georgetown University, Washington, DC
    Methods—*Deprivation Experiments*

Jane Goodall
Jane Goodall Institute, Silver Spring, Maryland
    Sanctuaries

Samuel D. Gosling
Assistant Professor of Psychology, University of Texas, Austin, Texas
    Personality and Temperament—*A Comparative Perspective*

James L. Gould
Professor of Biology, Princeton University, Princeton, New Jersey
    Griffin, Donald Redfield (1915–2003)

Patricia Adair Gowaty
Distinguished Research Professor, Institute of Ecology, University of Georgia, Athens, Georgia
    Reproductive Behavior—*Sex, Gender, and Sex Roles*
        Female–Female Sexual Selection

James W. Grau
Professor of Psychology, Texas A&M University, College Station, Texas
    Learning—*Instrumental Learning*

Jean Swingle Greek
DVM, Santa Barbara, California
    Emotions—*How Do We Know Animals Can Feel?*

Donald R. Griffin
Concord Field Station, Harvard University, Massachusetts (deceased)
    Communication—Auditory—*Bat Sonar*
    Nobel Prize—*The 1973 Nobel Prize for Physiology or Medicine*

Gary W. Guyot
Professor of Psychology, Regis University, Denver, Colorado
    Caregiving—*Attachment Behaviors*
    Cats—*Domestic Cats*

John Hadidian
Director, Urban Wildlife Programs, The Humane Society of the United States, Washington, DC
    Human (Anthropogenic) Effects—*Urban Wildlife Behavior*
    Margaret Morse Nice, Backyard Ornithologist

Jim Halfpenny
A Naturalist's World, Gardiner, Montana
    Careers—*Animal Tracking and Animal Behavior*

Mike Hansell
Division of Environmental and Evolutionary Biology, University of Glasgow, Glasgow, Scotland
    Burrowing Behavior

Brian Hare
Max Planck Institute of Evolutionary Anthropology, Leipzig, Germany
    Cognition—*Domestic Dogs Use Humans as Tools*

Benjamin Hart
Professor, University of California School of Veterinary Medicine, Davis, California
    Communication—Auditory—*Long Distance Calling, the Elephant's Way*
    Tools—*Tool Use by Elephants*

Lynette Hart
Professor, University of California School of Veterinary Medicine, Davis, California
    Communication—Auditory—*Long Distance Calling, the Elephant's Way*
    Tools—*Tool Use by Elephants*

Ethan Hay
Independent Scholar, New Xings Training and Development, Sausalito, California
    Sharks

Bernd Heinrich
Professor Zoology, University of Vermont, Burlington, Vermont
    Corvids—*The Crow Family*

Stephen Herrero
Professor Emeritus of Environmental Science, Universty of Calgary, Alberta, Canada
    Human (Anthropogenic) Effects—*Bears: Understanding, Respecting, and Being Safe around Them*

Adolf Heschl
Behavior and Evolution Research Group, Deer Park, Poellau, Austria
    Nature and Nature—*Nature Explains Nurture: Animals, Genes, and the Environment How to Solve the Nature/Nurture Dichotomy*

Heather M. Hill
Graduate Student in Experimental Psychology, University of Southern Mississippi, Hattiesburg, Mississippi
    Caregiving—*How Animals Care for Their Young*

Peggy Hill
Associate Professor of Biological Science, University of Tulsa, Tulsa, Oklahoma
    Communication—Tactile—*Vibrational Communication*

John L. Hoogland
Professor of Biology, Appalachian Laboratory, University of Maryland, Frostburg, Maryland
    Infanticide

Alexandra Horowitz
Department of Psychology, Hunter College, New York, New York
    Play—*Dog Minds and Dog Play*

Katherine A. Houpt
Professor of Physiology, Cornell University, Ithaca, New York
    Horses—*Behavior*
        Sleeping Standing Up

Sue Howell
Primate Foundation of Arizona, Mesa, Arizona
    Education—Classroom Activities—*All About Chimpanzees!*

Michael A. Huffman
Associate Professor of Primate Behavioral Ecology, Primate Research Institute,
    Kyoto University, Inuyam, Japan
    Self-Medication

Stuart H. Hurlbert
Professor of Biology, San Diego State University, San Diego, California
    Methods—*Research Methodology*
        Experimental Design
        Pseudoreplication

Rudolf Jander
Professor of Ecology and Evolutionary Biology, University of Kansas, Lawrence,
    Kansas
    Navigation—*Wayfinding*

Michael Jennions
Research Fellow, School of Botany & Zoology, Canberra, Australia
    Reproductive Behavior—*Female Multiple Mating*
    Reproductive Behavior—*Mate Choice*
    Reproductive Behavior—*Mate Desertion*
    Reproductive Behavior—*Sexual Selection*

Mary Lee Jensvold
Chimpanzee and Human Communication Institute, Central Washington University, Ellensburg,
    Washington
    Cognition—*Talking Chimpanzees*

Jörgen I. Johnsson
Associate Professor, Göteborg University, Sweden
Human (Anthropogenic) Effects—*Genetically Modified Fish*

R. Bryan Jones
Head of Welfare Biology Group, Roslin Institute, Scotland
Welfare, Well-Being, and Pain—*Enrichment for Chickens*

Alex Kacelnik
Professor of Behavioural Ecology, University of Oxford, Oxford, United Kingdom
Tools—*Tool Use and Manufacture by Birds*

Gisela Kaplan
Professor of Biological Sciences, University of New England, Armidale, New South Wales
Australia
Culture—*Orangutan Culture*
Darwin, Charles (1809–1882)
Mimicry—*Magpies*

Colleen Reichmuth Kastak
Long Marine Laboratory, University of California Santa Cruz, Santa Cruz, California
Cognition—*Equivalence Relations*
Learning—*Insight*

Ben Kenward
Ph.D. candidate, University of Oxford, Oxford, United Kingdom
Tools—*Tool Use and Manufacture by Birds*

Lesley A. King
Consultant, Humane Society of the United States, Linacre College, Oxford University,
United Kingdom
Welfare, Well-Being, and Pain—*Behavioral Correlates of Animal Pain and Distress*

Devra G. Kleiman
Adjunct Professor of Biology, University of Maryland, College Park, Maryland, and
Independent Consultant, Chevy Chase, Maryland
Conservation and Behavior—*Species Reintroduction*

Peter H. Klopfer
Professor Emeritus of Biology, Duke University, Durham, North Carolina
Lemurs—*Learning from Lemurs*

Janelle Knox
Research Assistant, University of Colorado, Boulder, Colorado
Caregiving—*Brood Parasitism in Freshwater Fish*

S. Mechiel Korte
Senior Scientist, Wageningen University and Research Center, The Netherlands
Welfare, Well-Being, and Pain—*Feather Pecking in Birds*

Kurt Kotrschal
Konrad-Lorenz-Forschungsstelle für Ethologie, Grünau, Austria
Feeding Behavior—*Scrounging*
Hormones and Behavior

Bernie Krause
Wild Sanctuary, Inc., Glen Ellen, California
    Careers—*Recording Animal Behavior Sounds: The Voice of the Natural World*

Egle H. Krosniunas
Assistant Professor of Biology, University of Wisconsin—Barron County, Rice Lake, Wisconsin
    Behavioral Physiology—*Turtle Behavior and Physiology*

Stan A. Kuczaj
Professor of Psychology, University of Southern Mississippi, Hattiesburg, Mississippi
    Caregiving—*How Animals Care for Their Young*

Tomás Landete-Castillejos
Researcher in Game Species and Lecturer in Zoology, University of Castilla–La Mancha, Albacete, Spain
    Caregiving—*Parental Investment*

Bob Landis
Bob Landis Wildlife Films, Gardiner, Montana
    Careers—*Wildlife Filmmaking*

Gary Landsberg
DVM, DACVB. President, American College of Veterinary Behaviorists, Doncaster Animal Clinic, Thornhill, Ontario, Canada
    Careers—*Veterinary Practice Opportunities for Ethologists*

Tom A. Langen
Departments of Biology and Psychology, Clarkson University, Potsdam, New York
    Cognition—*Food Storing*

Susan E. Lewis
Associate Professor and Chair of Biology, Carroll College, Waukesha, Wisconsin
    Caregiving—*Non-Offspring Nursing*
    Caregiving—*Parental Care*

Susan U. Linville
Undergraduate Internship Coordinator, Center for the Integrative Study of Animal Behavior, Indiana University, Bloomington, Indiana
    Behavioral Physiology—*Plumage Color and Vision in Birds*
    Methods—*DNA Fingerprinting*

Dingzhen Liu
Associate Professor of Animal Behavior, Beijing Normal University, Beijing, China
    Zoos and Aquariums—*Giant Pandas in Captivity*

Celia M. Lombardi
Researcher (CONICET), Museo Argentino de Ciencias Naturales, Buenos Aires, Argentina Experimental Design
    Methods—*Research Methodology*
      Experimental Design
      Pseudoreplication

Elba Muñoz Lopez
Centro de Rescate Rehabilitión de Primates, Peñaflor, Chile
Welfare, Well-Being, and Pain—*Primate Rescue Groups*

Cara Blessley Lowe
Writer and Filmmaker, Los Angeles, California
Careers—*Mapping Their Minds: Animals on the Other Side of the Lens*

Paula MacKay
Conservationist, Starksboro, Vermont
Applied Animal Behavior—*Scat-Sniffing Dogs*

Thomas D. Mangelsen
Images of Nature, Jackson, Wyoming
Careers—*Wildlife Photography*

Heidi Marcum
Senior Lecturer, Department of Environmental Studies, Baylor University,
Waco, Texas
Welfare, Well-Being, and Pain—*Back Scratching and Enrichment in Sea Turtles*
Welfare, Well-Being, and Pain—*Experiments in Enriching Captive Snakes*

Sue Margulis
Behavioral Research Manager, Brookfield Zoo, Committee on Evolutionary Biology,
University of Chicago, Chicago, Illinois
Education—*Classroom Activities in Behavior*
Classroom Research

P. Marler
Professor Emeritus, University of California, Davis, California
Nobel Prize—*The 1973 Nobel Prize for Physiology or Medicine*

Emília P. Martins
Associate Professor of Biology, Director, Center for the Integrative Study
of Animal Behavior, Indiana University, Bloomington, Indiana
Behavioral Phylogeny—*The Evolutionary Origins of Behavior*

Georgia Mason
Animal & Poultry Science, University of Guelph, Guelph, Ontario, Canada
Welfare, Well-Being, and Pain—*Carnivores in Captivity*

Jennifer A. Mather
Professor, Psychology & Neuroscience, University of Lethbridge, Lethbridge,
Canada
Behavioral Physiology—*Color Vision in Animals*
Colors—*How do Flowers and Bees Match?*
Do Squid Make a Language on Their Skin?
Careers—*Careers in Animal Behavior Science*
Cephalopods—*Octopuses, Squid, and Other Mollusks*

Nicolas Mathevon
Associate Professor, Jean Monnet University, Saint-Etienne, France
　　Communication—Auditory—*Acoustic Communication in Extreme Environments*

Tetsuro Matsuzawa
Professor, Primate Research Institute, Kyoto University, Kyoto, Japan
　　Caregiving—*Mother–Infant Relations in Chimpanzees*

Michael R. Maxwell
Research Associate, University of San Diego, California
　　Reproductive Behavior—*Sexual Cannibalism*

Paul McGreevy
Senior Lecturer in Animal Behaviour, Faculty of Veterinary Science, University of Sydney,
　　Sydney, Australia
　　Behavior Physiology
　　　　Vision, Skull Shape, and Behavior in Dogs

Charlene McIver
Department of Psychology, University of New Mexico, Albuquerque, New Mexico
　　Methods—*Deprivation Experiments*

Ian G. McLean
Research Analyst, Geneva International Centre for Humanitarian Demining, Geneva,
　　Switzerland
　　Applied Animal Behavior—*Mine-Sniffing Dogs*

Franklin D. McMillan
Veterinarian, Veterinary Centers of America, Los Angeles, California
　　Welfare, Well-Being, and Pain—*Psychological Well-Being*

Scott P. McRobert
Professor of Biology and Director of the Environmental Science Program, Saint Joseph's
　　University, Philadelphia, Pennsylvania
　　Reproductive Behavior—*Sexual Behavior in Fruit Flies*—Drosophila
　　Social Organization—*Shoaling Behavior in Fish*

Dwayne Meadows
Pacific Whale Foundation, Maalaea, Hawaii
　　Human (Anthropogenic) Effects—*The Effect of Roads and Trails on Animal Movement*

Carron A. Meaney
Research Associate, Denver Museum of Nature and Science, Denver, Colorado
　　Conservation and Behavior
　　　　Preble's Meadow Jumping Mouse and Culverts
　　Social Organization—*Pika Behavior Studies*

Rita Mehta
Ph.D. Candidate, Department of Psychology, University of Tennessee, Knoxville, Tennessee
　　Communication—*Modal Action Patterns*

Daniela G. Meissner
Department of Neurobiology, International University—Bremen, Bremen, Germany
  Communication—*Electrocommunication*
    Electric Fish

William W. Merkle
Wildlife Ecologist, National Park Service, Golden Gate National Recreation Area, San
    Francisco, California
  Human (Anthropogenic) Effects—*Edge Effects and Behavior*

Gail R. Michener
Professor of Biological Sciences, University of Lethbridge, Lethbridge, Canada
  Hibernation

Ádam Miklósi
Assistant Professor, Eötvös University, Budapest, Hungary
  Hibernation
  Cognition—*Imitation*

Myrna Milani
Independent Scholar, New Hampshire
  Cats—*Wild versus Domestic Behaviors: When Normal Behaviors Lead to Problems*

Brian J. Miller
Conservation Biologist, Denver Zoological Foundation, Denver, Colorado
  Conservation and Behavior—*Species Reintroduction*

Cory T. Miller
Postdoctoral Fellow, Johns Hopkins University School of Medicine, Baltimore,
    Maryland
  Communication—Auditory—*Audition*

Patrice Marie Miller
Department of Psychology, Salem State College, Salem, Massachusetts
  Development—*Behavioral Stages*

Robert W. Mitchell
Professor of Psychology, Eastern Kentucky University, Richmond, Kentucky
  Cognition—*Mirror Self-Recognition and Kinesthetic–Visual Matching*

Douglas Mock
Professor of Zoology, University of Oklahoma, Norman, Oklahoma
  Siblicide

Janice Moore
Professor of Biology, Colorado State University, Fort Collins, Colorado
  Parasite-Induced Behaviors

David B. Morton
Centre for Biomedical Ethics, University of Birmingham, Birmingham, United Kingdom
  Welfare, Well-Being, and Pain—*Behavior and Animal Suffering*

Miho Nakamura
Producer, ANC Corporation, Tokyo, Japan
Caregiving—*Mother–Infant Relations in Captive Chimpanzees*

Lucas P. J. J. Noldus
Managing Director, Noldus Information Technology bv, Wageningen, The Netherlands
Methods—*Computer Tools for Measurement and Analysis of Behavior*

Michael Noonan
Professor of Biology and Psychology, Canisius College, Buffalo, New York
Laterality—*Laterality in the Lives of Animals*
Zoos and Aquariums—*Animal Behavior Research in Zoos*

Janette Nystrom
Associate Director, Wallace Stegner Center for Land, Resources and the Environment,
S.J. Quinney College of Law, University of Utah, Salt Lake City, Utah
Domestication and Behavior—*The Border Collie, a Wolf in Sheep's Clothing*
Social Organization—*Social Order and Communication in Dogs*

Caitlin O'Connell-Rodwell
Researcher in Geophysics and Biology, Stanford University, California
Communication—Auditory—*Long Distance Calling, the Elephant's Way*

Terry J. Ord
Research Associate, Indiana University, Bloomington, Indiana
Behavioral Phylogeny—*The Evolutionary Origins of Behavior*

Catherine P. Ortega
Associate Professor of Biology, Fort Lewis College, Durango, Colorado
Caregiving—*Brood Parasitism among Birds*

Donald H. Owings
Professor of Psychology and Animal Behavior, University of California, Davis,
California
Antipredatory Behavior—*Predator–Prey Communication*

Jane M. Packard
Associate Professor of Wildlife and Fisheries Sciences, Texas A&M University,
College Station, Texas
Feeding Behavior—*Grizzly Foraging*

Brian Palestis
Assistant Professor of Biology, Wagner College, Staten Island, New York
Recognition—*Kin Recognition*
Four Instructive Examples of Kin Recognition
Social Facilitation

Jaak Panksepp
Distinguished Research Professor, Bowling Green State University, Bowling Green, Ohio;
Head, Affective Neuroscience Research, Falk Center for Molecular Therapeutics,
Northwestern University, Evanston, Illinois
Emotions—*Emotions and Affective Experiences*

Mauricio R. Papini
Professor of Psychology, Texas Christian University, Fort Worth, Texas
Comparative Psychology
Learning—*Evolution of Learning Mechanisms*

Sarah Partan
Assistant Professor of Psychology, University of South Florida, St. Petersburg, Florida
Robotics—*Animal Robots*

Penny Patterson
The Gorilla Foundation/Koko.org, Woodside, California
Gorillas—*Gorillas/Koko*

Scott Pawlowski
Research Assistant, University of Colorado, Boulder, Colorado
Caregiving—*Brood Parasitism in Freshwater Fish*

Irene M. Pepperberg
Research Associate Professor of Psychology, Brandeis University, Waltham, Massachusetts;
Research Scientist, MIT School of Architecture and Planning, Cambridge, Massachusetts
Cognition—*Grey Parrot Cognition and Communication*
What Use Does a Wild Parrot Make of Its Ability to Talk?

Michael E. Pereira
Scholar-in-Residence, The Latin School of Chicago, Chicago
Development—*Adaptive Behavior and Physical Development*

Anna L. Peterson
Professor of Religion, University of Florida, Gainesville, Florida
Religion and Animal Behavior

Brenda Peterson
Author and Independent Scholar, Seattle, Washington
Careers—*Writing about Animal Behavior: The Animals Are Also Watching Us*

Wojciech Pisula
Warsaw School of Social Psychology and Institute of Psychology, Polish Academy of
Sciences, Warsaw, Poland
Exploratory Behavior—*Inquisitiveness in Animals*

Stephanie D. Preston
Postdoctoral Fellow, Department of Neurology, University of Iowa, Iowa City, Iowa
Empathy

Frederick R. Prete
President, Visuo Technologies, LLC, Morton Grove, Illinois, and Adjunct
Professor of Psychology, DePaul University, Chicago, Illinois
Predatory Behavior—*Praying Mantids*
"Sexual Cannibalism": Is It Really "Sexual" or Is It Just Predatory Behavior?

Edward O. Price
Emeritus Professor of Animal Science, University of California, Davis, California
Domestication and Behavior

Sindhu Radhakrishna
Adjunct Associate Fellow, National Institute of Advanced Studies, Bangalore, India
Social Evolution—*Social Evolution in Bonnet Macaques*
Social Organization—*Social Knowledge in Wild Bonnet Macaques*

Hayley Randle
Senior Lecturer in Animal Science (Behaviour & Welfare), University of Plymouth,
Plymouth, UK
Personality and Temperament—*Personality, Temperament, and Behavioral Assessment in Animals*

Richard P. Reading
Director of Conservation Biology, Denver Zoological Foundation, Denver, Colorado
Conservation and Behavior—*Species Reintroduction*

Drew Rendall
Associate Professor of Psychology, University of Lethbridge, Lethbridge,
Alberta, Canada
Cognition—*Animal Languages, Animal Minds*

Michael J. Renner
Dean of the College of Arts and Sciences and Professor of Psychology, Nazareth College,
Rochester, New York
Curiosity

Erich K. Ritter
Chief Scientist, Global Shark Attack File, Shark Research Institute, Princeton,
New Jersey
Sharks—Mistaken Identity?

Lesley J. Rogers
Professor, University of New England, Armidale, Australia
Darwin, Charles (1809–1882)
Laterality

Antonio Rolando
Dipartimento di Biologia Animale Università degli studi di Torino, Torino, Italy
Social Organization—*Home Range*

Bernard E. Rollin
Professor of Philosophy, Colorado State University, Fort Collins, Colorado
Welfare, Well-Being, and Pain—*Veterinary Ethics and Behavior*

Barry Rosenbaum
Research Associate, Ecology and Evolutionary Biology, University of Colorado, Boulder,
Colorado
Human (Anthropogenic) Effects—*Logging, Behavior, and the Conservation of Primates*

Matthew P. Rowe
Professor of Biology, Appalachian State University, Boone, North Carolina
Mimicry

Daniel I. Rubenstein
Professor, Department of Ecology and Evolutionary Biology, Princeton University, Princeton, New Jersey
Social Organization—*Herd Dynamics: Why Aggregations Vary in Size and Complexity*

Anne E. Russon
Professor of Psychology, Glendon College of York University, Toronto, Canada
Culture—*Culture in our Relatives, the Orangutans*
History—*Human Views of the Great Apes from Ancient Greece to the Present*

Allen T. Rutberg
Research Assistant Professor, Tufts University School of Veterinary Medicine, North Grafton, Massachusetts
Wildlife Management and Behavior

Kenneth M. D. Rutherford
Postgraduate student, Roslin Institute, Edinburgh, United Kingdom
Welfare, Well-Being, and Pain—*Behavioral Assessment of Animal Pain*

Ronald L. Rutowski
Professor, School of Life Sciences, Arizona State University, Tempe, Arizona
Behavioral Physiology—*Insect Vision and Behavior*

Laurie R. Santos
Assistant Professor of Psychology, Yale University, New Haven, Connecticut
Cognition—*Theory of Mind*
Tools—*Tool Use*

Robert M. Sapolsky
Professor of Biology Sciences, Stanford University, Stanford, California
Personality and Temperament—*Stress, Social Rank, and Personality*

Cathy Schaeff
American University, Washington, DC
Caregiving—*Fostering Behavior*

Ingo Schlupp
Animal Behaviour Group, University of Hamburg, Germany
Unisexual Vertebrates

Ronald J. Schusterman
Long Marine Laboratory, University of California Santa Cruz, Santa Cruz, California
Cognition—*Equivalence Relations*

Abby L. Schwarz
Instructor, Biology Dept, Langara College, Vancouver, British Columbia, Canada
Reproductive Behavior—*Sex Change in Fish*

Michelle Pellissier Scott
Professor of Zoology, University of New Hampshire, Durham, New Hampshire
Caregiving—*Parental Care by Insects*

Laura Sewall
Independent Scholar and Land Use Consultant, Phippsburg, Maine
Ecopsychology—*Human–Nature Interconnections*

Robert M. Seyfarth
Professor of Psychology, University of Pennsylvania, Philadelphia, Pennsylvania
Cognition—*Social Cognition in Primates and Other Animals*

Kartik Shanker
Fellow, Ashoka Trust for Research in Ecology and the Environment,
Bangalore, India
Navigation—*Natal Homing and Mass Nesting in Marine Turtles*

Kenneth Shapiro
Executive Director, Psychologists for the Ethical Treatment of Animals (PsyETA),
Washington Grove, Maryland
Animal Abuse—*Animal and Human Abuse*
Animal Models of Human Psychology—*An Assessment of Their Effectiveness*

Rupert Sheldrake
Biologist and Author, London, England
Telepathy and Animal Behavior

Paul W. Sherman
Professor of Neurobiology and Behavior, Cornell University, Ithaca, New York
Levels of Analysis in Animal Behavior
Naked Mole-Rats
Reproductive Behavior—*Morning Sickness*

Robert W. Shumaker
Director of Orangutan Research, Iowa Primate Learning Sanctuary,
Des Moines, Iowa
Cognition—*Mirror Self-Recognition*
Individual Variation and MSR: The Response of Two Orangutans upon Seeing Their Reflection

Stephen M. Shuster
Professor of Invertebrate Zoology, Northern Arizona University, Flagstaff
Reproductive Behavior—*Mating Strategies in Marine Isopods*

Robert S. Sikes
Associate Professor, Department of Biology, University of Arkansas at Little Rock, Little
Rock, Arkansas
Communication—Vocal—*Vocalizations of Northern Grasshopper Mice*

Brandi Sima
Audubon Aquarium of the Americas, New Orleans, Louisiana
Welfare, Well-Being, and Pain—*Rehabilitation of Marine Mammals*

Patricia Simonet
Cognitive Ethology and Animal Behavior Consultant, Spokane, Washington
Emotions—*Laughter in Animals*

Anindya Sinha
Fellow, National Institute of Advanced Studies, Bangalore, India
Cognition—*Tactical Deception in Wild Bonnet Macaques*
Social Evolution—*Social Evolution in Bonnet Macaques*
Social Organization—*Social Knowledge in Wild Bonnet Macaques*
Tools—*Tool Manufacture by a Wild Bonnet Macaque*

Hans Slabbekoorn
Assistant Professor, Institute of Biology, Leiden University, Leiden,
The Netherlands
Communication—Vocal—*Singing Birds: From Communication to Speciation*
Dove Coo Code Decoded
Echoes in the Forest
Portraits of Sound

Con Slobodchikoff
Professor of Biological Sciences, Northern Arizona University, Flagstaff, Arizona
Communication—Vocal—*Referential Communication in Prairie Dogs*

Andrew T. Smith
Professor of Life Sciences, Arizona State University, Tempe, Arizona
Social Organization—*Social Dynamics in the Plateau Pika*

Douglas W. Smith
Yellowstone Wolf Project Leader, National Park Service, Yellowstone National Park,
Wyoming
Animals in Myth and Lore
The Wolf in Fairy Tales
Wolf Behavior—*Learning to Live in Life or Death Situations*

Rachel Smolker
Department of Biology, University of Vermont, Burlington, Vermont
Tools—*Tool Use by Dolphins*

Barbara Smuts
Professor of Psychology, University of Michigan, Ann Arbor, Michigan
Friendship in Animals

Charles T. Snowdon
Hilldale Professor of Psychology and Zoology, University of Wisconsin, Madison,
Wisconsin
Careers—*Significance of Animal Behavior Research*
Social Organization—*Monkey Families and Human Families*

Julia Sommerfeld
Primate Foundation of Arizona, Mesa, Arizona
Education—Classroom Activities—*All About Chimpanzees!*

Charles Southwick
Professor Emeritus, Ecology and Evolutionary Biology, University of Colorado, Boulder, Colorado
Communication—Visual—*Fish Display Behavior*

David A. Spector
Professor of Biology, Central Connecticut State University, New Britain, Connecticut
Communication—Vocal—*Variation in Bird Song*

Berry M. Spruijt
Professor of Ethology and Animal Welfare, University of Utrecht, Utrecht, The Netherlands
Welfare, Well-Being, and Pain—*Animal Welfare and Reward*

Judy Stamps
Professor of Biology, University of California, Davis, California
Social Organization—*Territoriality*
The Prior Residency Advantage

Martin L. Stephens
Vice President, Animal Research Issues, The Humane Society of the United States, Washington, DC
Welfare, Well-Being, and Pain—*Behavioral Correlates of Animal Pain and Distress*

L. Fredrik Sundström
Department of Zoology, Göteborg University, Sweden
Human (Anthropogenic) Effects—*Genetically Modified Fish*

Robert W. Sussman
Professor of Anthropology, Washington University, St. Louis, Missouri
Social Evolution—*Flowering Plants and the Origins of Primates*
Social Organization—*Cooperation and Aggression among Primates*

Ronald R. Swaisgood
Center for Reproduction of Endangered Species, Zoological Society of San Diego, California
Reproductive Behavior—*Captive Breeding*

Karyl B. Swartz
Professor Psychology, Lehman College of The City University of New York, New York
Cognition—*Mirror Self-Recognition*
Individual Variation and MSR: The Response of Two Orangutans upon Seeing Their Reflection

Daniel R. Tardona
Resource Interpretive Specialist, Timucuan Ecological & Historic Preserve, Jacksonville, Florida, and Adjunct, Department of Psychology, University of North Florida, Jacksonville, Florida
Conservation and Behavior—*Wildlife Behavior as a Management Tool in United States National Parks*

Inge Teblick
Belgian Horsemanship Association, Duffel, Belgium
Horses—*Horse Training*

Oakleigh Thorne II
Founder & Honorary President, Thorne Ecological Institute, Boulder, Colorado
  Reproductive Behavior—*Shredding Behavior of Rodents*

Elizabeth Tibbetts
Postdoctoral Fellow, University of Arizona, Tucson, Arizona
  Recognition—*Individual Recognition*

Ethel Tobach
Curator Emerita, American Museum of Natural History and Adjunct Professor, Psychology
    and Biology, The City University of New York, New York
  Zoos and Aquariums—*Studying Animal Behavior in Zoos and Aquariums*

Susan E. Townsend
Wildlife Ecologist/Consultant, Oakland, California
  Cognition—*The Audience Effect in Wolves*

Adrian Treves
Conservation Ecologist, Living Landscapes Program of the Wildlife Conservation Society,
    Bronx, New York
  Antipredatory Behavior—*Vigilance*

Rolan Tripp
DVM, Founder: www.AnimalBehavior.Net, La Mirada, California
  Careers—*Veterinary Practice Opportunities for Ethologists*

Walter R. Tschinkel
Professor of Biological Science, Florida State University
  Animal Architecture—*Subterranean Ant Nests*

Ruud van den Bos
Assistant Professor of Ethology and Welfare, Utrecht University, Utrecht,
    The Netherlands
  Emotions—*Emotions and Cognition*

Yvonne M. van Hierden
Animal Sciences Group, Wageningen University and Research Centre, Wageningen,
    The Netherlands
  Welfare, Well-Being, and Pain—*Feather Pecking in Birds*

Stephen B. Vander Wall
Associate Professor, University of Nevada, Reno, Nevada
  Cognition—*Caching Behavior*

Dirk H. Van Vuren
Professor of Wildlife Biology, University of California, Davis, California
  Social Organization—*Dispersal*
    Dispersal in Marmots

Simine Vazire
Graduate Student, Department of Psychology, University of Texas, Austin, Texas
  Personality and Temperament—*A Comparative Perspective*

Carmen Viera
Professor of Entomology, Departemento Biologa Animal, Facultad de Ciencias, Universidad de la República, Montevideo, Uruguary
Predatory Behavior—*Orb-Web Spiders*

Elisabetta Visalberghi
Istituto di Scienze e Tecnologie della Cognizione, CNR Consiglio Nazionale delle Ricerche, Rome, Italy
Feeding Behavior—*Social Learning: Food*

Augusto Vitale
Researcher in Animal Behaviour, Istituto Superiore di Sanità, Rome, Italy
Caregiving—*Helpers in Common Marmosets*
Welfare, Well-Being, and Pain—*Enrichment for New World Monkeys*

Paul Waldau
Center for Animals and Public Policy, Tufts University School of Veterinary Medicine, North Grafton, Massachusetts
Careers—*Animal Behavior and the Law*

Ralf Wanker
Research Assistant, Hamburg, Germany
Communication—Vocal—*Social System and Acoustic Communication in Spectacled Parrotlets*

Edward A. Wasserman
Stuit Professor of Experimental Psychology, The University of Iowa, Iowa City, Iowa
Behavioral Physiology—*Visual Perception Mechanisms*
Behaviorism
Amusing Tales of Animal Mind
Concept Learning in Pigeons

Paul Watson
Research Assistant Professor of Biology, University of New Mexico, Albuquerque, New Mexico
Aggressive Behavior—*Ritualized Fighting*

Ann Weaver
Assistant Professor of Psychology and Behavioral Sciences, Argosy University, Sarasota, Florida
Animals in Myth and Lore—*Fairy Tales and Myths of Animal Behavior*
Methods—*Ethograms*

Alex A.S. Weir
Ph.D. student, University of Oxford, Oxford, United Kingdom
Tools—*Tool Use and Manufacture by Birds*

Hal Whitehead
Killam Professor of Biology, Dalhousie University, Halifax, Nova Scotia, Canada
Culture—*Whale Culture and Conservation*

Howard H. Whiteman
Associate Professor of Biological Sciences, Murray State University,
Murray, Kentucky
Human (Anthropogenic) Effects—*Environmentally Induced Behavioral Polymorphisms*

Fred Whoriskey
Vice-President of Research & Environment, Atlantic Salmon Federation, St. Andrews, Canada
Methods—*Sonic Tracking of Endangered Atlantic Salmon*

Ben Wilson
Researcher, Behaviour@Sea Project, University of British Columbia, Vancouver, Canada
Communication—Auditory
Noisy Herring

Thomas Wynn
Professor of Anthropology, University of Colorado, Colorado Springs, Colorado
Cognition—*Behavior, Archaeology, and Cognitive Evolution*

Stephen Zawistowski
Senior Vice President and National Program Director, Science Advisor, The American Society for the Prevention of Cruelty to Animals, New York
Careers—*Applied Animal Behavior*

Günther K. H. Zupanc
Professor of Neurobiology, International University Bremen, Bremen, Germany
Communication—*Electrocommunication*
Electric Fish

◼ Index